Basic Concepts for Simple and Complex Liquids

This book presents a unified approach to the liquid state, focussing on the concepts and theoretical methods that are necessary for an understanding of the physics and chemistry of fluids.

The authors do not attempt to cover the whole field in an encyclopedic manner. Instead, important ideas are presented in a concise and rigorous manner, illustrated with examples from both simple molecular liquids and more complex soft condensed matter systems such as polymers, colloids or liquid crystals. After a general chapter introducing the liquid state, the book is in four parts devoted to thermodynamics, structure and fluctuations, phase transitions, interfaces and inhomogeneous fluids, and, finally, transport and dynamics. Each chapter introduces a new set of closely related concepts and theoretical methods, which are then illustrated by a number of specific applications covering a broad range of physical situations.

Accessible to graduate students with a basic knowledge of statistical physics or physical chemistry, this advanced text will be useful to physicists, chemists or chemical engineers seeking a broad-scope introduction to the liquid state of matter, from thermodynamics and structure to transport properties and activated dynamics.

JEAN-LOUIS BARRAT graduated from ENS Paris, and obtained his Ph.D. from Université Paris 6 in 1987. He was a postdoc at the TU in Munich and at UCSB, and worked as a CNRS associate at ENS Lyon. He is now a professor at the University of Lyon, where he heads the condensed matter theory group. His research interests cover application of statistical physics and computer simulation to various aspects of material sciences, e.g. interfaces, glasses and amorphous systems, clusters or polymers.

JEAN-PIERRE HANSEN obtained a Ph.D. in Physics at the Université de Paris-Sud in 1969, and was appointed Professor of Physics at Université Pierre and Marie Curie. In 1987 he moved to the newly created Ecole Normale Supérieure de Lyon, where he headed the Physics Department until 1997, when he was appointed to the Chair of Theoretical Chemistry at Cambridge University. He was elected Fellow of the Royal Society in 2002. J. P. Hansen has made important contributions in Statistical Mechanics of simple and complex liquids, of quantum fluids and of dense plasmas.

Basic Concepts for Simple and Complex Liquids

Jean-Louis Barrat and Jean-Pierre Hansen

CAMBRIDGE
UNIVERSITY PRESS

PUBLISHED BY THE PRESS SYNDICATE OF THE UNIVERSITY OF CAMBRIDGE
The Pitt Building, Trumpington Street, Cambridge, United Kingdom

CAMBRIDGE UNIVERSITY PRESS
The Edinburgh Building, Cambridge CB2 2RU, UK
40 West 20th Street, New York, NY 10011-4211, USA
477 Williamstown Road, Port Melbourne, VIC 3207, Australia
Ruiz de Alarcón 13, 28014 Madrid, Spain
Dock House, The Waterfront, Cape Town 8001, South Africa

http://www.cambridge.org

First published 2003

Printed in the United Kingdom at the University Press, Cambridge

Typefaces Times New Roman 10/13 pt and Univers   *System* LaTeX 2_ε [TB]

A catalogue record for this book is available from the British Library

Library of Congress Cataloguing in Publication data

Barrat, Jean-Louis, 1964–
Basic concepts for simple and complex liquids / Jean-Louis Barrat and Jean-Pierre Hansen.
    p.   cm.
Includes bibliographical references and index.
ISBN 0-521-78344-5 – ISBN 0-521-78953-2 (pb.)
1. Liquids.   2. Complex fluids.   I. Hansen, Jean Pierre.   II. Title.
QC145.2 .B35   2003
530.4′2–dc21   2002031051

ISBN 0 521 78344 5 hardback
ISBN 0 521 78953 2 paperback

Contents

Preface

While the solid state of matter is mostly associated with the mineral world and much of modern technology, liquids are more closely related to living matter and biological processes. In fact life is generally believed to have emerged in the primordial ocean which was formed when the right temperature conditions came to prevail on the young Earth, providing a striking illustration of the 'marginal' character of the liquid state, compared to the solid and gaseous phases of the same substances, which exist over much wider ranges of temperature and pressure. The liquid state arises from a delicate balance between 'packing' of molecules and cohesive forces or, more formally, between entropy and energy, which renders a statistical description very difficult, due to the absence of any obvious 'small parameter'. This may explain that, while the gaseous and crystalline states of matter were well understood by the 1950s, significant theoretical progress on the liquid state only just started around that time, and was then speeded up by early neutron scattering and computer simulation data.

From the start, the exploration of the liquid state was interdisciplinary *par excellence*, thanks to the combined efforts of physicists, physical chemists and chemical engineers. However, early theoretical work evolved along two lines of thought. 'Simple' liquids were studied on a molecular scale, using statistical mechanics and computer simulations as basic tools, while 'complex' fluids (sometimes referred to as 'soft matter' following P.G. de Gennes) were mostly examined on a more coarse-grained level, epitomized in the scaling approach to the theory of polymer solutions. Over the last ten to fifteen years the two complementary viewpoints have moved closer, mostly through detailed experimental and theoretical studies of concentrated colloidal dispersions, where micrometre-sized particles in many respects mimic the behaviour of atoms or molecules in simple liquids, apart from the obvious change of scale. A unified picture of the liquid state is gradually emerging, based on common concepts and approximation schemes which are applicable to both 'simple' and 'complex' liquids. Depending on the problem at hand, a proper balance must be struck between molecular detail and phenomenological coarse-graining, in particular when widely different length and time scales are involved.

The present book is an attempt to present such a unified approach to the liquid state, focussing on general concepts and theoretical methods applicable to a broad range of physical situations. The book does not intend to provide a comprehensive catalogue of the overwhelming variety of simple or complex fluids, but rather investigates in some detail selected aspects of fluid systems as applications of the general theoretical framework, to illustrate the wide-ranging applicability of concepts and approximation schemes, like mean field theory, the random phase approximation, scaling arguments, density functional theory, entropic forces, stochastic equations of motion or mode-coupling ideas. It is hence not a substitute for some of the excellent texts covering in much more detail specific classes of liquids, like simple liquids, ionic liquids, liquid crystals, polymer solutions or colloidal dispersions, but refers the reader to such specialized monographs under suggestions for 'further reading' at the end of every chapter, while remaining self-contained as far as possible.

After a general introductory chapter to the liquid state, the book spans eleven chapters grouped into four parts devoted to thermodynamics, structure and fluctuations, phase transitions, interfaces and inhomogeneous fluids, and finally dynamics. Each chapter introduces a new set of closely related concepts and theoretical methods, which are then immediately illustrated, in most chapters, by a number of specific applications. Although some of the topics in these chapters are relatively standard, and are covered in many other texts, they have been included on purpose for the sake of clarity, completeness and self-containment. Other subjects are very rarely, if ever, presented in existing books in any detail or with sufficient clarity. A small number of relatively advanced exercises are included in several chapters. Apart from the suggestions for 'further reading', which almost exclusively refer to relatively recent books, the chapters contain only very few references to original articles in the footnotes. The bibliography on liquids has grown enormously over recent years, and the references given in the text are only to a very small number of key articles, selected for their importance and their pedagogical value; they are not a prerequisite for understanding the present text.

This book grew out of advanced undergraduate and postgraduate lecture courses taught by the authors in Cambridge and Lyon, together in particular with our colleagues Mark Warner and Lydéric Bocquet. The general level of most chapters is aimed at postgraduate students wishing to learn the basic theory of simple and complex fluids. The only prerequisites are some fluency with thermodynamics and statistical mechanics (although brief reminders to these are provided in sections 2.1 and 2.2), and with basic mathematical techniques, including Fourier transformation and functional differentiation. It is hoped that academic and industrial researchers in the field may gain additional insight from the unified approach, and that the book will allow experimentalists to familiarize themselves with basic theory, without the need of excessive formalism and technicalities.

The authors are most grateful to Elisabeth Charlaix, Walter Kob, Ard Louis and Peter Holdsworth for comments, suggestions and criticisms, to Christiane Alba, Peter Bolhuis, Eckhardt Bartsch, John Crocker, Marjolein Dijkstra, David Goulding, François Lequeux, Ed Thomas, Virgile Viasnoff, David Wales for providing figures or data and giving permission to reproduce them, to Henk Lekkerkerker for providing the magnificent cover photograph, to Sue Harding for patiently typing large parts of the book, to Toby White for skilfully preparing most of the figures, and last, but not least, to Elisabeth and Martine for their patience and support.

1 An introduction to liquid matter

One of the most remarkable observations in physical sciences or, for that matter, of everyday life, is that most substances, with a well defined chemical composition, can exist in one of several states, exhibiting very different physical properties on the macroscopic scale; moreover one can transform the substance from one state (or phase) to another, simply by varying thermodynamic conditions, like temperature or pressure. In other words, a collection of N molecules, where N is typically of the order of Avogadro's number $\mathcal{N}_A$, will spontaneously assemble into macroscopic states of different symmetry and physical behaviour, depending on a limited number of thermodynamic parameters. The most common states are either solid or fluid in character, and are characterized by qualitatively different responses to an applied stress. At ambient temperatures, the solid states of matter are generally associated with the mineral world, while 'soft' matter, and in particular the liquid state, are more intimately related to life sciences. In fact it is generally accepted that life took its origin in the primordial oceans, thus underlining the importance of a full quantitative understanding of liquids. However, even for the simplest substances, there are at least two different fluid states, namely a low density 'volatile' gas (or vapour) phase, which condenses into a liquid phase of much higher density upon compression or cooling. For more complex substances, generally made up of highly anisotropic molecules or of flexible macromolecules, the liquid state itself exhibits a rich variety of structures and phases, often referred to as 'complex fluids'.

The present book deals with some of the more generic aspects and concepts of the liquid state of matter. Rather than attempting a systematic description of the many classes of liquids, this monograph intends to illustrate generic statistical concepts and theoretical tools on a number of examples, covering a wide range of structural, dynamic and phase behaviour. This introductory chapter offers general background material and basic facts of the liquid state. It is intended to provide the indispensable link between subsequent chapters devoted to more specific aspects and examples.

1.1 Fluid states of simple substances

Consider a sample of a simple, pure substance containing a large number ($N \simeq \mathcal{N}_A = 6.02 \times 10^{23}$) of identical molecules, say water or methane. At sufficiently high temperature T, and for not too high pressures P, the substance will be in its vapour or gas phase. Under ambient pressure, gases constitute a low density phase, where molecules are far apart, and undergo only occasional binary collisions, a regime well described by the ideal gas model and the Boltzmann kinetic equation. Molecular configurations are highly disordered, as signalled by a large entropy per molecule, S/N. All physical properties of this high temperature phase are invariant under arbitrary rotations and translations, i.e. the gas phase has full rotational and translational symmetry.

When the temperature is lowered, the vapour will generally condense into droplets of a much denser liquid phase, which has a greatly reduced entropy per molecule, indicating some degree of molecular order. In fact, if $v = V/N$ is the mean volume per molecule in the liquid, $\rho = 1/v$ is the number density and v_0 is the volume of one molecule, then the packing fraction

$$\phi = \frac{v_0}{v} = \rho v_0 \tag{1.1}$$

is typically of the order of 0.3–0.5 in most liquids. At such high densities neighbouring molecules almost touch, and form well defined shells of nearest neighbours around any given molecule, characteristic of short-range order. This order is, however, lost beyond a few intermolecular distances, so that liquids still preserve, in general, full rotational and translational symmetry, both locally and on macroscopic scales.

Upon further gradual cooling, the liquid samples will generally freeze into a crystalline solid phase, which is characterized by the appearance of long-range order, embodied in a periodic crystal lattice. The regularly spaced crystal planes will Bragg reflect X-rays, and the observed diffraction patterns obtained upon varying the crystal orientation in the X-ray beam allow an unambiguous characterization of the crystal structure. The appearance of long-range order leads to a reduction in symmetry: physical properties of the crystalline solid are now invariant only under a discrete set of reflections, rotations and translations constituting one of 230 possible space groups. The full rotational and translational symmetry of the fluid phases is said to be broken at the freezing transition to the low temperature crystal phase.

The scenario just described, which is generic for simple molecular systems, is summarized in the phase diagrams of figure 1.1, which represent three cuts through the surfaces in (P, T, ρ) space, bounding the gas, liquid and solid phases. It should be noted that the liquid state only occupies a relatively small portion in the three orthogonal planes, over a range of temperatures limited to $T_t \leq T \leq T_c$, where T_t denotes the triple point temperature, where all three phases coexist,

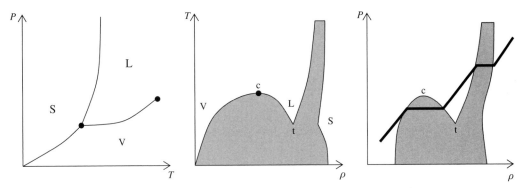

Figure 1.1. Schematic phase diagrams of a simple one-component substance, in the temperature (T)–pressure (P), density (ρ)–T and ρ–P planes. The shaded areas in the middle and right panels indicate regions of two-phase coexistence. The bold line in the right panel is an isotherm while t and c show the locations of the triple and critical points.

and T_c the critical point temperature, above which the liquid and the gas merge into a single fluid phase. The shaded areas in the (ρ, T) and (ρ, P) diagrams in figure 1.1 are two-phase regions: the corresponding thermodynamic states are either metastable or unstable, and will eventually lead to a separation into two coexisting stable phases. Examples of such states are superheated solids, supercooled liquids or supersaturated vapours. Depending on the degree of supersaturation, the latter will either form liquid droplets by a process of nucleation and growth (to be discussed in section 10.6), or undergo rapid spinodal decomposition (considered in section 9.3). Glasses constitute another particularly important class of metastable materials, often obtained by rapid cooling (or 'quenching') of a liquid well below its freezing temperature. Most substances form glasses only under rather extreme cooling conditions, but silicate melts, for instance, are easy and excellent glass-formers, as may be readily observed in the glass-blower's workshop. Glasses are amorphous ('structureless') solids, which maintain the disordered structure and rotational/translational symmetry of liquids on the molecular scale, while exhibiting the rigidity (or resistance to shear deformation) of crystalline solids at the macroscopic level, although glasses may eventually flow over extremely long time scales [1].

The distinction between the spatial arrangements of molecules in gases, liquids and solids is illustrated very schematically in figure 1.2, which shows 'snapshots' of typical configurations of disc-like molecules in two-dimensional counterparts of the three phases. A quantitative measure of the local order on the molecular

---

[1] This slow flowing is often illustrated by medieval stained-glass windows, which should tend to be thicker at the bottom. Whether this is actually the case remains however a controversial question, see the discussion by E.D. Zanotto, *Am. J. Phys.* **66**, 392 (1998).

Figure 1.2. Typical atomic
configurations in a gas
(left), liquid (middle) and
crystalline solid (right).

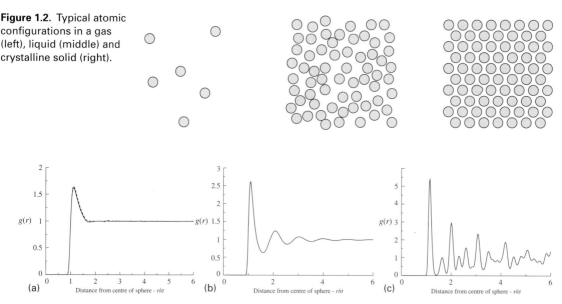

Figure 1.3. Typical pair distribution functions for (a) a gas, (b) a liquid and (c) a solid.
These functions have been generated using a molecular dynamics simulation (see
section 1.6) of atoms interacting through a Lennard-Jones potential (equation (1.9),
figure 1.4). The Boltzmann factor for the Lennard-Jones potential has been
superimposed on the gas phase distribution function (dashed curve). The
thermodynamic states are: $T = 2\epsilon / k_B$, $\rho\sigma^3 = 0.05$ (gas), $T = \epsilon / k_B$, $\rho\sigma^3 = 0.8$ (liquid),
$T = 0.2\epsilon / k_B$, $\rho\sigma^3 = 0.9$ (solid).

scale is provided by the radial (or pair) distribution function $g(r)$, which
characterizes the modulation of the local density $\rho(r)$ around a given molecule,
as a function of the distance r from that molecule.

In the limit of an ideal gas of non-interacting (point) molecules, the local
density, as seen from any one fixed molecule, is everywhere equal to its average ρ.
In reality, molecules interact via a pair potential v, which, for spherical molecules,
will only depend on the centre-to-centre distance r; even in a dilute gas, the local
density around a fixed molecule will rapidly vanish for r less than the molecular
diameter, and will be modulated by the Boltzmann factor $\exp(-\beta v(r))$, where
$\beta = 1/k_B T$. In other words

$$\rho(r) \equiv \rho g(r) = \rho \exp(-\beta v(r)) \tag{1.2}$$

In the liquid, however, the positions of neighbouring molecules are strongly
correlated, leading to a modulation of $\rho(r)$ extending over a few molecular diameters, so that the radial distribution would behave as shown schematically in
figure 1.3(b); the maxima may be associated with shells of neighbours, but the
oscillations are rapidly damped, showing the gradual smearing out of short-range

order; for $r \gg \sigma$, $g(r)$ goes to one, i.e. the local density around a fixed molecule tends rapidly to its macroscopic value ρ characteristic of a uniform (translationally invariant) fluid.

At the transition to the crystal phase the short-range order of the liquid grows spontaneously into full long-range order, signalled by well defined intermolecular distances and coordination numbers extending to macroscopic scales, which are characteristic of the periodic pattern of the crystal lattice. At zero temperature, this would reduce $g(r)$ to a sequence of δ-functions located at distances r dictated by the lattice geometry; at finite temperatures the δ-peaks are broadened by the thermal vibrations of molecules around their equilibrium lattice sites $\mathbf{R}_i$, as shown in figure 1.3(c). Note that since the crystal is anisotropic, its radial distribution function involves an average over all orientations of the lattice. An amorphous solid, however, is again isotropic, and its pair distribution function is generally difficult to distinguish from that of a liquid.

The two key macroscopic properties which distinguish liquids from their vapour, on the one hand, and fluids in general from solids, on the other hand, are cohesion and fluidity. Cohesion, shared by liquids and solids, is a consequence of intermolecular attractions, which lead to a significant lowering of the internal energy due to molecular clustering, compared to the gas phase; in the latter the thermal kinetic energy of the molecules is sufficient to overcome the short-range attraction, so that the molecules in the gas tend to occupy all the available volume. The fluidity of liquids and gases distinguishes them from the rigidity of crystalline solids. Consider the response to an applied stress (or force per unit area) characterized by the components of a macroscopic stress tensor. In the elastic regime of solids, corresponding to small strains or deformations, the components $\sigma_{\alpha\beta}$ of the stress tensor are proportional to those of the dimensionless symmetric strain tensor u, which measures the gradient of the displacement field $\mathbf{u}(\mathbf{r})$

$$u_{\alpha\beta} = \frac{1}{2}\left(\frac{\partial u_\alpha}{\partial x_\beta} + \frac{\partial u_\beta}{\partial x_\alpha}\right) \qquad \alpha, \beta = x, y, z \qquad (1.3)$$

The coefficients of the linear relation between the components of the tensors σ and u are the components of a fourth rank tensor of elastic constants. The number of independent elastic constants is strongly reduced by symmetry considerations. In particular, for an isotropic solid, like a glass or a polycrystalline sample, the stress–strain relation reduces to

$$\sigma_{\alpha\beta} = B\delta_{\alpha\beta}u_{\zeta\zeta} + 2G\left(u_{\alpha\beta} - \frac{1}{3}\delta_{\alpha\beta}u_{\zeta\zeta}\right) \qquad (1.4)$$

where B and G are the bulk and shear moduli, δ refers to the usual Kronecker symbol, and the Einstein convention of summation over repeated indices has been adopted. In particular, for a shear stress, the relation reads

$$\sigma_{xy} = 2Gu_{xy} \qquad (1.5)$$

In fluids, however, the application of an external stress will result in flow characterized by a fluid velocity field $\mathbf{v}(\mathbf{r})$, rather than by an elastic displacement field $\mathbf{u}(\mathbf{r})$. The phenomenological linear relationship valid for the most common Newtonian fluids is now between the stress tensor and the rate of strain tensor

$$\gamma_{\alpha\beta} = \frac{1}{2}\left(\frac{\partial v_\alpha}{\partial x_\beta} + \frac{\partial v_\beta}{\partial x_\alpha}\right) \tag{1.6}$$

Since the velocity field is the time derivative of the displacement field, $\gamma_{\alpha\beta}$ has the dimension of inverse time or frequency. In the case of a shear stress, the elastic relation (1.5) is now replaced by

$$\sigma_{xy} = 2\eta\gamma_{xy} \tag{1.7}$$

where η is the shear viscosity, which characterizes the internal friction of the fluid. In contrast to the elastic response of a solid, the viscous flow in a liquid dissipates energy. In water, at room temperature, $\eta = 10^{-2}$ poise (1 poise = 0.1 kg/m s). On the molecular scale, the distinction between the rigidity of solids and the fluidity of liquids reflects itself in molecular diffusion. While in a solid, molecules remain localized in the vicinity of equilibrium lattice positions, except for occasional very rare jump events between neighbouring sites, the same molecules will gradually drift away from their initial positions in a liquid. The mean square displacement at time t of a molecule from its initial ($t = 0$) position is characterized, for times long compared to molecular time scales (which are typically of the order of femtoseconds), by Einstein's relation

$$\langle |\mathbf{r(t)} - \mathbf{r(0)}|^2 \rangle = 6Dt \qquad t \gg \tau \tag{1.8}$$

where the angular brackets denote a statistical average over initial conditions and D is the self-diffusion coefficient. For simple liquids, D is of the order of $10^{-9}\,\mathrm{m^2/s}$ while in solids, D is several orders of magnitude smaller. Molecular diffusion will be studied in more detail in subsequent chapters. It is worth stressing already at this stage that, contrarily to viscous flow, molecular diffusion occurs spontaneously, in the absence of any externally applied stress, and is a consequence of thermal fluctuations and a signature of irreversible behaviour on mesoscopic and macroscopic scales. To conclude this introductory section it is important to underline that the perfect elastic behaviour of solids, embodied e.g. in equation (1.5), and the ideal Newtonian behaviour of liquids, described by equation (1.7), constitute ideal limits, and that many materials, either solid or liquid, exhibit macroscopic behaviour which may deviate very significantly from the above simple phenomenological laws. This is particularly true of complex fluids and soft condensed matter, as will become clearer in the course of this presentation.

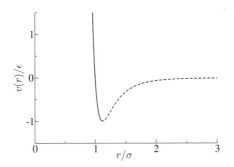

Figure 1.4. The Lennard-Jones potential, equation (1.9), as a function of interatomic distance.

1.2 From simple to complex fluids

The simple phase diagrams shown in figure 1.1, and the generic fluid behaviour sketched in section 1.1, are typical of 'simple' materials or substances composed of a single species of small, quasi-spherical molecules, say argon (Ar), nitrogen (N_2), methane (CH_4) or ammonia (NH_3). Argon atoms are strictly spherical and interact via a short-range repulsion, originating in the Pauli principle which opposes the overlap of electronic orbitals on neighbouring atoms, and longer ranged van der Waals–London dispersion forces, which are attractive; the dominant dispersion interaction decays like $1/r^6$. A convenient, semi-empirical representation incorporating these essential repulsive and attractive contributions is provided by the Lennard-Jones potential, shown in figure 1.4

$$v(r) = 4\epsilon \left(\left(\frac{\sigma}{r} \right)^{12} - \left(\frac{\sigma}{r} \right)^6 \right) \tag{1.9}$$

where σ is the atomic diameter, and ϵ the depth of the attractive well; these two parameters are generally determined by fitting certain properties derived from the potential, like the second virial coefficient, to experimental data. For Ar, $\sigma \simeq 0.34$ nm and $\epsilon/k_B \simeq 120$ K [2]. Note that the simple $1/r^{12}$ form of the Lennard-Jones repulsion is for convenience; an exponential repulsion would be more realistic, but lacks the simplicity of an inverse power law.

The potential energy of interaction between two small polyatomic molecules is generally split into v^2 spherically symmetric pair interactions between v sites (e.g. the atomic nuclei) associated with each molecule. For highly polar molecules, the interaction sites carry electric charges, chosen such as to reproduce the known multipole moments of the molecule. An example illustrated in figure 1.5 is the simple point charge (SPC) model for water, involving $v = 3$ sites on each molecule.

An alternative, useful for quasi-spherical molecules, is to supplement a spherically symmetric potential, like the Lennard-Jones potential (1.9), with the

---

[2] Interaction energies are often expressed in temperature units, which gives a simple order of magnitude estimate for the triple point temperature.

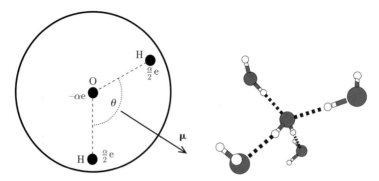

Figure 1.5. The simple point charge model of water (left): the oxygen atom carries an electric charge $-\alpha e$, and is the centre of a Lennard-Jones potential (cf. equation (1.9)); each hydrogen atom carries a charge $\alpha e/2$, with $\alpha = 0.8476$, and is situated at a distance $d = 0.1$ nm from the O atom; the HOH bond angle is 109.5°; this charge distribution gives rise to a dipole moment of magnitude $\mu = 2.3$ debye (H.J.C. Berendsen, J.R. Grigera and T.P. Straatsma, *J. Phys. Chem.* **91**, 6269 (1987)). The right-hand frame shows a typical hydrogen-bonded tetrahedral configuration of four molecules around a central water molecule. (Courtesy of D. Wales.)

anisotropic potential energy of electrostatic point multipoles. Molecules lacking a centre of symmetry, like NH_3 or CO, carry an electric dipole $\boldsymbol{\mu}$. The potential energy of two point dipoles $\boldsymbol{\mu}_1$ and $\boldsymbol{\mu}_2$ placed at a relative position $\mathbf{r} = \mathbf{r}_{12}$ is

$$v(\boldsymbol{\mu}_1, \boldsymbol{\mu}_2, \mathbf{r}) = -\sum_{\alpha,\beta} \mu_{1\alpha} T_{\alpha\beta}(\mathbf{r})\mu_{2\beta} \tag{1.10}$$

where T is the dipole–dipole interaction tensor, with components

$$T_{\alpha\beta} = \nabla_\alpha \nabla_\beta \left(\frac{1}{4\pi\epsilon_0 r}\right) = \frac{1}{4\pi\epsilon_0 r^3}\left(\frac{r_\alpha r_\beta}{r^2} - \delta_{\alpha\beta}\right) \tag{1.11}$$

The dipole moment of a small molecule is of the order of the product of the elementary (proton) charge by a typical intramolecular length scale, say 0.1 nm, i.e. 1.6×10^{-29} C m in standard units. A convenient unit is the debye $(1\,\text{D} = (1/3)10^{-29}\,\text{C m})$; typical values for isolated molecules are $\mu = 1.08$ D for HCl, $\mu = 1.47$ D for NH_3 and $\mu = 1.85$ D for H_2O. The resulting dipolar interaction energy between two molecules is strong, $v/k_B \sim 10^4$ K for separations comparable to the molecular diameter; this is significantly larger than the van der Waals dispersion interaction. The dipolar interaction can be attractive (e.g. for head-to-tail configurations) or repulsive, depending on the mutual orientations of the vectors $\boldsymbol{\mu}_1$, $\boldsymbol{\mu}_2$ and $\mathbf{r}$. Dipolar interactions play a dominant part in some colloidal dispersions called ferrofluids, where colloidal particles with diameters $\sigma \simeq 10$ nm carry *magnetic* dipole moments [3]; the strong dipolar interactions lead

---

[3] For magnetic dipole moments the $1/4\pi\epsilon_0$ in equation (1.11) has to be replaced by its magnetic counterpart μ_0.

then to alignment of dipoles and chain formation, giving rise to very peculiar rheological properties. On the molecular scale, a particular form of chain or network formation is observed for some strongly polar molecules involving H^+ groups, the most prominent examples being HF, H_2O and H_3N. Because of the small size of the H atom, the electronegative atoms F, O or N of neighbouring molecules can approach very close to H^+ groups of a given molecule, such that the approximate point dipole picture to describe the charge distribution on the molecules breaks down. The true Coulomb interaction between two such molecules is stronger and highly directional, leading to the formation of hydrogen bonds with 'bond' energies intermediate between the usual intermolecular energies, and chemical bond energies (which are of the order of a few electronvolts). In the most important case of water, the intermolecular H-bond energy of a water dimer is typically of the order of $\epsilon_H/k_B = -3000$ K, i.e. much larger than thermal energies. It is also significantly larger, in absolute value, than the energy predicted by the point dipole approximation. The H-bond is highly directional, with the lowest energy achieved when the chemical (intramolecular) OH bond is aligned with the H–O 'bond' between the H on one molecule and the O on the neighbouring one; the corresponding optimal OH–O distance of the dimer is about 0.28 nm. This strong directionality, and the fact that the intramolecular HOH bond angle of water is 105°, very close to the tetrahedral value of 109°, leads to a network-like organization of H_2O molecules in ice, whereby each O atom is linked to four O atoms of neighbouring molecules via four H-bonds, as illustrated in figure 1.5. This characteristic network, which implies a low coordination number of 4, essentially survives upon melting, at least locally, and explains many of the unusual properties of liquid water [4]. The example of the H-bond network in crystalline or liquid water shows that as simple a molecule as H_2O can already give rise to complex behaviour. More generally complex behaviour of liquid matter may often be traced back to one or several of the following characteristics of a given substance or material.

- Multicomponent systems, including mixtures of several molecular species and solutions, may give rise to compositional ordering or disordering and hence to fluid–fluid phase separation or demixing. The corresponding phase diagrams, which now involve the additional thermodynamic concentration variables characterizing the chemical composition, can rapidly become very complicated. In particular there may be a competition between condensation and demixing. However the step from pure, one-component simple fluids to mixtures of simple molecular species does not involve any new fundamental concepts. This is no longer true of mixtures involving charged (ionic) species, which are generally referred to as solutions.
- Ionic fluids involve at least two chemical species carrying electric charges of opposite sign, to ensure overall charge neutrality. The corresponding attractive and repulsive

---

[4] For example, the fact that the viscosity of liquid water *decreases* upon application of pressure, which disrupts the H-bond network and makes the system more fluid.

Coulomb interactions are of infinite $(1/r)$ range, and give rise to highly collective behaviour like the Debye screening of the bare Coulomb forces, which will be discussed in section 3.10. The simplest ionic fluids are molten monovalent salts, which involve only oppositely charged anions (e.g. Cl^-) and cations (e.g. Na^+). The high melting temperature of most salts (e.g. $T_m = 1073$ K for NaCl) reflects the strength of the bare Coulomb attraction $e^2/4\pi\epsilon_0 r$ of an anion–cation pair. When a salt, or more generally an electrolyte, is dissolved in water, the anions and cations dissociate because their Coulomb interaction is reduced by a factor ϵ, where ϵ is the dielectric constant of water ($\epsilon \simeq 78$ at room temperature). The gain in entropy due to dissociation more than compensates for the reduction in the electrostatic attraction. Ionic solutions play a vital role in many chemical and biological processes. The strong coupling between ionic charges and the water dipoles leads to the hydration of individual ions by long-lived shells of water molecules. The strong electric field of the ions, or of strongly polar molecules, leads to polarization effects responsible for many-body induction forces. Microscopic ions also form electric double-layers near the highly charged surfaces of colloidal particles or other mesoscopic objects, which will be examined in section 7.6. Metals are another class of ionic systems, where the role of the anion is played by degenerate valence (or conduction) electrons. In many situations the metallic cations (e.g. Na^+, Ca^{2+} or Al^{3+}) and the neutralizing fraction of polarized valence electrons surrounding each cation, may be treated as 'pseudoatoms' interacting via short-range (screened), density-dependent pair potentials. The problem of liquid metal structure and thermodynamics is then similar to that of simple atomic liquids, apart from some additional twists due to the valence electron component, in particular as regards electronic transport properties (electric and thermal conductivities).

- Mesogenic substances. As long as it is not too far from spherical, molecular shape generally has no qualitative influence on phase behaviour of substances. However, sufficiently elongated (rod-like) or flat (plate-like) molecules may lead to additional anisotropic fluid phases, called liquid crystal phases. Such molecules are called meso-genic, and the new anisotropic phases, which are in some sense 'intermediate' between the isotropic liquid and the fully periodic crystal, are referred to as mesophases. The anisotropy gives rise to optical birefringence, as well as to peculiar elastic behaviour of liquid crystal materials. Some of the more common mesophases are shown schemat-ically in figure 1.6. For rod-like molecules, these include the nematic, smectic A, smectic C and cholesteric phases, while plate-like molecules can form nematic and columnar phases. In the nematic phase, molecules are preferentially aligned along one direction, embodied in a unit vector $\mathbf{n}$, called the director, thus breaking the rotational invariance of the isotropic phase, while translational invariance of the centres of mass is preserved. The director is an *order parameter* which plays a key role in the study of phase transitions (cf. chapter 4). Smectic (or lamellar) phases are characterized by orientational order, and the rods are organized in layers, with a one-dimensional period-icity in the direction perpendicular to the layers, thus breaking translational invariance in that direction; translational invariance is preserved inside each layer, along the two remaining directions. The distinction between smectic A and C phases is that in the former the director is along the normal to the layers, while it is tilted in the latter. In the cholesteric phase, the director rotates along an axis, thus leading to a helicoidal

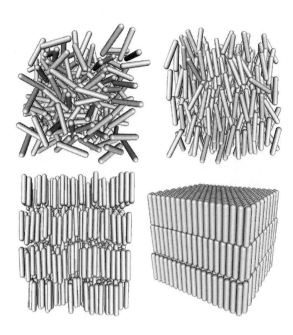

Figure 1.6. Typical configurations of the isotropic (upper left), nematic (upper right), smectic (lower right) and crystalline phases of a system of mesogenic molecules modelled as hard spherocylinders (i.e. cylinders capped by hemispheres of the same diameter). (Courtesy of P. Bolhuis.)

arrangement of rod-like molecules, with a pitch of typically a few hundred nm. Finally, in the columnar phase, flat (discotic) molecules are organized in parallel stacks or columns, the traces of which are ordered on a periodic two-dimensional hexagonal lattice, while the positions of the centres of the discotic molecules are disordered along the axes of the columns. The spontaneous shape-induced partial ordering of mesogenic molecules (see section 4.5) thus leads to much richer phase behaviour than shown in the standard phase diagrams in figure 1.1.

- Macromolecular systems. Whilst the mesogenic molecules considered above are fairly rigid, macromolecules of very high molecular weight (10^5 g/mol or more) are highly flexible objects. Linear polymers are chemically bonded chains of thousands of monomers, which can take on enormous numbers of different intramolecular conformations not unlike the many possible trajectories of an N-step random walk. A single long polymer chain is thus in itself a statistical object, a 'coil' which exhibits self-similarity, or scale invariance. This new type of symmetry means that macromolecular conformations are invariant under dilation, or change of scale, as illustrated in figure 1.7; this in turn implies that many characteristic properties of polymer chains may be expressed as simple power laws of the number N of monomers, provided $N \gg 1$. Simple models of polymer chains will be discussed in section 1.5. The study of a single polymer chain is relevant for the description of very dilute polymer solutions. Collective behaviour sets in in semi-dilute solutions, when neighbouring polymer coils begin to overlap, and even more in concentrated solutions or in solvent-free polymer melts, which exhibit complex dynamics. Chemical or physical cross-linking between different polymer chains can lead to the formation of a space-filling network called a gel, with peculiar viscoelastic properties intermediate between the elastic behaviour of a solid and the viscous flow of a simple liquid. The wealth of conformational,

Figure 1.7. A typical
conformation of a
polymer coil, with two
successive magnifications
(cf. circular insets),
illustrating scale
invariance.

structural and dynamic behaviour is enhanced when one considers branched polymers, copolymers made up of several types of monomeric units, or polyelectrolytes,
where monomers carry electric charges. Important examples are proteins, made up
of random assemblies of 20 different amino-acid residues, or DNA, whose famous
double-helix structure is an example of a particularly 'stiff' polyelectrolyte. These
two macromolecules are key components of living cells, and significant ingredients of
biological complexity.

- Self-assembly. Some large molecules have multiple chemical functionalities which
  strongly favour the spontaneous formation of supramolecular aggregates, a phenomenon often referred to as self-assembly. A first example are diblock copolymers,
  made up of two incompatible polymer chains, say A and B, linked together by a
  chemical bond. Since A and B chains do not mix, a solution or melt of diblock copolymers will lead to local phase separation into A-rich and B-rich microdomains, organized such as to minimize the contact between A and B segments. The second
  very important example is provided by amphiphilic molecules, or surfactants, like
  lipids made up of a polar head-group and one or several hydrocarbon chains. The
  head-group is hydrophilic, while the hydrocarbon tails are hydrophobic. When dissolved in water, such amphiphilic molecules will spontaneously form supramolecular aggregates called micelles, i.e. quasi-spherical assemblies with the hydrophilic
  heads at the surface, in contact with the solvent, while the hydrocarbon chains avoid
  being exposed to the water by clustering inside the micelle. The thermodynamics
  of micelle formation will be examined in section 2.6. Depending on their shape,
  lipids will minimize the contact between their hydrocarbon tails and water not by
  forming micelles, but rather by forming lipid bilayers, which are lamellar structures,

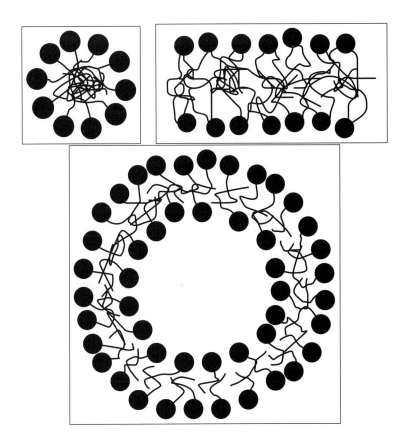

Figure 1.8. Schematic configurations of amphiphilic (surfactant) molecules, self-assembled into a micelle (upper left frame), a lamellar bilayer (upper right frame), and a vesicle (lower frame). The hydrophilic head-groups are represented by black circles, while the wiggly lines represent the hydrophobic tails; note that in the bilayer, each surfactant molecule has two tails.

as shown in figure 1.8. Contact of the hydrophobic alkane chains with water is again avoided by a spontaneous alignment of the tails inside the bilayer, while the polar heads point towards the solvent on both sides. The resulting flexible thin films are the simplest examples of biological membranes, the curvature elasticity of which is controlled by bending moduli (section 8.2). Depending on lipid concentration, and the energy cost of curvature, edge effects of planar bilayers may be avoided by the spontaneous formation of closed surfaces called vesicles, which are prototypes of living cells. The morphology of vesicles is essentially determined by their surface to volume ratio. Surfactant molecules will also spontaneously aggregate at the interface between immiscible water and oil, where they will tend to form monolayers, with the polar heads pointing to the water side and the hydrophobic alkane chain tails intruding into the oil phase. The surfactant molecules in fact reduce the strong surface tension of the liquid–oil interface, thus favouring the stability of microemulsions, which are mesoscopic two-phase systems of finely divided droplets of oil in water, or of water in oil, depending on the concentrations of the two liquids. Upon varying the concentration of surfactant, a large variety of microphases may be generated, including smectic lamellar phases, with alternating layers of oil and water, as well as more complex structures like bicontinuous phases where oil and water form interpenetrating domains separated by fully connected monolayer membranes spanning the whole volume of the sample.

- Suspensions and dispersions are highly heterogeneous two-phase (or multi-phase) systems, of which the previously mentioned microemulsions are a good example involving two liquid phases. Fog is another example involving finely divided water droplets suspended in air. More generally, *colloids* are dispersions of mesoscopic solid or liquid particles, with sizes ranging typically from tens to thousands of nanometres, in a suspending liquid. Colloidal dispersions, like ink, paints, lubricants, cosmetics, or milk, are ubiquitous in everyday life and play a key role in many industrial processes. Due to the highly divided nature, or high degree of heterogeneity, of a colloidal dispersion, the total surface separating the two constituent phases is very large, so that colloidal dispersions are dominated by interfacial properties. For example, 1 litre of a fairly concentrated suspension of spherical colloidal particles with a diameter of 10^2 nm, and a packing fraction $\phi = 0.1$, contains 2×10^{17} particles, and the total internal surface is about 1500 m^2! A widely studied category of colloidal dispersions involves mesoscopic solid particles, which may be crystallites, like naturally occurring silicates, or polymeric amorphous particles, like synthetic polystyrene or latex spheres, or small biological organisms, like viruses. Such solid colloids come in different shapes: spherical like the synthetic latexes, lamellar like clay platelets, or long rods, of which the tobacco mosaic virus (TMV) is a good example. It is then tempting to seek analogies between large assemblies of microscopic molecules forming various phases discussed earlier in this chapter, and suspensions of mesoscopic colloidal particles in a liquid. And indeed, colloidal dispersions are observed to give rise to various phases, including colloidal crystals and, in the case of rod-like particles, nematic phases, embedded in the suspending fluid. Quite apart from the difference in scale there are, however, a number of rather fundamental differences between molecular and colloidal assemblies. First of all, contrarily to molecules of a given species, which are all identical, colloidal particles have a size distribution, i.e. they are *polydisperse*; careful synthesis can limit relative differences in size to a few per cent in favourable cases, but size polydispersity is an intrinsic property of colloidal dispersions. A second, fundamental difference is the nature of particle dynamics: while the motions of interacting molecules are governed by reversible Newtonian dynamics, colloidal particles undergo irreversible Brownian dynamics due to their coupling to the suspending fluid; due to the very different time scales between the rapid thermal motions of the molecules, and the considerably slower motion of the much heavier colloidal (or Brownian) particles, this coupling can only be described in stochastic terms, as in the framework of the Langevin equation examined in section 10.1. A final, very significant feature of colloidal systems is that within a coarse-grained statistical description, where microscopic degrees of freedom are integrated out, the interaction between colloidal particles is an effective one, including an entropic component. In other words, the effective interaction energy between colloidal particles in a bath of microscopic molecules and ions, or in the presence of macromolecules, is a free energy, rather than a mere potential energy, and hence generally depends on the thermodynamic state variables (temperature, concentrations, etc...) of the suspension. Examples of such effective interactions include depletion forces, to be discussed in section 2.7, and the interactions between electric double-layers, associated with charged colloidal particles (cf. section 7.6). Finally, flow properties of concentrated dispersions generally deviate very significantly from the simple Newtonian behaviour of simple fluids, as embodied e.g. in equation (1.7) for a simple shear flow. The study of such non-Newtonian flow is the realm of rheology, and its many technological implications.

1.3 Exploring the liquid state

The present book describes some of the key concepts and models which have been developed to understand and predict the macroscopic and microscopic properties of liquids and their interfaces, in particular their structure and dynamical behaviour on molecular or mesoscopic scales. Ultimately the validity of theoretical predictions must be gauged against experimental measurements of such properties. Although later chapters will not dwell on experimental techniques, it is important to mention, at least briefly, some of the more widely used techniques, and the physical quantities to which they give access. Broadly speaking, the properties that are most frequently measured, and the corresponding experimental probes, fall into four categories, two of which are of macroscopic nature, while the other two relate more directly to the molecular or macromolecular constituents, on microscopic or mesoscopic scales.

- Thermodynamics characterizes macroscopic samples in equilibrium. The key bulk thermodynamic properties are the equation of state, compressibility and specific heat (at constant volume or pressure); they involve the measurement of the fluid density as a function of applied pressure, as well as various calorimetric techniques which also give access to latent heats at phase transitions. Free energies follow from thermodynamic integration, using basic thermodynamic relations which will be summarized in chapter 2. Solutions are characterized by their osmotic properties. In particular, the osmotic pressure Π exerted by the solute is traditionally measured by monitoring the rise of the solution in a vertical tube in osmotic equilibrium with the pure solvent across a semi-permeable membrane which allows the exchange of solvent molecules, but not of the solute. If h is the height to which the solution rises in the capillary, ρ_s the mass density of the solution, and g the acceleration of gravity, the osmotic pressure is $\Pi = \rho_s g h$. Osmometry is a direct application of osmotic equilibrium which allows the determination of the molar mass of macromolecules by measuring the osmotic pressure of a macromolecular solution.

  The static dielectric constant (or permittivity) ϵ is another important macroscopic property of a fluid, which determines the response of a sample, in the form of the electric polarization $\mathbf{P}$, to an externally applied electric field, or corresponding displacement field $\mathbf{D}$. The field $\mathbf{E}$ inside the sample is related to $\mathbf{D}$ via the relation

$$\mathbf{D} = \epsilon_0 \mathbf{E} + \mathbf{P} = \epsilon \epsilon_0 \mathbf{E} \tag{1.12}$$

where the second equality is valid in the linear regime. ϵ is a scalar if the fluid sample is isotropic, and ϵ_0 is the permittivity of empty space. ϵ is thus directly determined by measuring the electric field inside a condenser in the absence and in the presence of the fluid sample. The dielectric constant is state dependent, and a high value will ensure strong dissociation of electrolytes in highly polar solvents, like water. Another important application of equation (1.12) is the propagation of light, since ϵ (which depends on the frequency) is related to the index of refraction. In anisotropic, complex fluids, birefingence (i.e. the dependence of the index of refraction on the polarization of light) yields important information on molecular orientations.

Interfacial phenomena are dominated by the surface tension, a force per unit length (or energy per unit area), which may be measured by a number of techniques, based on the Young–Laplace equation for the pressure drop across a curved interface (cf. section 6.4)

$$\Delta P = \gamma \left(\frac{1}{R_1} + \frac{1}{R_2} \right) \tag{1.13}$$

where R_1 and R_2 are the curvature radii of the dividing surface (see section 6.2). A classic technique is based on the measurement of the angle made by the meniscus of the liquid–gas interface in a capillary rise experiment. Related methods are based on digital image analysis of the shape of drops or bubbles under gravity (sessile drop or bubble method). The thickness of interfaces or adsorbed thin liquid films is routinely measured by ellipsometric methods analysing the elliptical polarization of light beams reflected by interfaces or films. Similar information may be derived from X-ray reflectivity measurements.

- Macroscopic transport coefficients determine the dissipative (irreversible) behaviour and the flow properties of fluids. The thermal and electrical conductivities determine the heat and charge currents through a fluid subjected to gradients in temperature or electric potential, while the shear and bulk viscosities, η and ζ, characterize the internal friction in a flowing Newtonian fluid, obeying a linear relationship between stress and strain rate (cf. equation (1.7)). Standard viscosimeters measure η, either by monitoring fluid flow through capillaries or by using rotating coaxial cylinders; in the latter experiment, the viscosity of a fluid placed between an outer and an inner cylinder is directly determined by measuring the torque exerted on the inner cylinder, when the outer cylinder is rotated at constant angular velocity. A combination of η and ζ determines the attenuation of sound waves propagating though a liquid (cf. section 11.5). The vast field of rheology explores the flow properties of complex fluids, including polymer solutions and melts, and colloidal dispersions or various other suspensions. Such fluids are generally non-Newtonian, i.e. the stress versus strain rate relation is no longer linear. In particular, Bingham fluids flow only beyond a threshold of applied stress (the yield stress). Suspensions often exhibit thixotropy, i.e. the reduction of viscosity at higher flow rates ('shear thinning'). A well known illustration is the reduction of the viscosity of paints by brushing. Despite its obvious technological importance, the rheological behaviour of complex fluids will not be considered further in this book.
- Shifting the attention to mesoscopic and molecular scales, a key objective is to characterize the static local structure of fluids, i.e. the average spatial organization of the basic constituents (molecules, macromolecules or supramolecular aggregates). The spatial arrangement may be described by a set of static correlation functions, of which the pair distribution function $g(r)$, already introduced in section 1.1 (cf. equation (1.2))

is the simplest and most studied. In fact, as will be shown in more detail in section 3.8, the spatial Fourier transform of $g(r)$, called the static structure factor $S(k)$ (where the wavenumber k is the variable conjugate to r), is directly accessible to radiation diffraction experiments. If molecular scales (of the order of a nanometre or less) are to be probed, the natural radiations are X-rays or thermal neutrons from reactors or spallation sources, because their wavelength is of the order of angstroms. The availability of very intense X-ray sources from synchrotron radiation makes it possible to obtain time-resolved diffraction patterns, allowing e.g. the evolution of non-equilibrium and metastable structures (like those of glassy materials) to be monitored on time scales of a fraction of a second. If mesoscopic structures are to be explored, as in colloidal suspensions, X-rays and neutrons may still be used, provided that the diffractometers can resolve very small wavenumbers; as will become clear in section 3.8, this corresponds to the regime of small angle scattering. Alternatively one may use radiations of longer wavelengths, like visible light which is diffracted by assemblies of particles of colloidal size, or one can resort to direct visualization using a powerful microscope; this is combined with video recordings in the widely used technique of video microscopy. Diffraction measurements are generally applied to local order determinations of bulk fluids, on the assumption of homogeneity or translational invariance. Inhomogeneous fluids and interfacial regions break translational invariance, and are hence characterized by a spatially varying local density (or density profile), rather than by a mere constant bulk number density ρ. Such density profiles $\rho(\mathbf{r})$ are accessible in X-ray or neutron reflectivity experiments, and by grazing incidence X-ray diffraction. The interfacial structure, and in particular the layering of molecules in strongly confined fluids, e.g. near a solid wall, may also be measured by a surface force apparatus, capable of measuring forces between plates induced by very thin fluid films, on the nanonewton scale, with a spatial resolution of the order of an angstrom. The same apparatus is also well adapted to investigate lubrication forces and capillary condensation, as well as various aspects of wetting phenomena (sections 6.3, 6.4).

• Experimental investigations of individual or collective motions of molecules or particles require the use of dynamical, time or frequency-dependent probes. One of the most widely used dynamical diagnostics is inelastic scattering of neutrons or photons. Since thermal neutrons have energies comparable to the kinetic energy of molecules in fluids, the inelastic scattering cross-section contains detailed information on molecular motions (cf. section 11.6). Inelastic or quasi-elastic scattering of light probes length scales of the order of the wavelength of light, and is hence well adapted to examine collective dynamical fluctuations in molecular fluids (as in Rayleigh–Brillouin scattering), or the motion of mesoscopic colloidal particles. Dynamics of the latter may also be explored by photon correlation spectroscopy, an interferometric method well adapted to slow relaxation processes, typically in the range of milliseconds. Slow diffusion of molecules and macromolecules, on the scale of microseconds, can be resolved by nuclear magnetic resonance (NMR, section 11.7), while reorientational motions of molecular dipoles are conveniently measured by dielectric relaxation (section 10.3). Detailed information of how intermolecular forces affect the rotational and vibrational motions of molecules in liquids, as compared to their gas-phase behaviour, may be gained from various spectroscopic techniques, including infrared (IR) and Raman spectroscopy, which are beyond the scope of this presentation.

To conclude this introductory chapter it seems worthwhile to illustrate two of the key concepts in complex fluids, namely free volume and scale invariance, by simple examples, before embarking on a more systematic presentation in the following chapters. As a third example, we shall briefly show how computer simulations of atomistic models can provide extremely valuable information on the structure, phase behaviour and dynamics of simple and complex fluids, supplementing the data obtained from experimental probes.

1.4   Application 1: excluded and free volume

The single most important feature of the interaction between molecules or colloidal particles is their strong mutual repulsion whenever their centres come within a distance of the order of the particle diameter. For non-spherical particles, this shape-dependent repulsion will obviously be a function of the mutual orientation of the molecules, and this will be a very important factor e.g. in the case of liquid crystals. For the sake of clarity, we shall first restrict the discussion to spherical particles of diameter $\sigma = 2R$. In view of the steepness of the repulsive potential between two such particles, the latter may very often, to a good approximation, be regarded as non-interpenetrating hard spheres, such that the interaction 'potential' is simply $v(r) = \infty, r < \sigma$ and $v(r) = 0, r > \sigma$. The corresponding Boltzmann factor reduces to a Heaviside step function

$$\exp(-\beta v(r)) = \theta(r - \sigma) \tag{1.14}$$

where $\theta(x) = 1(0)$ if $x > 0(< 0)$. Note that the Boltzmann factor for a pair of hard spheres is independent of temperature; this is a direct consequence of the absence of any energy scale in a hard sphere fluid, and is a common characteristic of athermal systems made up of particles having only excluded volume interactions. In the case of hard spheres, the excluded volume around the centre of any particle is that of a sphere centred on that particle and of radius equal to the particle diameter σ, so that the excluded volume $v_{ex} = 4\pi\sigma^3/3$ is eight times the volume $v_0 = 4\pi R^3/3$ of the particle itself. Each particle has such an exclusion sphere associated with it, as shown in the equivalent two-dimensional situation in figure 1.9. The centre of any additional (or test) particle inserted into the fluid cannot come closer than σ to the centres of the existing spheres, and hence cannot penetrate into the exclusion spheres around the latter. It is important to realize that, while the excluded volume associated with each individual sphere is v_{ex}, the total volume V_{ex} from which the centre of a test sphere is excluded is less than $N \times v_{ex}$, since the exclusion spheres of neighbouring particles can overlap, as illustrated in figure 1.9. The volume accessible to a test particle is simply the total volume of the system minus the excluded volume, $V' = V - V_{ex}$. V_{ex}, and hence V', depend on the instantaneous configuration of the N non-overlapping particles; V_{ex} fluctuates as the positions $\mathbf{r}_i$ $(1 < i < N)$ of the spheres vary in

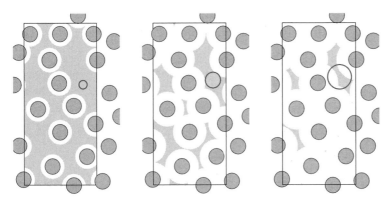

Figure 1.9. Schematic two-dimensional representation of free (or accessible) 'volume' (area in this case) for typical fluid-like configurations of hard discs. The test particle is shown as a heavy-lined circle; the hard discs are the dark grey circles; the concentric white areas represent the excluded area around each disc. As the diameter of the test particle increases (from left to right), so does the diameter of each exclusion circle; the light grey area represents the free area, accessible to the centre of the test particle without causing overlap with any of the discs; the free area is shown only within the rectangular frames. (Courtesy of D. Goulding.)

time, due to their thermal motion. The value of the accessible volume V' averaged over all allowed configurations of spheres is called the free volume. Allowed configurations are such that $|\mathbf{r}_i - \mathbf{r}_j| > \sigma$ for all $N(N - 1)/2$ pairs of particles. Bearing in mind that the test particle is identical to the spheres in the system, it will be shown later (section 2.5) that V_{ex} is intimately related to the chemical potential of a hard sphere fluid. Meanwhile we consider two limiting cases. At very low densities, the volume per particle $v \gg v_{\mathrm{ex}}$, so that the overlap of the exclusion volumes associated with different particles is very unlikely. In that case the volume accessible to the test particle, which may be identified with any one of the spheres, is simply $V' = V - Nv_{\mathrm{ex}}$. The phase space volume for one sphere is $\omega = (V - Nv_{\mathrm{ex}})/\Lambda^3$, where

$$\Lambda = h/\sqrt{2\pi m k_\mathrm{B} T} \tag{1.15}$$

is the de Broglie thermal wavelength, stemming from the integration over a Maxwell distribution of momenta for particles of mass m, and h is Planck's constant. The total phase space volume, or partition function, for hard spheres is then approximately given by

$$\Omega = \frac{(V - Nv_{\mathrm{ex}}/2)^N}{N! \Lambda^{3N}} \tag{1.16}$$

The factor $1/2$ occurs to avoid double-counting of the excluded volume for pairs of spheres. The entropy finally follows from Boltzmann's relation

$$S = k_\mathrm{B} \ln \Omega = N k_\mathrm{B} \ln \left(\frac{v - v_{\mathrm{ex}}/2}{\Lambda^3} \right) \tag{1.17}$$

where Stirling's formula $\ln N! \simeq N \ln N - N$ has been used. In an athermal system, the Helmholtz free energy reduces to $-TS$, and the pressure follows from

$$P = -\left(\frac{\partial F}{\partial V}\right)_{N,T} = T\left(\frac{\partial S}{\partial V}\right)_{N,T} = \frac{k_B T}{v - v_{ex}/2} \qquad (1.18)$$

The dimensionless compressibility factor (or equation of state) reduces to

$$Z = \frac{PV}{Nk_B T} = \frac{1}{1 - 4\phi} \qquad (1.19)$$

where $\phi = \pi\sigma^3/6v$ is the packing fraction for hard spheres. Note that (1.19) can only be expected to be valid for $\phi \ll 1$, so that to lowest order in ϕ, $Z = 1 + 4\phi + \mathcal{O}(\phi^2)$; the term linear in ϕ is the leading correction to the ideal gas law in a *virial* expansion of Z in powers of the density $\rho = 1/v$. In the opposite limit of high densities, which is much more relevant for the condensed (liquid or solid) states of matter, calculation of the free volume is a much more difficult task, due to the numerous overlaps of the excluded volumes associated with each particle. On average each particle is trapped in a cage of neighbouring spheres, and the free volume per particle $v' = V'/N$ is much less than the volume per particle $v = V/N$.

The free volume v' accessible to any one particle trapped in its cage may then be calculated with a *cell model*, whereby the nearest neighbours of the particle are assumed to be fixed at some favourable average position. In a crystal, the average positions naturally coincide with the lattice positions of the first coordination shell of nearest neighbours around the trapped particle. The very concept of an average position of molecules is meaningless in fluids, since the molecules diffuse away from any initial position, according to Einstein's law (1.8). However, on a sufficiently short time scale, say 1 ps, a small molecule will have moved typically less than 0.1 nm in a dense fluid, so that the neighbours forming the cage may be considered as effectively 'frozen' for the purpose of calculating the free volume v'. For the sake of an easier graphical free volume representation, we consider in figure 1.10 the two-dimensional case of hard discs. The most compact packing is achieved by the triangular lattice, where each atom is surrounded by six nearest neighbours placed at the vertices of a hexagonal cell, at a distance $d = (2/\sqrt{3}\rho)^{1/2}$ from the centre, where ρ is the number of atoms per unit area (number density of the system). The free area available to the atom within the hexagonal cage is the shaded area in figure 1.10, resulting from the intersection of the exclusion discs of radius $\sigma = 2R$ centred on each of the six vertices. For dense fluids, d is only slightly larger than σ, and an elementary calculation leads to a simple expression for the free area, valid to second order in $(d/\sigma - 1)$, namely $a' = 2\sqrt{3}(d - \sigma)^2$. The resulting entropy is $S = Nk_B \ln[2\sqrt{3}(d - \sigma)^2/\Lambda^2]$ and

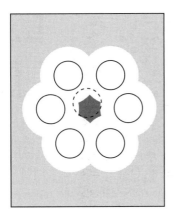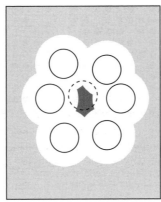

Figure 1.10. Free area (in dark grey), accessible to the centre of a disc inside the hexagonal cage of six nearest-neighbour discs fixed at the equilibrium positions of a triangular lattice (left-hand frame). The right-hand frame shows how the free area is distorted when the six discs of the cage move away from the equilibrium lattice positions in the course of thermal vibrations.

the corresponding two-dimensional equation of state reduces to:

$$Z = \frac{P}{\rho k_B T} = \frac{1}{1 - (\sigma/d)} = \frac{1}{1 - (\phi/\phi_0)^{1/2}} \tag{1.20}$$

where $\phi = \pi \rho \sigma^2/4$ is the two-dimensional packing fraction and $\phi_0 = \pi/2\sqrt{3} = 0.907$ is the value of ϕ at close packing (i.e. when $d = \sigma$). The corresponding free volume result in three dimensions would be:

$$Z = \frac{1}{1 - (\phi/\phi_0)^{1/3}} \tag{1.21}$$

where now $\phi_0 = \pi\sqrt{2}/6 = 0.740$, corresponding to close packing on a face-centred cubic (FCC) lattice. Note the completely different functional form of (1.21) compared to the low density result (1.19). In particular (1.21) diverges at close-packing, and provides a reasonable equation of state of the hard sphere solid. It provides only a very rough estimate of the pressure of the hard sphere fluid near freezing, i.e. for $\phi = 0.5$. Better agreement with 'exact' computer simulation data (to be discussed in section 1.6) for the equation of state of the hard sphere fluid is achieved upon replacing the close-packed value of ϕ_0 pertaining to a regular FCC lattice, by the value ϕ_0' appropriate for *random close-packing*. The value of ϕ_0' corresponding to a completely jammed (or 'glassy') fluid is not known exactly nor even rigorously defined. It is however believed to be of the order of 0.64, from measurements on random packings of small macroscopic steel spheres, achieved by J.D. Bernal in the early 1960s.

The main deficiency of the above free volume approximation is that it is a mean field theory which oversimplifies the effect of the strong positional correlations in dense fluids. Such excluded volume correlations will be examined in more detail in chapter 3.

A highly coarse-grained model of fluids, accounting for excluded volume effects, is the lattice gas with single occupancy constraint. In this model the total volume V is subdivided into M identical cubic cells of volume $v_0 = V/M$ of

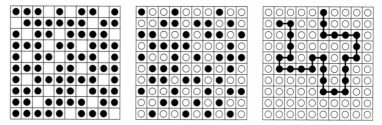

Figure 1.11. Two-dimensional lattice gas model of a one-component fluid (left), of a binary mixture (middle) and of a linear polymer in solution. The binary mixture model is incompressible, since the two species, represented by full or open circles, occupy all available cells; the same is true of the polymer in solution, where the full, connected circles represent the monomers, while the white circles represent solvent molecules.

the order of the volume $\pi\sigma^3/6$. The N molecules of the fluid are distributed over the M cells, such that each cell contains either 0 or 1 molecule; a typical configuration of a two-dimensional lattice gas is shown in figure 1.11. In this discretized representation of configuration space, the total number of independent configurations is simply

$$\Omega = \frac{M!}{N!(M-N)!} \tag{1.22}$$

Exercise: Using Boltzmann's expression (1.17) for the entropy, and Stirling's formula, show that the equation of state of the lattice gas reduces to

$$\frac{PV}{Nk_{\mathrm{B}}T} = -\frac{1}{f}\ln(1-f) \tag{1.23}$$

where f is the fraction of occupied sites, which plays the same role as the packing fraction in a continuous model. Note that the result is independent of the space dimension d.

The crude lattice gas model may be generalized in many ways, to include additional physical features, or model different physical situations. Short-range attractions between molecules may be incorporated by attributing an energy $-\epsilon$ to pairs of occupied nearest-neighbour cells. The resulting model is highly non-trivial, and may be shown to be isomorphous to an Ising spin model (cf. section 4.1); this lattice model with nearest-neighbour attraction exhibits a liquid–gas transition below a critical temperature, the lattice gas equivalent of the Curie temperature of the Ising model, which signals the transition from paramagnetic to ferromagnetic behaviour. The lattice gas may also be generalized to model binary mixtures of molecules of species A and B. In this case each of the M cells is occupied either by an A or a B particle (no empty cells), and energies are associated with nearest-neighbour cells, depending on whether they contain two A molecules, an A and a B molecule, or two B molecules. Such mixtures will be considered in more detail in section 4.1; depending on the sign of the combination $\Delta\epsilon = \epsilon_{AA} + \epsilon_{BB} - 2\epsilon_{AB}$, the lattice mixture model may exhibit a demixing transition below a critical temperature. By linking together the molecules of species A (corresponding each to a monomer), as shown in figure 1.11, one obtains a coarse-grained model for a linear polymer chain in a solvent of B molecules.

In the simplest version of the model, the polymer chain maps out a random walk on the underlying lattice. However, since monomers repel each other at short range, a more realistic model is a *self-avoiding walk* (SAW), whereby the chain is not allowed to intersect; this is the fundamental model of a polymer chain in a 'good solvent', to which we shall return in the following section, and in chapter 4.

Further generalizations of the basic lattice gas picture have been developed in recent years to model the action of surfactant (amphiphilic) molecules at the interface of two immiscible liquids (like oil and water), and to study various forms of self-assembly in such ternary systems.

Lattice gas models deal only with the configurational part of the statistical mechanics of fluids, in a highly simplified way. Such coarse-grained representations, which ignore molecular details of fluids, may only be expected to provide insight into large-scale, collective properties of simple and complex fluids, which are insensitive to the precise chemical nature of the basic constituents. This is particularly true of polymers, made up of very large numbers M of monomeric units, and which exhibit the important property of scale invariance in the limit where $M \to \infty$ (called the 'scaling limit'), a characteristic to be discussed in the following section. A similar scale invariance holds for fluids near a liquid–gas or near a demixing critical point; the critical behaviour is independent of chemical detail, and hence the highly simplified lattice gas model is particularly well suited to the study of critical phenomena.

1.5 Application 2: polymer chains and scale invariance

Polymeric liquids (polymer solutions in which the polymer is mixed with a solvent, or polymer melts in which the polymer is the only component) have unique physical properties. For example, adding a small amount of polymer to a solvent will increase the viscosity of the latter by a much larger amount than upon dissolving the same component in its monomeric form. Many properties of polymeric liquids are therefore related not to the specific chemical nature of the polymer, but rather to the fact that the monomers are *connected* to form long chains. This implies that many properties of the polymeric liquid will be determined mainly by the length of these chains, and the length scales they introduce, such as the size (radius of gyration) of a coil in the solution. For long chains, these length scales are much larger than molecular sizes. Hence, in modelling polymeric systems, some freedom will be permissible in the choice of the model at the microscopic level, provided this model gives rise to the correct length and time scales on the mesoscopic level of the whole chain. This idea underlies the so-called 'scaling' approach to the physics of polymer systems, which was already introduced in figure 1.7. In this section we present several models for polymer chains, and show how these models can be mapped onto each other by an appropriate redefinition of the microscopic parameters to give rise to the same mesoscopic quantities. We consider only ideal chains, i.e. chains for which the only interaction between monomers is the connectivity between neighbours along the chain. The role of interactions between monomers will be considered in sections 3.10 and 5.5.

The freely jointed chain

A first approach in modelling a polymer chain, with some relationship to the chemical reality, consists in assuming that it is composed of M freely jointed segments of length ℓ, the 'monomers'. The whole chain has total extension $M\ell$, and M will generally be referred to as the molecular weight. In the model, M is actually a dimensionless quantity, but it is easily mapped onto the actual molecular weight expressed in grams per mole [5]. As indicated in section 1.2, the polymer chain is a complex, statistical object. Hence, we are interested in the average value and probability distribution of global quantities such as the end to end distance $\mathbf{R}$ or the radius of gyration R_G. The latter is defined as the average distance from the centre of mass, i.e.

$$R_G^2 = \frac{1}{M} \left\langle \sum_{i=1}^{M} (\mathbf{r}_i - \mathbf{r}_{cm})^2 \right\rangle \tag{1.24}$$

where $\mathbf{r}_i$ is the position of monomer i, $\mathbf{r}_{cm} = \frac{1}{M} \sum_{i=1}^{M} \mathbf{r}_i$, and the angular brackets denote an average over all possible conformations of the chain.

Exercise: Show that

$$R_G^2 = \frac{1}{2M^2} \left\langle \sum_{i=1}^{M} \sum_{j=1}^{M} (\mathbf{r}_i - \mathbf{r}_j)^2 \right\rangle \tag{1.25}$$

The end-to-end vector is $\mathbf{R} = \mathbf{r}_M - \mathbf{r}_1 = \sum_{i=1}^{M-1} \mathbf{u}_i$, where $\mathbf{u}_i = \mathbf{r}_{i+1} - \mathbf{r}_i$ is the vector that defines the ith chain segment. Since the chain is 'freely jointed', there are no angular restrictions on the orientations of successive segments, so that $\langle \mathbf{u}_i \cdot \mathbf{u}_{i+1} \rangle = 0$. Hence $\mathbf{R}$ is the sum of $M - 1$ independent variables, each of these variables being a vector of fixed length ℓ and random orientation. The *central limit theorem* of probability theory states that the probability distribution

---

[5] For polystyrene, the appropriate factor is close to 100 g. Chains with a molecular weight 10^6 g/mol are made of typically 10 000 monomers.

of the sum of a large number of independent variables converges towards a Gaussian distribution, which is entirely defined by its average and variance. Since

$$\langle \mathbf{R} \rangle = 0 \quad \text{and} \quad \langle \mathbf{R}^2 \rangle = (M - 1)\ell^2 \tag{1.26}$$

the probability distribution of $\mathbf{R}$ follows as

$$p(\mathbf{R}) = \left(\frac{3}{2\pi(M - 1)\ell^2} \right)^{3/2} \exp\left(-\frac{3\mathbf{R}^2}{2(M - 1)\ell^2} \right) \tag{1.27}$$

A detailed numerical study shows that for $M > 10$, (1.27) provides an excellent approximation to the exact probability distribution, except of course for the largest (and very unlikely) extensions, as seen in the exercise below.

Equation (1.27) implies that the conformational entropy $S(\mathbf{R})$ associated with a given extension of the chain, is of the form

$$S(\mathbf{R}) = k_B \ln P(\mathbf{R}) = \text{constant} - \frac{3\mathbf{R}^2}{2(M - 1)\ell^2} \tag{1.28}$$

The associated free energy, $F(\mathbf{R}) = -TS(\mathbf{R})$, is therefore quadratic, similar to the energy of a Hookean spring. The large number of conformations accessible to the polymer chains is responsible for its *elasticity*, often described as *entropic elasticity*. This finding underlies the formulation of the Gaussian chain model, considered next.

Exercise: Entropic elasticity of a polymer chain, small and large extension limits.

Consider a *one-dimensional* 'freely jointed polymer chain', made of M consecutive segments of length ℓ. Each segment can take any one of the two possible orientations on the Ox axis. Show that the total number of configurations that result in an end-to-end vector $R = N\ell$ is

$$\Omega(R) = \frac{M!}{N_+!N_-!} \tag{1.29}$$

with N_+ (N_-) the number of segments oriented to the right (left). For non-interacting segments, the free energy of the system is purely entropic, $F(R) = -k_B T S(R) = -k_B T \ln \Omega(R)$. The relation between the force f and the extension R is obtained using $f = -\partial F/\partial R$. Show that this simple model results in the non-linear force–extension relation

$$f = \frac{k_B T}{2\ell} \ln \frac{1 + R/M\ell}{1 - R/M\ell} \tag{1.30}$$

Discuss the small and large extension limits of this relation. In the limit of small extensions ($R \ll M\ell$), show that the quadratic expression for the free energy and the corresponding Hookean behaviour are recovered. Note that the corresponding elastic stiffness *increases* with increasing temperature.

The Gaussian chain model

In this model, the configuration of the chain is specified by $N + 1$ monomer positions, $\mathbf{R}_1 \ldots \mathbf{R}_N$ [6], and the energy associated with a given configuration is

$$H = \frac{3k_B T}{2b^2} \sum_{i=1}^{N-1} (\mathbf{R}_{i+1} - \mathbf{R}_i)^2 \tag{1.31}$$

where b has the dimension of a length. As is obvious from the explicit temperature factor, H is actually a free energy rather than an energy, and the correct interpretation of equation (1.31) is that the statistical weight of a configuration $\mathbf{R}_i$ is

$$W(\mathbf{R}_1, \ldots, \mathbf{R}_N) = \left(\frac{3}{2\pi b^2} \right)^{3N-1/2} \exp\left(-\frac{H}{k_B T} \right) \tag{1.32}$$

In this model, all the segments $\mathbf{R}_{i+1} - \mathbf{R}_i$ are independent Gaussian variables. A simple calculation shows that the end-to-end distance and radius of gyration are given, respectively, by

$$\langle \mathbf{R}^2 \rangle = (N - 1)b^2 \qquad R_G^2 = (N - 1)b^2/6 \tag{1.33}$$

Obviously this Gaussian model [7] is logically interpreted as a coarse-grained version of the freely jointed model. Each Gaussian monomer represents a subchain of $M_0 \geq 10$ freely jointed segments, and the two models will lead to identical results for the end-to-end distance and its probability distribution provided

$$N = M/M_0 \qquad b^2 = M_0 \ell^2 \tag{1.34}$$

Equation (1.34) illustrates the philosophy behind the physical modelling of polymers. The 'microscopic' parameters M and ℓ, or N and b^2, are phenomenological parameters of a model. The model is used to determine the large-scale quantities characteristic of the polymer chain. Models can be mapped onto each other by imposing that they yield the same large-scale quantities, which will to a large extent determine the physical properties of the melt and solutions. *Scaling* relations, such as the proportionality between the end-to-end distance and the square root of the molecular weight, $\langle \mathbf{R}^2 \rangle \propto M$, are independent of the precise underlying microscopic model.

Finally, let us mention that the Gaussian model is conveniently transformed into a continuous model by treating the monomer number as a continuous variable

---

[6] The reason for the change of notation in the monomer position will become clear below, when it is realized that the Gaussian chain monomer is an *effective* one.

[7] This model is also described as the 'bead spring' model or sometimes as Edwards' model, after S.F. Edwards who promoted its use in polymer science.

$s, 0 < s < N$. The free energy (1.31) is then rewritten as

$$H = \frac{3k_B T}{2b^2} \int_0^N ds \left(\frac{\partial \mathbf{R}(s)}{\partial s} \right)^2 \tag{1.35}$$

The Kratky–Porod chain

The freely jointed chain is a reasonable *microscopic* description of very flexible polymers, such as polystyrene (PS) or poly-dimethyl-siloxane (PDMS). Many polymer chains, however, have strong orientation constraints between successive segments which are not accounted for in this model. A simple way of approaching such constraints is to assume that the angle between two successive segments of length ℓ (monomers) is fixed, equal to α. Using the same notation as in the freely jointed case, we now have

$$\langle \mathbf{u}_{i+1} \cdot \mathbf{u}_i \rangle = \ell^2 \cos \alpha \tag{1.36}$$

Proceeding by induction, it is easily seen [8] that

$$\langle \mathbf{u}_{i+k} \cdot \mathbf{u}_i \rangle = \ell^2 (\cos \alpha)^k = \ell^2 \exp \left(-\frac{k\ell}{\ell_p} \right) \tag{1.37}$$

where $\ell_p = -\ell / \ln(\cos \alpha) \simeq 2\ell/\alpha^2$ is the *persistence length*, and the second equality is valid for small α. ℓ_p is the length scale above which the correlation between orientations along the chain becomes negligible.

Again, the Kratky–Porod chain can be 'coarse-grained' into a freely jointed model with identical large-scale properties. If the chain is divided into subunits of length ℓ_p, it is clear from equation (1.37) that the orientations of successive subunits are essentially uncorrelated. Hence the chain can be described as a succession of rigid subunits, each of length ℓ_p, with random relative orientations. More precisely, the mapping is carried out by computing the end-to-end distance

$$\langle \mathbf{R}^2 \rangle = \langle (\mathbf{u}_1 + \cdots + \mathbf{u}_{M-1})^2 \rangle = (M-1)\ell^2 + 2\ell^2 \sum_{i=1}^{M-1} \sum_{j=i+1}^{M-1} (\cos \alpha)^{j-i} \tag{1.38}$$

Summing the geometric series in (1.38), one eventually obtains

$$\langle \mathbf{R}^2 \rangle = (M-1)\ell^2 \left[\frac{1 + \cos \alpha}{1 - \cos \alpha} - \frac{2 \cos \alpha}{M-1} \frac{(1 - (\cos \alpha)^M)}{(1 - \cos \alpha)^2} \right] \tag{1.39}$$

In the long chain limit $M \gg 1$, and for small α this reduces to

$$\langle \mathbf{R}^2 \rangle = 2M\ell\ell_p = (M\ell/2\ell_p)(2\ell_p)^2 \tag{1.40}$$

---

[8] The proof goes by decomposing $\mathbf{u}_{i+k+1} = \cos \alpha \, \mathbf{u}_{i+k} + \mathbf{v}_{i+k}$. Free rotation implies that $\mathbf{v}_{i+k}$ is uniformly distributed on a circle, so that $\langle \mathbf{v}_{i+k} \cdot \mathbf{u}_i \rangle = 0$.

An introduction to liquid matter

where we have used $\ell_p \simeq 2\ell/\alpha^2$. The chain is therefore equivalent to $M' = M\ell/2\ell_p$ freely jointed segments of length $2\ell_p$.

Exercise: Like the Gaussian model, the Kratky–Porod model admits a continuous version, in which the configuration of the chain is described by a continuous curve $\mathbf{R}(s)$, where $0 < s < L$ is a curvilinear coordinate along the chain of total length $L = M\ell$. The free energy associated with a given configuration is written in the form

$$H = \frac{1}{2}k_B T \ell_p \int_0^L ds \left(\frac{\partial \mathbf{t}}{\partial s}\right)^2 \tag{1.41}$$

where $\mathbf{t} = \frac{\partial \mathbf{R}}{\partial s}$ is the unit vector tangent to the chain.

1. Discuss qualitatively the free energy (1.41).

2. Rewrite the free energy (1.41) in a discretized form involving the M tangent vectors $\mathbf{t}_1, \ldots, \mathbf{t}_M$ at points $0, L/M, 2L/M, \ldots, L$ along the chain, where M is a large number.

3. Using this discretized free energy, show that

$$\langle \mathbf{t}_{i+1} \cdot \mathbf{t}_i \rangle = -\frac{ds}{\ell_p} + \coth \frac{\ell_p}{ds} \tag{1.42}$$

where $ds = L/M$. Show that ℓ_p can therefore be identified with the persistence length defined in (1.37).

4. The continuous version of (1.37) can be written as

$$\langle \mathbf{t}(s) \cdot \mathbf{t}(s') \rangle = \exp - \left(\frac{|s - s'|}{\ell_p}\right) = f(s - s') \tag{1.43}$$

Show that the mean squared distance between two points along the chain is given by

$$\langle (\mathbf{R}(s) - \mathbf{R}(0))^2 \rangle = 2 \int_0^s (s - t) f(t) \, dt = 2 \left(s\ell_p - \ell_p^2 + \ell_p^2 \exp(-s/\ell_p)\right) \tag{1.44}$$

Discuss the small and large s limits of this expression.

1.6 Application 3: numerical experiments

The advent and rapid growth in the power of digital computers, beginning in the 1950s, led naturally to the idea of carrying out numerical 'experiments' on condensed matter systems, by simulating samples of hundreds or thousands of interacting atoms or molecules. The objective of such simulations is to estimate statistical averages of well defined functions of molecular coordinates and velocities, related to local or macroscopic properties of a given fluid or solid material, by an appropriate sampling of the phase space of the many-particle system. The statistical description may be carried out, broadly speaking, on one of three levels of detail, corresponding to different length and time scales.

On the most fundamental (or 'first principles') level, any material is made up of nuclei and electrons. In the liquid state, the former follow almost invariably classical statistical mechanics, since the corresponding thermal de Broglie wavelength (1.15) is generally a small fraction of an angstrom at room temperature, much less than typical internuclear distances in condensed matter, so that quantum diffraction effects, due to overlap of wavepackets associated with neighbouring nuclei, are negligible. The valence electrons, however, are highly degenerate since the Fermi temperature $T_F = \hbar^2(3\pi^2 n_e)^{2/3}/2m_e k_B$ (where n_e is the number of valence electrons per unit volume and m_e the electron mass) greatly exceeds room temperature at condensed matter densities. Electron valence states hence require a full quantum mechanical treatment. The joint ab initio description of a many-body system of degenerate valence electrons and classical nuclei 'dressed' with the tightly bound core electrons, poses a formidable theoretical challenge. The large difference in energy and associated time scales may be exploited within the adiabatic Born–Oppenheimer approximation, whereby the electronic structure problem is solved for 'frozen' configurations of the nuclei. This procedure is particularly well adapted to crystalline solids, where the nuclei may be assumed to be fixed at the sites of a regular lattice; the periodicity of the latter leads to considerable simplifications which are lost in disordered systems, like amorphous solids and fluids. In the latter, the nuclei undergo constant diffusive motion, making it impossible to define long-lived equilibrium positions similar to the sites of a crystal lattice. Invoking once more adiabaticity, the electronic structure must be calculated for successive, instantaneous configurations of the nuclei, and the resulting local valence electronic density may then be used to compute the forces between the nuclei. These forces may be divided into intramolecular forces between nuclei belonging to the same molecule, and intermolecular forces between different molecules; they determine the subsequent motion of the nuclei. Such a step-by-step procedure, alternating quantum electronic structure calculations and the solution of the classical equations of motion of the nuclei interacting through forces calculated 'on the fly', may be achieved by a clever combination of electronic density functional theory (DFT), and a classical molecular dynamics (MD) algorithm for the nuclear motion, to be described later on in this section. Such an ab initio MD scheme, first put forward by R. Car and M. Parrinello in 1985[9], is very computer intensive, but indispensable for a realistic description of fluids involving highly covalent intermolecular bonding, or strong hydrogen bonds between molecules. The Car–Parrinello method also allows molecular scale investigations of chemical reactions in liquids.

Fortunately, for fluids made up of atoms, ions or molecules with closed electronic shells, a simpler, semi-empirical procedure proves sufficient for most purposes. The highly localized electron orbitals give rise to rather weak

---

[9] R. Car and M. Parrinello, *Phys. Rev. Lett.* **55**, 2471 (1985).

intermolecular forces (except for the short-range repulsion resulting from the overlap of orbitals on different molecules), which have been described in section 1.2. The important simplification is that these interactions are, to a good approximation, pair-wise additive, i.e. depend essentially on the relative positions and orientations of two molecules, and not on the full configuration of neighbouring particles. This means that intermolecular forces may be calculated once and for all from a single quantum electronic structure calculation for a pair of molecules, and then transferred, after generally minor adjustments, to condensed states involving many such molecules. This considerable simplification allows the simulation of much larger systems, involving up to several million atoms, than in the more fundamental ab initio approach. This second level of description also allows numerical 'experiments' on increasingly complex fluids, like liquid crystals or self-assembling amphiphilic systems. At this level of detail, one must distinguish between intramolecular motions, like relative vibrations of the nuclei inside the same molecule, or torsional motions around intramolecular bonds, and overall molecular motions, including the translational motion of the centre of mass, and the rotational motion of the molecular axes. Translational and rotational energies are of the order of $k_B T$, and the associated time scales of the order of

$$\tau_t = \left(\frac{m\sigma^2}{k_B T}\right)^{1/2} \simeq \tau_r = \left(\frac{I}{k_B T}\right)^{1/2} \simeq 10^{-12} \text{ s} \tag{1.45}$$

for small molecules of diameter σ and moment of inertia $I \simeq m\sigma^2$ at room temperature. Vibrational periods are typically 2 to 3 orders of magnitude shorter, and hence it is, for many purposes, a good approximation to make the adiabatic assumption that the intramolecular bonds are rigid, with fixed bond lengths. The relative torsional motions of monomers in macromolecules, like polymers, are much slower and must in principle be treated on an equal footing as overall translations and rotations. In this kind of description, the two fundamental simulation schemes are the Monte-Carlo (MC) and molecular dynamics (MD) methods, to be described briefly later on.

If larger length and time scales are to be explored, as in complex fluids involving macromolecules, supramolecular aggregates or colloidal particles, a full atomic scale description becomes untractable, and some degree of 'coarse-graining' is required to allow practical simulations. Coarse-graining invariably amounts to a considerable reduction of the initial number of atomic degrees of freedom. This reduction may be formally achieved by integrating out microscopic degrees of freedom in the partition function, leaving only the degrees of freedom associated with mesoscopic entities, like colloidal particles; the latter then interact via effective forces, which derive from a free energy, rather than merely a potential energy, as would be the case at the atomic level. Such effective interactions have an entropic component, and depend on the thermodynamic state of the fluid, as will be illustrated in later chapters in the cases of depletion forces or interactions between electric double-layers. At this level of description, simulations are carried out on systems of mesoscopic particles interacting via such effective forces. An extreme example of such coarse-graining is provided by discretized lattice gas models introduced in section 1.4.

The two most widely used classical simulation methods, Monte-Carlo and molecular dynamics, correspond to two different ways of exploring the phase space of a sample of N particles interacting via given force laws. For the sake of simplicity, the subsequent discussion will be restricted to fluids of spherical, classical particles interacting via central forces deriving from a pair potential $v(r)$, where r is the distance between the centres of two particles. A microscopic state of the system is entirely specified by a point Γ_N in $6N$-dimensional phase space. The coordinates of Γ_N are the N position vectors $(\mathbf{r}_1, \ldots, \mathbf{r}_N)$ and N momenta $(\mathbf{p}_1, \ldots, \mathbf{p}_N)$ of the particles. According to Gibbs' ensemble theory, the probability of occurrence of such a state is given by a phase space probability density $P_N(\Gamma_N)$. If the system under consideration is closed (i.e. does not exchange particles with the outside world, so that their total number N is fixed) and in contact with a thermostat, which fixes the temperature T, the relevant Gibbs ensemble is the canonical ensemble, where

$$P_N(\Gamma_N) = \frac{1}{N! h^{3N}} \frac{\exp(-\beta H_N)}{Q_N} \qquad (1.46)$$

The probability of a microscopic state is proportional to the Boltzmann distribution involving the total energy $H_N = K_N(\{\mathbf{p}_i\}) + V_N(\{\mathbf{r}_i\})$ where

$$K_N(\{\mathbf{p}_i\}) = \sum_i \frac{\mathbf{p}_i^2}{2m} \qquad V_N(\{\mathbf{r}_i\}) = \sum_{i<j} v(|\mathbf{r}_i - \mathbf{r}_j|) \qquad (1.47)$$

and the canonical partition function

$$Q_N = \frac{1}{N! h^{3N}} \int d\Gamma_N \exp(-\beta H_N) \qquad (1.48)$$

ensures the proper normalization of the probability density (1.46). Note that the latter factorizes into momentum- and configuration-dependent parts, implying that in classical statistical mechanics, positions and momenta are uncorrelated at any given time. The factor $1/h^{3N}$ (where h is Planck's constant) makes Q_N dimensionless, while the factor $1/N!$ is required to ensure the extensivity of the resulting free energy (cf. section 2.2). Integration over the Maxwellian distribution of momenta is trivial, leaving

$$Q_N = \frac{1}{N! \Lambda^{3N}} \int d\mathbf{r}_1 \cdots d\mathbf{r}_N \exp(-\beta V_N(\mathbf{r}_1, \ldots, \mathbf{r}_N)) = \frac{1}{N! \Lambda^{3N}} Z_N \qquad (1.49)$$

where Λ is the thermal de Broglie wavelength (1.15) and the second equality defines the configuration integral Z_N. The corresponding configuration space probability density reduces to

$$P_N(\mathbf{r}_1, \ldots, \mathbf{r}_N) = \frac{\exp(-\beta V_N(\mathbf{r}_1, \ldots, \mathbf{r}_N))}{Z_N} \tag{1.50}$$

A purely random sampling of phase space would be most inefficient at condensed matter densities, since almost all randomly generated configurations would lead to overlap of particles, and hence to a very low probability (1.50), due to the highly repulsive excluded volume interaction, resulting in large positive values of V_N. The importance sampling algorithm, introduced by Metropolis et al. in 1953 [10], ensures that only configurations of high probability are generated in a MC simulation. Statistical averages of configuration-dependent variables $A(\mathbf{r}_1, \ldots, \mathbf{r}_N)$, in the form

$$\langle A \rangle = \int d\mathbf{r}_1 \cdots d\mathbf{r}_N \, P_N(\mathbf{r}_1, \ldots, \mathbf{r}_N) A(\mathbf{r}_1, \ldots, \mathbf{r}_N) \tag{1.51}$$

are then estimated from

$$\langle A \rangle = \frac{1}{N_c} \sum_{k=1}^{N_c} A\left\{ \mathbf{r}_i^{(k)} \right\} \tag{1.52}$$

where N_c is the total number of configurations generated in a MC run, and $\{\mathbf{r}_i^{(k)}\}$ denotes the set of coordinates of the N particles in the configuration k. Note that the MC algorithm allows the calculation of weighted averages of the form (1.51), but does not yield, in its simplest form, the normalization factor itself, the logarithm of which is directly related to the free energy, and hence to the entropy. The MC algorithm is conveniently illustrated in the simple example of hard spheres, of diameter σ, where the Boltzmann factor in equation (1.50) can only take two values: 0 if there is at least one overlap, i.e. if for at least one pair (i, j), $|\mathbf{r}_i - \mathbf{r}_j| < \sigma$; 1 if there is no overlap. The latter configuration is allowed, and all allowed configurations have the same probability, while the former configuration is of zero probability, i.e. forbidden. The corresponding MC algorithm is easily implemented. An initial allowed configuration is constructed, e.g. by placing N spheres on N sites of a face-centred cubic (FCC) lattice. New configurations are then generated by a series of trial moves of randomly chosen spheres. The centre of a sphere at $\mathbf{r}_i$ is moved to a new position $\mathbf{r}_i' = \mathbf{r}_i + \delta\mathbf{r}$, where $\delta\mathbf{r}$ is a vector chosen at random inside a cube of edge δ. The trial position $\mathbf{r}_i'$ is tested for possible overlaps by comparing $|\mathbf{r}_i' - \mathbf{r}_j|$ to σ. If any of these distances is less than σ, the trial configuration is rejected and the 'old' configuration (before the trial displacement) is retained. If no overlap occurs, the 'new' configuration is accepted, and a next trial move is attempted on another randomly chosen sphere. This basic step is repeated N_c times, and averages are taken over the

---

[10] N. Metropolis, A. Rosenbluth, M. Rosenbluth, A. Teller and E. Teller, *J. Chem. Phys.* **21**, 1087–1092 (1953).

N_c configurations, *a fraction of which are duplicated*, due to the rejection of trial moves leading to overlap. In practice the best convergence of the statistical averages is observed when δ is adjusted such that the rejection rate is about 50%. As one might expect, the optimum displacement δ decreases as the packing fraction ϕ increases. By making histograms of the interparticle distances one may, for instance, compute the pair distribution function $g(r)$ after appropriate normalization.

The molecular dynamics method exploits Boltzmann's prescription for calculating statistical averages, by mapping out the phase space trajectory of a single system over a sufficiently long time interval τ, and taking the time average of appropriate functions of the instantaneous positions and momenta $\{\mathbf{r}_i(t), \mathbf{p}_i(t)\}$ of all particles in the system. An estimate of the required average then follows from:

$$\langle A \rangle = \frac{1}{\tau} \int_0^\tau A\left(\{\mathbf{r}_i(t), \mathbf{p}_i(t)\}\right) dt \tag{1.53}$$

The main task is thus to determine the phase space trajectory, by starting from given initial conditions $\{\mathbf{r}_i(t=0), \mathbf{p}_i(t=0)\}$. This is achieved by solving the coupled classical equations of motion of the N particles interacting via a given force law; these are just Newton's equations of motion

$$m \frac{d^2 \mathbf{r}_i}{dt^2} = \mathbf{F}_i = -\nabla_i V_N\left(\{\mathbf{r}_j\}\right) \tag{1.54}$$

$\mathbf{F}_i$ is the total force acting on particle i, due to all other particles in the system. The N coupled vectorial equations of motion are solved in practice by a finite difference algorithm, using a finite time step δt. A very simple and stable algorithm, satisfying time reversal symmetry, is that due to Verlet (1967)

$$\mathbf{r}_i(t+\delta t) = -\mathbf{r}_i(t-\delta t) + 2\mathbf{r}_i(t) + \frac{\delta t^2}{2}\mathbf{F}_i(t) + \mathcal{O}(\delta t^4) \tag{1.55}$$

The time step δt is chosen such as to ensure a good conservation of the total energy H_N in the course of time; typically the optimum δt is of the order of 1% of τ_t in equation (1.45), i.e. of the order of 10^{-14} s. The finite difference algorithm is iterated for 10^4–10^7 time steps, which allows time scales of the order of 10 ns to be reached rather routinely. Statistical averages are now averages over time which are estimated over the finite time interval $\tau = N_t \, \delta t$, where N_t is the total number of iterations:

$$\langle A \rangle = \frac{1}{N_t} \sum_{k=1}^{N_t} A(k \, \delta t) \tag{1.56}$$

Such averages along a phase space trajectory correspond to the microcanonical (or constant energy) Gibbs ensemble if the system is ergodic. As an example, the instantaneous temperature $T(t)$, related to the kinetic energy of the atoms, fluctuates in the course of time, and the mean temperature may be estimated from the equipartition theorem:

$$\langle T \rangle = \frac{1}{N_t} \sum_{k=1}^{N_t} \frac{1}{3k_B N} \sum_i \frac{\mathbf{p}_i^2(k\,\delta t)}{m} \qquad (1.57)$$

Since MC and MD simulations are carried out on samples of $N = 10^3$–10^6 particles, which is very small compared to Avogadro's number, the question of the deviation of the statistical averages from their thermodynamic limit arises naturally. If bulk properties are to be investigated, undesirable surface effects are eliminated by the use of periodic boundary conditions. Extrapolation of averages computed for several values of N generally shows a rapid convergence of intensive properties towards the thermodynamic limit for $N \geq 10^2$, at least for particles interacting via short-range forces, in states not too close to phase transitions.

By construction, MC simulations can only probe static properties of thermodynamic equilibrium states, like the equation of state or pair distribution functions. The intrinsically stochastic MC method has the advantage over deterministic molecular dynamics that 'unphysical' particle moves may be generated which can speed up the convergence properties considerably. This is particularly true of macromolecular systems, with enormous numbers of molecular conformations. Thus trial moves can be attempted whereby monomers at one end of a polymer chain are removed and reattached to the opposite end, leading to a significant conformational change in a single move. The MC method can also be easily generalized to other ensembles, like the grand canonical ensemble corresponding to a fluctuating number N of particles; the latter is particularly well adapted to the study of phase transitions. A related extension of the MC method allows a direct and efficient computation of free energies, as will be illustrated in section 2.5.

Molecular dynamics, in contrast, allows one to follow the time evolution of microscopic variables, and to compute time-dependent correlation functions which will be discussed in part IV; such correlation functions are intimately related to various relaxation phenomena and frequency-dependent cross-sections characterizing inelastic scattering of radiation. They also relate directly to linear transport coefficients like viscosity or thermal conductivity of fluids. MD can cope both with equilibrium states and non-equilibrium situations, as is the case for systems subjected to external forces or gradients in thermodynamic variables (thermal or concentration gradients for instance). The term 'non-equilibrium molecular dynamics' (NEMD) has been coined for such simulations. MD has also been generalized to coarse-grained mesoscopic scales by replacing the Newtonian dynamics, obeying deterministic, time-reversible equations of motion, by irreversible dynamics incorporating random 'Brownian' forces, due to fast microscopic scales, and hydrodynamic interactions induced by the motion of colloidal particles through a fluid continuum. The corresponding, most common, algorithms are 'Brownian dynamics' and 'dissipative particle dynamics'.

Despite their limitations as to the length and time scales which they can simulate, numerical 'experiments' are a very useful complement of laboratory experiments, in that they can measure various quantities and correlation functions that are not experimentally accessible, but do provide additional physical insight.

Further reading

D. Tabor, *Liquids, Gases and Solids*, Cambridge University Press, Cambridge, 1991. Classic and very readable reference, at the undergraduate level. Liquids (both simple and complex) are covered at a mostly descriptive and qualitative level. Thermodynamics and phase diagrams of simple substances are also discussed.

R.G. Larson, *The Structure and Rheology of Complex Fluids*, Oxford University Press, Oxford, 1999. An extensive and very readable discussion of all kinds of complex fluids, with emphasis on their rheological properties; careful presentation of many recent and older experimental references. Discussion of theoretical models is also included, albeit in some cases in a somewhat allusive form.

D. Frenkel and B. Smit, *Understanding Molecular Simulation*, 2nd edn., Academic Press, New York 2001 (http://molsim.chem.uva.nl/frenkelsmit/) is a very nice exposition of the simulation techniques used in all types of liquids, with emphasis on sophisticated algorithms and their statistical mechanical foundations.

M.P. Allen and D. Tildesley, *Computer Simulation of Liquids*, Clarendon Press, Oxford, 1987. Contains a very useful practical description of computer simulation techniques and their implementation (http://www.ccp5.ac.uk/librar.shtml#ALLENTID)

J. Israelachvili, *Intermolecular and Surface Forces*, 2nd edn., Academic Press, New York, 1992. Gives a thorough discussion of interactions between various types of molecules and surfaces.

J.S. Rowlinson and F. Swinton, *Liquids and Liquid Mixtures*, 3rd edn., Butterworths, London, 1983. Contains a clear account of the physical chemistry of fluid mixtures and of their phase diagrams.

I Thermodynamics, structure and fluctuations

2 A reminder of thermodynamics

Thermodynamics is the branch of physical sciences developed in the 19th century to provide a systematic theoretical framework for the characterization of physical and chemical transformations of substances, involving exchange of heat, work and matter. It is an essentially macroscopic and phenomenological theory, which ignores the molecular nature of matter; in fact at the time the laws of thermodynamics were formulated, atomistic theories were far from being universally accepted. The link between macroscopic thermodynamics and the modern microscopic description of matter was provided later by statistical mechanics, which is the central theoretical tool for the study of complex fluids; this link will be briefly recalled in section 2.2.

A brief summary of macroscopic thermodynamics is provided in section 2.1. This reminder, which emphasizes key concepts based on the two fundamental laws of thermodynamics, more than specific thermodynamic relations, will serve as a constant reference throughout the book. In particular the summary will provide the basis for the phenomenology of phase transitions and interfacial phenomena (section 2.3), as well as for the macroscopic theory of fluctuations in chapter 3, which in turn can be generalized to mesoscopic and microscopic scales (chapter 3, sections 3.4 to 3.5).

2.1 State variables and thermodynamic equilibrium

Numerous measurements carried out on a broad variety of gaseous, liquid or solid substances, clearly show that the macroscopic equilibrium states of any substance can be fully characterized by specifying the values of a small number of state variables. In other words all measurable, macroscopic properties of a given substance, say its specific heat, its thermal expansivity or its viscosity, are well defined functions of these state variables. There is no unique set of state variables, and the precise choice of these variables is generally dictated by experimental considerations or convenience, but to characterize the bulk properties of a homogeneous substance in thermodynamic equilibrium, one must specify the values of $\nu + 2$ state variables, where ν is the number of (non-reacting) chemical species in the substance. Once the chemical composition has been specified, by giving e.g. the mass (in kilograms or moles) of each species, only two further variables are needed to characterize completely the state of that substance; in practice the most common choices are temperature T and pressure P or total volume V of the sample. This section will focus on translationally invariant, homogeneous substances, made up of a single phase. Inhomogeneous substances, involving several phases separated by interfaces, will be considered in sections 2.3 and 6.1.

The choice of state variables is largely dictated by the way in which the system under consideration is coupled to the external world, i.e. by thermodynamic boundary conditions and external constraints. Systems may be closed or open. A closed system has a fixed chemical composition, i.e. the numbers N_α of molecules of each species $1 \le \alpha \le \nu$ are held constant. An open system may exchange matter with the external world, considered as an infinite reservoir of molecules of the various species; in the open case, the relevant chemical composition variables are not the N_α, which may now vary, but rather the chemical potentials μ_α of the species, which are fixed by the reservoir. The 'chemical' variables N_α and μ_α are said to be conjugate; note that the N_α are *extensive* variables (i.e. proportional to the size of the system), while the μ_α are *intensive* variables.

Closed systems may either be thermally insulated, or exchange thermal energy with the external world which plays the role of a heat reservoir; when the system is in thermal equilibrium with the reservoir, it is at the same temperature as the latter. Finally, a system may produce or receive mechanical work, δW: for fluid systems, the most common form of work is the pressure work upon compression or expansion of the system, $-P \, dV$, where dV is the change in volume of the system at constant external pressure P.

To each of these situations and associated choice of state variables, there corresponds a *thermodynamic potential* (or characteristic function) with the dimension of energy, which satisfies an extremum condition at equilibrium, and which allows a systematic derivation of all known thermodynamic relations. Moreover, these various thermodynamic potentials are related by *Legendre transformations* which allow one to switch from one set of state variables to another.

We first consider closed systems. The first law of thermodynamics extends the mechanical principle of conservation of energy to thermodynamic systems, by stating that in any transformation of a system, the variation of its internal energy U is the sum of the heat exchanged with the external world, δQ, and of the work spent by or done on the system, δW:

$$dU = \delta Q + \delta W = \delta Q - P \, dV \qquad (2.1)$$

The internal energy U is a state function, implying that dU depends only on the initial and final states of the system, before and after the transformation.

The second law of thermodynamics introduces another state function, the entropy S, which restricts the class of thermodynamic transformations that are actually possible for a given system. For an infinitesimal transformation involving a heat exchange δQ with a reservoir at temperature T, the second law states that the entropy variation of the system obeys

$$dS \geq \frac{\delta Q}{T} \tag{2.2}$$

The equality is obtained only for a reversible (or quasistatic) transformation, i.e. one that is sufficiently slow that the system is, at each stage, in thermodynamic equilibrium. An irreversible transformation is one during which the system, initially removed from equilibrium, relaxes back towards thermodynamic equilibrium. In particular, the entropy of a thermally insulated system can only increase or be stationary:

$$dS^{\text{ins}} \geq 0 \tag{2.3}$$

which shows that thermodynamic equilibrium corresponds to a state of maximum entropy. If (2.2) is substituted into (2.1) we find that for a reversible transformation,

$$dU = T \, dS - P \, dV \tag{2.4}$$

Equation (2.4) is now an identity relating changes in state functions when the system undergoes an infinitesimal transformation between two equilibrium states. It can be used to compute the entropy $S(U, V)$ as a function of energy and volume, or the energy $U(S, V)$ as a function of entropy and volume. The expression (2.4) of the first law may be generalized to open systems, in the form

$$dU = T \, dS - P \, dV + \sum_{\alpha} \mu_{\alpha} \, dN_{\alpha} \tag{2.5}$$

where the chemical potential μ_{α} of species α is the increase in internal energy when one molecule of that species is transferred from the reservoir to the system. Equation (2.5) leads directly to the standard thermodynamic relations

$$T = \left(\frac{\partial U}{\partial S} \right)_{V, \{N_{\alpha}\}} \qquad P = - \left(\frac{\partial U}{\partial V} \right)_{S, \{N_{\alpha}\}} \qquad \mu_{\alpha} = \left(\frac{\partial U}{\partial N_{\alpha}} \right)_{S, V, \{N_{\beta}\}} \tag{2.6}$$

while the equality of second derivatives leads to Maxwell relations, like:

$$\left(\frac{\partial T}{\partial V} \right)_{S, \{N_{\alpha}\}} = - \left(\frac{\partial P}{\partial S} \right)_{V, \{N_{\alpha}\}} \tag{2.7}$$

To switch from the state variables S, V, $\{N_{\alpha}\}$ to the set S, P, $\{N_{\alpha}\}$ where, according to (2.4), the pressure P is the variable conjugate to the volume V, it suffices to carry out the following Legendre transformation from $U(S, V, \{N_{\alpha}\})$ to the enthalpy H

$$H(S, P, \{N_{\alpha}\}) = U(S, V, \{N_{\alpha}\}) - V \left(\frac{\partial U}{\partial V} \right)_{S, \{N_{\alpha}\}} = U(S, V, \{N_{\alpha}\}) + PV \tag{2.8}$$

Clearly:

$$dH = T\,dS + V\,dP + \sum_\alpha \mu_\alpha\,dN_\alpha \tag{2.9}$$

which immediately yields thermodynamic relations similar to (2.6). In fact the entropy S is not a convenient state variable, since it is not easily controlled experimentally. It can be eliminated in favour of its conjugate variable T by the following Legendre transformation

$$F(T, V, \{N_\alpha\}) = U(S, V, \{N_\alpha\}) - TS \tag{2.10}$$

where F is the Helmholtz free energy, the differential of which is

$$dF = -S\,dT - P\,dV + \sum_\alpha \mu_\alpha\,dN_\alpha \tag{2.11}$$

which leads to the widely used thermodynamic relations:

$$S = -\left(\frac{\partial F}{\partial T}\right)_{V,\{N_\alpha\}} \qquad P = -\left(\frac{\partial F}{\partial V}\right)_{T,\{N_\alpha\}} \qquad \mu_\alpha = \left(\frac{\partial F}{\partial N_\alpha}\right)_{T,V,\{N_\beta\}} \tag{2.12}$$

and to new Maxwell relations such as

$$\left(\frac{\partial S}{\partial N_\alpha}\right)_{T,V,\{N_\beta\}} = -\left(\frac{\partial \mu_\alpha}{\partial T}\right)_{V,\{N_\alpha\}} \tag{2.13}$$

Similarly, one may switch from the enthalpy H to the free enthalpy, or Gibbs free energy G, by

$$G(T, P, \{N_\alpha\}) = H(S, P, \{N_\alpha\}) - TS = F(T, V, \{N_\alpha\}) + PV \tag{2.14}$$

such that

$$dG = -S\,dT + V\,dP + \sum_\alpha \mu_\alpha\,dN_\alpha \tag{2.15}$$

Since G and the N_α are extensive quantities, G is necessarily a first-order homogeneous function of the N_α, i.e.

$$G(T, P, \{\lambda N_\alpha\}) = \lambda G(T, P, \{N_\alpha\}) \tag{2.16}$$

Then, according to Euler's theorem for first-order homogeneous functions, it follows that

$$G(T, P, \{N_\alpha\}) = \sum_\alpha N_\alpha \left(\frac{\partial G}{\partial N_\alpha}\right)_{T,P,\{N_\beta\}} = \sum_\alpha \mu_\alpha N_\alpha \tag{2.17}$$

Differentiating both sides of this equation and comparing to (2.15), one arrives directly at the Gibbs–Duhem relation

$$S\,dT - V\,dP + \sum_\alpha N_\alpha\,d\mu_\alpha = 0 \tag{2.18}$$

A final Legendre transformation allows one to switch from the chemical composition variables $\{N_\alpha\}$ to the chemical potentials $\{\mu_\alpha\}$ in the set of independent state variables, by introducing the following thermodynamic potential, called the grand potential,

$$\Omega(T, V, \{\mu_\alpha\}) = F(T, V, \{N_\alpha\}) - \sum_\alpha \mu_\alpha N_\alpha \tag{2.19}$$

$$d\Omega = -S\,dT - P\,dV - \sum_\alpha N_\alpha d\mu_\alpha \tag{2.20}$$

Recalling equations (2.14) and (2.17), the grand potential is seen to be directly related to the equation of state, namely

$$\Omega = F - G = -PV \tag{2.21}$$

A very important property of all thermodynamic potentials, U, H, F, G and Ω, which is a direct consequence of the second law in its form (2.3), is that they reach a minimum, with respect to all possible internal changes in the system, at thermodynamic equilibrium; they decrease during any irreversible transformation, and are stationary at equilibrium, e.g.

$$dF \leq 0 \tag{2.22}$$

The stationarity property (2.22) is easily proved in the simple case of a closed system ($dN_\alpha = 0$), in contact with a much larger external heat reservoir, which imposes the temperature T_0 (thermostat, referred to by the index 0). If the system and the thermostat are thermally insulated from the rest of the world, then for any transformation the total energy is conserved, while the total entropy obeys (2.3):

$$dU + dU_0 = 0 \qquad dS + dS_0 \geq 0 \tag{2.23}$$

Since the much larger thermostat is not significantly perturbed by the system during the transformation, it remains in thermodynamic equilibrium; if δQ is the amount of heat exchanged between system and thermostat during a transformation at constant volume, then,

$$dS_0 = \frac{\delta Q}{T_0} = \frac{dU_0}{T_0} = \frac{-dU}{T_0} \tag{2.24}$$

Substitution into (2.23) leads to the desired result:

$$dS - \frac{dU}{T_0} \geq 0 \Rightarrow dF = d(U - T_0 S) \leq 0 \tag{2.25}$$

Consider now the system as the sum of its two halves. Assuming a one-component system for simplicity, the extensivity of the free energy implies that

$$F(N, V, T) = 2F(N/2, V/2, T) \tag{2.26}$$

If, by some external constraint, e.g. the positioning of a piston, the volume of one part (containing $N/2$ molecules) is increased by an amount δV at the expense of the other part (containing the same number of molecules), then clearly the system is removed from its state of thermodynamic equilibrium, and according to (2.22):

$$F(N/2, V/2 + \delta V, T) + F(N/2, V/2 - \delta V, T) \geq 2F(N/2, V/2, T) \qquad (2.27)$$

This inequality, valid for any δV, shows that the free energy is a convex function of the extensive variable V. This implies the stability condition:

$$\left(\frac{\partial^2 F}{\partial V^2}\right)_{N,T} = -\left(\frac{\partial P}{\partial V}\right)_{N,T} = \frac{1}{V \chi_T} \geq 0 \qquad (2.28)$$

where χ_T is the isothermal compressibility. Similarly one can show that F is a convex function of N.

This simple example illustrates the fact that, while three variables are sufficient to characterize the equilibrium state of a one-component system, further variables or parameters (δV in the present example) are needed to describe a non-equilibrium state. These additional parameters generally specify some external constraint, which keeps the system away from thermodynamic equilibrium. When the constraint is removed, the system will spontaneously relax to the state of lowest free energy corresponding to thermodynamic equilibrium.

Variations of free energy in reversible transformations determine the maximum amount of mechanical work, W, which can be extracted from an isothermal transformation. In other words, the work produced by the system ($W < 0$) is determined by the loss in free energy during a reversible transformation:

$$W = \Delta F = F_{\text{final}} - F_{\text{initial}} \qquad (2.29)$$

As a direct consequence of the second law (2.2), the work produced by the system during an irreversible transformation is less than the corresponding variation in free energy, i.e. $|W| < |\Delta F|$.

The thermodynamic relations considered so far must be generalized in the case where the system is a dielectric medium subjected to an external electric field, or a magnetic medium placed in a magnetic field. In the case of a dielectric medium characterized by a dielectric permittivity, the work carried out on the medium by an external generator is

$$\delta W = \int_{\text{space}} \mathbf{E} \cdot \delta \mathbf{D} \, d\mathbf{r} \qquad (2.30)$$

where $\mathbf{E}$ is the electric field inside the dielectric and $\mathbf{D}$ is the corresponding displacement field, $\mathbf{D} = \epsilon_0 \mathbf{E} + \mathbf{P}$. ϵ_0 is the permittivity of free space, $\mathbf{P}$ is the electric polarization vector, and the integration in equation (2.30) is over all space. If $\mathbf{E}_0$ denotes the electric field produced by the same generators in the absence of the dielectric medium, e.g. inside an empty condenser, then (2.30) may be rewritten as

$$\delta W = \delta \int_{\text{space}} \epsilon_0 \frac{\mathbf{E_0}^2}{2} \, d\mathbf{r} - \int_{\text{dielectric}} \mathbf{P} \cdot \delta \mathbf{E_0} \, d^3\mathbf{r} \tag{2.31}$$

where the second integral is now taken over the volume occupied by the dielectric.

Substituting (2.30) into (2.1), we find that for a dielectric fluid inside a capacitor of volume V_c, the differential of the free energy generalizes to

$$dU = T \, dS - P \, dV + V_c \mathbf{E} \cdot d\mathbf{D} + \sum_\alpha \mu_\alpha \, dN_\alpha \tag{2.32}$$

2.2 Link with statistical mechanics

Statistical mechanics provides a systematic link between the macroscopic level of phenomenological thermodynamics, where the state of a system in equilibrium is fully characterized by a small number of experimentally controlled state variables, and a microscopic level of description, which requires in principle the specification of $2 \sum_\alpha N_\alpha f_\alpha$ coordinate and momentum variables (where N_α is the total number of particles, and f_α the number of degrees of freedom per particle of species α) at any instant of time. The reduction of the detailed microscopic description involving an untractable number of rapidly varying degrees of freedom, to the macroscopic characterization of the system requiring only a small number of variables, which are, moreover, constant in time for equilibrium states, is achieved by a process of statistical averaging. Formally, the link is provided by partition functions, which are multi-dimensional integrals of probability distribution functions, like the Boltzmann factor, over phase space (see e.g. equation (1.48)), or multiple sums over discrete states, as in the case of lattice models.

As already mentioned, a microstate is entirely specified by a point in $2fN$-dimensional phase space, Γ_N (where $N = \sum_\alpha N_\alpha$, and Nf is short for $\sum_\alpha N_\alpha f_\alpha$). The usual Gibbs ensembles correspond to different sets of macroscopic constraints imposed on the systems of the ensemble. These constraints amount to choosing well defined sets of macroscopic state variables, and hence correspond to given thermodynamic potentials. Statistical mechanics provides a one-to-one correspondence between thermodynamic potentials and the well defined phase-space integrals, referred to as partition functions. We briefly recall here the three most commonly used Gibbs ensembles, and the corresponding relations between partition functions and thermodynamic potentials.

- The *microcanonical* ensemble is appropriate for the statistical description of fully isolated systems, i.e. systems that exchange neither heat (i.e. are thermally insulated) nor mechanical work with the external world. Their total energy E is then conserved within an experimental uncertainty ΔE, and their chemical composition $\{N_\alpha\}$ and volume V are fixed. Let $\Omega(E, V, \{N_\alpha\})$ be the total number of microstates accessible to a system subjected to these constraints. This number is the ratio of the accessible phase space 'volume' over some elementary phase space volume conventionally chosen to be h^{fN}, where h is Planck's constant (which has the dimension of a length times a momentum). The link with thermodynamics is then provided by Boltzmann's relation which incorporates the fundamental postulate of equal probabilities in phase space

$$S(E, V, \{N_\alpha\}) = k_B \ln \Omega(E, V, \{N_\alpha\}) \qquad (2.33)$$

This relation was already used in the example of the 'free volume' in section 1.4. Equation (2.33) clearly points to a statistical interpretation of entropy as a measure of the number of different microstates accessible for given chemical composition, volume and energy. It is, in a certain sense, a measure of 'disorder', although it will be seen later that this intuitive interpretation must be used with caution, since the degree of disorder of a system is a largely subjective concept. The interpretation of the second law is now much clearer: any additional constraint which maintains the system in a non-equilibrium state will reduce the number of accessible microstates; when the constraint is released, additional microstates become available, and the entropy increases towards its maximum value, corresponding to full thermodynamic equilibrium. The total energy E must of course be identified with the internal energy U, and the mean temperature of the system follows from equation (2.6). Thus, if we are capable of computing Ω (and hence S) as a function of $\{N_\alpha\}$, E and V (in general a formidable task in the case of dense fluids!), all thermodynamic properties follow from the usual relations adapted to this choice of variables.

As mentioned in section 1.6, the temperature of an isolated system fluctuates, due to the constant interchange between the potential and kinetic components of the (conserved) total energy of the system; the mean temperature is determined by the microcanonical average of the total kinetic energy according to equation (1.57).

• Closed systems with a well defined chemical composition $\{N_\alpha\}$, occupying a given volume V and in thermal equilibrium with a heat reservoir, or thermostat, which fixes the temperature T, are conveniently considered as belonging to a *canonical* ensemble. If $H_N = K_N + V_N$ denotes the total energy which depends on all coordinates q_i and as many momenta p_i, then the microstates are distributed in phase space according to the normalized Boltzmann factor

$$P(\Gamma_N) = \frac{1}{\prod_\alpha N_\alpha! h^{N_\alpha f_\alpha}} \frac{\exp(-\beta H_N)}{Q_N} \qquad (2.34)$$

where the canonical partition function is the dimensionless phase space integral

$$Q_N = \frac{1}{\prod_\alpha N_\alpha! h^{N_\alpha f_\alpha}} \int_{\Gamma_N} \exp(-\beta H_N)\, d\Gamma_N \qquad (2.35)$$

Equations (2.34) and (2.35) generalize equations (1.46) and (1.48), valid for one-component systems of spherical particles. The link with thermodynamics is provided by the following statistical definition of the Helmholtz free energy:

$$F(V, T, \{N_\alpha\}) = -k_B T \ln Q_N(V, T, \{N_\alpha\}) \qquad (2.36)$$

which is a function of the state variables $\{N_\alpha\}$, V and T, and satisfies all the thermodynamic relations recalled in section 2.1. In particular, it is easily verified from equations (2.35) and (2.36), that the internal energy U follows from

$$U \equiv \int H_N P_N \, d\Gamma_N = -\left(\frac{\partial \ln Q_N}{\partial \beta}\right)_{V,\{N_\alpha\}} = \left(\frac{\partial \beta F}{\partial \beta}\right)_{V,\{N_\alpha\}} \tag{2.37}$$

which agrees with the thermodynamic relations (2.11). Reverting, for the sake of simplicity, to point-like particles, for which $f_\alpha = 3$ (centre of mass coordinates and momenta), the integration of the Boltzmann factor $\exp(-\beta K_N)$ over momenta can be carried out immediately (product of $3N$ Gaussian integrals), resulting in

$$Q_N = \frac{1}{\prod_\alpha N_\alpha! \Lambda_\alpha^{3N_\alpha}} Z_N \tag{2.38}$$

The configuration integral Z_N can be calculated explicitly only in the 'ideal' low density or concentration limit, as in a very dilute solution, where the interactions between particles can be neglected, i.e. $V_N = 0$, and:

$$Z_N^{(\mathrm{id})} = V^N = V^{\sum_\alpha N_\alpha} \tag{2.39}$$

In that case the free energy of a multicomponent system of point particles reduces to

$$F_{\mathrm{id}} = -k_\mathrm{B} T \ln \prod_\alpha^N \frac{V_\alpha^N}{N_\alpha! \Lambda_\alpha^{3N_\alpha}} = k_\mathrm{B} T \sum_\alpha N_\alpha \left[\ln(\rho_\alpha \Lambda_\alpha^3) - 1\right] \tag{2.40}$$

where $\rho_\alpha = N_\alpha/V$ is the number density of species α, and Stirling's formula has been used; Λ_α is the thermal de Broglie wavelength (1.15) associated with species α. The thermodynamic relations (2.12) lead directly back to the familiar ideal gas or ideal solution results for monoatomic species

$$U_{\mathrm{id}} = \frac{3}{2} N k_\mathrm{B} T$$

$$P_{\mathrm{id}} = k_\mathrm{B} T \sum_\alpha \rho_\alpha \tag{2.41}$$

$$\mu_{\mathrm{id},\alpha} = k_\mathrm{B} T \ln\left(\rho_\alpha \Lambda_\alpha^3\right)$$

If the particles have internal degrees of freedom, linked e.g. to the vibrational and torsional motions of large molecules, the partition function for non-interacting particles factorizes into:

$$Q_N^{\mathrm{id}} = \prod_\alpha \frac{1}{N_\alpha!} \left(\frac{V q_\alpha^{\mathrm{int}}}{\Lambda_\alpha^3}\right) \tag{2.42}$$

where q_α^{int} is the internal partition function of a single particle, while V/Λ_α^3 is again the contribution of the centre of mass translational motion. Note that in the case of macromolecules, the internal partition function itself is a complicated quantity involving the many coupled degrees of freedom of the numerous monomers. Since each q_α^{int} is independent of the total volume accessible to the molecules, due to the intramolecular constraints, the pressure of an ideal gas or solution is still given by (2.41), while the relations (2.41) for the internal energy and chemical potentials generalize to

$$U_{id} = \frac{3}{2} N k_B T - \sum_\alpha N_\alpha \frac{\partial \ln q_\alpha^{int}}{\partial \beta}$$

$$\mu_{id,\alpha} = k_B T \ln \left(\frac{\rho_\alpha \Lambda_\alpha^3}{q_\alpha^{int}} \right)$$

(2.43)

Returning to the physically more relevant case of fluids of interacting particles requires a calculation of the configuration integral Z_N, which can only be achieved by making approximations, or by resorting to computer simulations. It is convenient to factorize Q_N into a product of its ideal part (2.42), and an *excess* part, due to interactions between particles. From equation (2.36) it then follows that the free energy splits into two terms, $F = F_{id} + F_{ex}$, and that all its derivatives separate in a similar fashion, i.e. $U = U_{id} + U_{ex}$, $P = P_{id} + P_{ex}, \ldots$. Clearly U_{ex} is the statistical (or ensemble) average of the total potential energy, $U_{ex} = \langle V_N \rangle$. P_{ex} can be expressed in terms of the total potential energy via the virial theorem [1]

$$P_{ex} = -\frac{1}{3V} \left\langle \sum_{i=1}^{N} \mathbf{r}_i \cdot \nabla_i V_N(\{\mathbf{r}_j\}) \right\rangle$$

(2.44)

where the sum is over all N particles in the fluid (regardless of their chemical species) and $\mathbf{r}_i$ is the centre of mass position vector of the ith particle.

The total energy of systems belonging to a canonical ensemble fluctuates, due to thermal exchanges with the thermostat; the mean square deviation of the total energy from its mean value is directly related to the specific heat at constant volume. The latter is defined by the thermodynamic relation:

$$C_v = \left(\frac{\partial U}{\partial T} \right)_{V, \{N_\alpha\}} = \left(\frac{\partial \langle H_N \rangle}{\partial T} \right)_{V, \{N_\alpha\}}$$

(2.45)

From the definition of the canonical average, it is then easily verified that

$$C_v = \frac{1}{k_B T^2} \left(\langle H_N^2 \rangle - \langle H_N \rangle^2 \right)$$

(2.46)

Note that the mean square fluctuation of an extensive quantity (H_N in the present case) is also an extensive quantity.

• Open systems, capable of exchanging energy and particles with a reservoir which fixes the temperature T and the chemical potentials μ_α of the various species, are most conveniently treated within a *grand canonical* ensemble. The numbers N_α are now allowed to fluctuate and the probability of finding the system with a given composition inside a volume V at temperature T must be weighted by the activities $z_\alpha = \exp(\beta \mu_\alpha)$ according to:

$$P(V, T, \{\mu_\alpha\}) = \frac{1}{\Xi} \left(\prod_\alpha z_\alpha^{N_\alpha} \right) Q_N(V, T, \{N_\alpha\})$$

(2.47)

---

[1] See e.g. H. Goldstein, *Classical Mechanics*, 2nd edn., Addison-Wesley, Reading, MA, 1990.

where the normalization is provided by the *grand partition function*:

$$\Xi = \sum_{N_1=0}^{\infty} \cdots \sum_{N_\nu=0}^{\infty} \left(\prod_\alpha z_\alpha^{N_\alpha} \right) Q_N(V, T, \{N_\alpha\}) \qquad (2.48)$$

The summations over numbers of particles are formally taken to $N_\alpha = \infty$, but in practice natural upper bounds arise due to excluded volume considerations. In particular,
for the simple lattice gas model introduced at the end of section 1.4, the number N
of particles (in a one-component system) cannot exceed the total number M of cells,
because of the single occupancy constraint. The grand potential (not to be confused
with the phase space volume appearing e.g. in (2.33)) is related to Ξ via:

$$\Omega(V, T, \{\mu_\alpha\}) = -k_B T \ln \Xi(V, T, \{\mu_\alpha\}) \qquad (2.49)$$

Remembering equation (2.47), it is easily verified that the mean number of particles
of species α is given by

$$\langle N_\alpha \rangle = \sum_{N_1=0}^{\infty} \cdots \sum_{N_\nu=0}^{\infty} N_\alpha P(V, T, \{\mu_\alpha\}) = -\left(\frac{\partial \Omega}{\partial \mu_\alpha} \right)_{V,T,\{\mu_\beta\}} \qquad (2.50)$$

in agreement with the thermodynamic relation (2.20). Taking second derivatives of Ω
leads to the fluctuation relations:

$$\langle N_\alpha N_\beta \rangle - \langle N_\alpha \rangle \langle N_\beta \rangle = -k_B T \left(\frac{\partial^2 \Omega}{\partial \mu_\alpha \partial \mu_\beta} \right)_{V,T,\{\mu_\gamma\}} \qquad (2.51)$$

In the special case of a one-component system, (2.51) reduces to

$$\frac{\langle N^2 \rangle - \langle N \rangle^2}{\langle N \rangle} = \frac{k_B T}{\langle N \rangle} \left(\frac{\partial \langle N \rangle}{\partial \mu} \right)_{V,T} = \rho k_B T \chi_T \qquad (2.52)$$

where the second equality follows from the isothermal ($dT = 0$) version of the Gibbs–
Duhem relation (2.18), and χ_T is the isothermal compressibility defined by equation
(2.28). The fluctuation is seen to be extensive and positive definite, confirming the
stability condition $\chi_T > 0$ derived on purely thermodynamic grounds in section (2.1).
    Solutions and dispersions involving osmotic equilibria are often more conveniently
handled within a semi-grand canonical ensemble, where the numbers of colloidal
particles or macromolecules are fixed, while the microscopic molecules of the solvent
are considered at fixed chemical potential, since they can be freely exchanged with a
reservoir across a membrane which is impermeable to the larger particles, as shown
schematically in figure 2.1. The *osmotic pressure* exerted by the solute or dispersed
particles is equal to the difference in total pressure across the membrane. Restricting the
discussion to a solvent (species A) and a single solute (species B), it proves convenient
to introduce the following 'semi-grand potential':

$$\Omega_s(T, V, \mu_A, N_B) = F(T, V, N_A, N_B) - \mu_A N_A \qquad (2.53)$$

such that

$$d\Omega_s = -S\,dT - P\,dV - N_A\,d\mu_A + \mu_B\,dN_B \qquad (2.54)$$

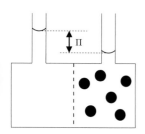

Figure 2.1. Schematic
representation of osmotic
equilibrium between a
pure solvent (in the left
half of the recipient) and a
solution or dispersion of
solute molecules or
particles represented by
full circles in the right half;
the two compartments are
separated by a semi-
permeable membrane.
The rise of the column of
solvent in the left-hand
side tube yields a
measure of the osmotic
pressure Π.

and a straightforward application of Euler's theorem shows that:

$$\Omega_s(T, V, \mu_A, N_B) = -PV + \mu_B N_B \qquad (2.55)$$

In the reservoir, where $N_B = 0$, the pressure is P', while in the solution, the pressure is P. The osmotic pressure is the difference:

$$\Pi = P - P' = \mu_A(\rho_A - \rho_A') + \mu_B \rho_B + f(T, \rho_A', 0) - f(T, \rho_A, \rho_B) \qquad (2.56)$$

where $f = F/V$ is the Helmholtz free energy density, while ρ_A and ρ_A' are the solvent densities in the solution and in the reservoir, which are determined by the chemical potential μ_A, according to:

$$\frac{\partial f(T, \rho_A', 0)}{\partial \rho_A'} = \frac{\partial f(T, \rho_A, \rho_B)}{\partial \rho_A} = \mu_A \qquad (2.57)$$

If $\phi = \rho_B/\rho$ denotes the mole fraction of solute, the free energy density may be expressed as a function of the total density ρ and of ϕ, rather than of ρ_A and ρ_B. If, moreover, the solution may for all practical purposes be considered as incompressible (i.e. ρ is regarded as fixed), the chemical potentials simplify to:

$$\mu_A = \left(\frac{\partial f}{\partial \rho_A}\right)_{\rho_B} = -\frac{\phi}{\rho}\left(\frac{\partial f}{\partial \phi}\right)_{\rho}, \qquad \mu_B = \left(\frac{\partial f}{\partial \rho_B}\right)_{\rho_A} = \frac{1-\phi}{\rho}\left(\frac{\partial f}{\partial \phi}\right)_{\rho} \qquad (2.58)$$

and the expression for the osmotic pressure reduces to:

$$\Pi = \phi\left(\frac{\partial f}{\partial \phi}\right)_{\rho} - f(\phi) + f(\phi = 0) \qquad (2.59)$$

which will prove useful in the calculation of the osmotic pressure of polymer solutions (section 4.4).

Exercise: Derive from equation (2.59) van't Hoff's law for 'ideal' solutions, valid in the limit of very low solute concentrations.

Hint: Separate f into its ideal and excess parts; the latter is an analytic function of ϕ in the limit $\phi \to 0$.

2.3 Phase coexistence and interfaces

In this section we turn our attention from homogeneous bulk behaviour of fluids, to inhomogeneous multi-phase systems involving interfaces separating coexisting homogeneous phases. Consider two phases, say a and b, of a substance containing ν chemical species. The usual equilibrium conditions for the two phases to coexist are the equality of temperatures (thermal equilibrium), pressures (mechanical equilibrium) and chemical potentials of each species (chemical equilibrium), i.e.

$$T^{(a)} = T^{(b)} \qquad P^{(a)} = P^{(b)} \qquad \mu_\alpha^{(a)} = \mu_\alpha^{(b)} \qquad (\alpha = 1 \ldots \nu) \qquad (2.60)$$

Strictly speaking these conditions hold in the thermodynamic limit of infinitely large systems, separated by a planar interface. For finite systems, like small droplets, surface effects become important, and mechanical equilibrium does not imply the equality of the pressures on both sides of a *curved* interface. Such effects will be considered in section 6.4.

The chemical potentials are functions of the $\nu + 1$ intensive variables P, T and $\{x_\alpha\}$ $(1 \leq \alpha \leq \nu - 1)$, where the $\{x_\alpha\}$ are the concentrations or mole fractions N_α/N. If ϕ phases are present, there will be $\nu(\phi - 1)$ chemical equilibrium constraints, while the total number of intensive variables is $2 + \phi(\nu - 1)$, i.e. P, T and the $\nu - 1$ composition variables in each of the ϕ phases. The total number of intensive variables which may be varied independently is hence given by Gibbs' phase rule

$$\xi = 2 + (\nu - 1)\phi - \nu(\phi - 1) = 2 + \nu - \phi \tag{2.61}$$

In the simplest case of a one-component system $(\nu = 1)$, the coexistence of two phases $(\phi = 2)$ leaves one independent variable, e.g. the temperature. The two phases are in equilibrium along a coexistence line $P(T)$ in the pressure–temperature plane, as illustrated in figure 1.1. This line may be calculated from the single chemical equilibrium condition

$$\mu^{(a)}(P, T) = \mu^{(b)}(P, T) \tag{2.62}$$

provided the chemical potentials are known as functions of P and T in both phases. The situation is pictured schematically in figure 2.2, where phase a is the low-temperature phase (e.g. the solid phase). The volume and entropy per particle,

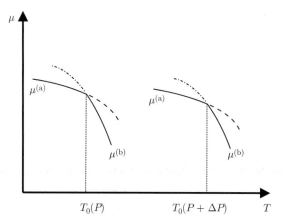

Figure 2.2. Chemical potentials $\mu^{(a)}$ and $\mu^{(b)}$ of two phases as functions of temperature T, for two pressures P and $P + \Delta P$. The intersection of the two curves $\mu^{(a)}$ and $\mu^{(b)}$ marks the phase coexistence temperature T_0, a function of pressure; the full curves correspond to the stable phase below and above T_0, while the dashed and dash-dotted parts of the intersecting curves correspond to the metastable phase.

$v = V/N$ and $s = S/N$ are generally different in two coexisting phases, separated by a first-order phase transition. The changes in entropy $\delta s = s^{(a)} - s^{(b)}$ and in volume $\delta v = v^{(a)} - v^{(b)}$, determine the slope of the coexistence curve. This is easily established by rewriting the equilibrium condition (2.62) for an infinitesimally close thermodynamic state $(P + dP, T + dT)$ along the coexistence line and subtracting (2.62)

$$\mu^{(a)}(P + dP, T + dT) - \mu^{(a)}(P, T) = \mu^{(b)}(P + dP, T + dT) - \mu^{(b)}(P, T) \quad (2.63)$$

Expanding the differences to first order in dP and dT leads to

$$\left(\frac{\partial \mu^{(a)}}{\partial P}\right)_T dP + \left(\frac{\partial \mu^{(a)}}{\partial T}\right)_P dT = v^{(a)} dP - s^{(a)} dT = v^{(b)} dP - s^{(b)} dT \quad (2.64)$$

and hence to the Clausius–Clapeyron relation

$$\frac{dP}{dT} = \frac{s^{(b)} - s^{(a)}}{v^{(b)} - v^{(a)}} = \frac{L}{T\Delta V} \quad (2.65)$$

where $L = T\Delta S = T\mathcal{N}_A(s^{(b)} - s^{(a)})$ is the latent heat per mole and $\Delta V = \mathcal{N}_A(v^{(b)} - v^{(a)})$ is the molar volume change at the transition. If a is the more ordered low temperature phase, L is always positive, and the same is generally true of ΔV, so that $dP/dT > 0$ in most situations. A notable exception is the freezing line of water, where $\Delta V < 0$ (water expands upon freezing!) and $dP/dT < 0$.

Three-phase coexistence is also possible for a one-component substance, but only at discrete points in the P–T plane, since the two chemical equilibrium conditions, namely $\mu^{(a)}(P, T) = \mu^{(b)}(P, T) = \mu^{(c)}(P, T)$ provide a system of two equations for two unknowns. This mirrors the Gibbs phase rule (2.61) which predicts $\xi = 0$ independent variables for $\nu = 1$ and $\phi = 3$.

In the case of a binary mixture, the coexistence between two phases implies the constraints

$$\mu_1^{(a)}(P, T, x_1^{(a)}) = \mu_1^{(b)}(P, T, x_1^{(b)}) \qquad \mu_2^{(a)}(P, T, x_1^{(a)}) = \mu_2^{(b)}(P, T, x_1^{(b)}) \quad (2.66)$$

where $x_1^{(a)}$ and $x_1^{(b)}$ are the concentrations of species 1 in the two phases. For any given pressure and temperature, the values of these two concentrations are entirely determined by the two equations (2.66), i.e. for a given total number of particles of the two species, the chemical equilibrium conditions control the *partitioning* of the two species between the two phases. Three-phase coexistence is possible at any temperature (below some critical temperature if the phases involved are fluid), for well defined compositions of the two species in the three phases, and a well defined pressure $P(T)$.

Phase coexistence means that two phases in contact, which generally have different overall densities, are separated by a surface or interface. In the presence of a gravitational field, the surface which separates bulk phases is a horizontal plane, with the denser phase below. Common examples are the planar interface between a liquid and its vapour, or the interface between the two phases, with different chemical compositions, of two partially miscible liquids. At the macroscopic level the planar interface appears to be sharp, but on the molecular scale, the interfacial *density profile* $\rho(z)$ (where z is the vertical coordinate) drops continuously from the bulk density of the liquid (phase a) to that of the vapour (phase b), over several molecular diameters. The description of interfacial density profiles at the molecular level will be considered in great detail in part III. At the macroscopic level, the presence of an interface means that a new extensive variable, the interfacial area A, has to be considered in the thermodynamic description of the system. The differential of the free energy (2.11) is then modified into

$$dF = -S\,dT - P\,dV + \gamma\,dA + \sum_\alpha \mu_\alpha\,dN_\alpha \qquad (2.67)$$

The coefficient γ, which has dimensions of an energy per unit area, is the surface (or interfacial) tension or surface free energy. $\gamma\,dA$ is equal to the work required to increase the surface area by an amount dA; the surface tension is hence expressed in J/m^2. At a molecular level, the surface tension results from the imbalance of the forces exerted by surrounding molecules on molecules situated at the interface. γ is also the force per unit length which has to be applied to one edge of an interface to increase its area by dA, as is illustrated in the case of a thin liquid film attached to a frame, shown in figure 2.3. One side of the frame is mobile; the work produced by the force F pulling on this mobile side during a displacement is related to the increase in surface energy associated with the areas of the two sides of the film,

$$\delta W = F\,dx = \gamma\,dA = 2\gamma L\,dx \qquad (2.68)$$

Hence γ can also be expressed in N/m, or more conveniently in mN/m.

From (2.67), the surface tension is given by the thermodynamic derivative:

$$\gamma = \left(\frac{\partial F}{\partial A}\right)_{V,T,\{N_\alpha\}} \qquad (2.69)$$

Figure 2.3. Schematic representation of the measurement of the interfacial tension of a liquid film spanning a rectangular frame of width L. The work done in pulling the mobile part of the frame a distance dx by applying a constant force F yields the interfacial tension via equation (2.68).

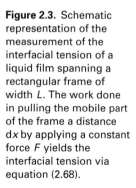
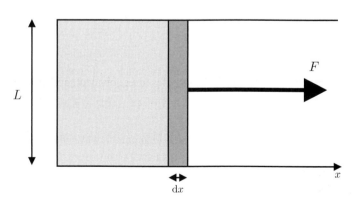

Similarly, for an open system:

$$\gamma = \left(\frac{\partial \Omega}{\partial A}\right)_{V,T,\{\mu_\alpha\}} \tag{2.70}$$

and since the extensive thermodynamic potentials F and Ω are homogeneous functions of first order in the extensive variables V, A and $\{N_\alpha\}$ (in the case of F), it is immediately clear by integration that:

$$F = -PV + \gamma A + \sum_\alpha \mu_\alpha N_\alpha$$
$$\Omega = -PV + \gamma A \tag{2.71}$$

The surface tension is always positive, i.e. intermolecular forces tend to reduce the interfacial area as much as possible. This means, in particular, that in the absence of gravity, a spherical interface will be favoured, since a sphere provides the smallest area for a given volume. For that reason liquid drops, or vapour bubbles inside a liquid, will tend to be spherical if deformations due to gravitational forces can be neglected.

2.4 Application 1: scaled particle theory

Consider an athermal fluid of hard core particles, like hard spheres or hard rods and hard ellipsoids, which provide basic models for simple fluids of spherical molecules, or for liquid crystals. At high densities, such fluids are dominated by excluded volume effects, and scaled particle theory provides a semi-macroscopic route to their equation of state. This presentation is restricted to the simplest case of a hard sphere fluid. Starting from the virial theorem (2.44), we first derive an expression for the pressure of the fluid in terms of the radial distribution function $g(r)$, and the pair potential $v(r)$ in the important case where the total potential energy V_N is pair-wise additive, as in equation (1.47). Substituting the latter into equation (2.44), and using the radial symmetry of $v(r)$, valid for spherical particles, one arrives, for a fluid of average density $\rho = N/V$ at:

$$P = \rho k_B T - \frac{1}{6V}\left\langle \sum_{i\neq j} \mathbf{r}_{ij} \cdot \nabla v(\mathbf{r}_{ij})\right\rangle = \rho k_B T - \frac{1}{6V}\left\langle \sum_{i\neq j} r_{ij} v'(r_{ij})\right\rangle \tag{2.72}$$

where $v'(r) = dv(r)/dr$. Since all N particles are equivalent, and exploiting the translational invariance of a homogeneous fluid, (2.72) may be re-expressed as:

$$P = \rho k_B T - \frac{N}{6V}\left\langle \sum_{j>1} r_{ij} v'(r_{ij})\right\rangle = \rho k_B T \left(1 - \frac{1}{6k_B T}\int r v'(r)\rho g(r)\,d\mathbf{r}\right) \tag{2.73}$$

where, according to equation (1.2), $\rho g(r)\,d\mathbf{r}$ is the mean number of molecules in a volume element $d\mathbf{r}$ at position $\mathbf{r}$ from any molecule (e.g. molecule 1) fixed at the origin; in view of the spherical symmetry, $4\pi\rho g(r)r^2\,dr$ is the mean number of molecules in a spherical shell around one molecule. Gathering results:

$$Z \equiv \frac{P}{\rho k_B T} = 1 - \frac{2\pi\rho}{3k_B T}\int_0^\infty g(r)v'(r)r^3\,dr \tag{2.74}$$

Equation (2.74) clearly shows how the interaction potential modifies the equation of state compared to the ideal gas result. Attractive potentials (or attractive parts in the $v(r)$ function, with $v'(r) > 0$) tend to make P smaller, while repulsive contributions ($v'(r) < 0$) increase the pressure. This equation, valid for continuous pair potentials $v(r)$, is not directly applicable to the case of hard spheres, where $v(r)$ is singular. Equation (2.74) may be rewritten in a more suitable form, by re-expressing $g(r)$ as:

$$g(r) = y(r)\exp(-\beta v(r)) \tag{2.75}$$

The Boltzmann factor $\exp(-\beta v(r))$ takes care of the rapid variation of the pair distribution function $g(r)$ when two particles approach very close. In the low density limit, $y(r)$ reduces to unity according to equation (1.2), and this function is expected to vary smoothly with r at higher density when two particles are near contact. Substitution of (2.75) into (2.74) leads immediately to

$$\frac{P}{\rho k_B T} = 1 + \frac{2\pi\rho}{3}\int_0^\infty \left[\frac{d}{dr}\exp(-\beta v(r))\right]y(r)r^3\,dr \tag{2.76}$$

This equation is suitable for going to the hard sphere limit, where $\exp(-\beta v(r)) = \theta(r - \sigma)$, θ denoting the Heaviside step function; the derivative of θ is the Dirac δ function. Since $y(r)$ reduces to $g(r)$ for $r > \sigma$, equation (2.76) finally yields the following virial expression for the pressure of a hard sphere fluid:

$$Z = \frac{P}{\rho k_B T} = 1 + \frac{2\pi\rho}{3}\sigma^3 g(\sigma) = 1 + 4\phi g(\sigma) \tag{2.77}$$

where $g(\sigma)$ is the $r \to \sigma^+$ limit, or contact value, of the pair distribution function of hard spheres at packing fraction ϕ. For sufficiently low densities or packing fractions, $g(r)$ reduces to the step function $\theta(r - \sigma)$, so that $g(\sigma) = 1$, and the equation of state reduces to the second virial coefficient approximation,

$$Z = 1 + 4\phi + \mathcal{O}(\phi^2) \tag{2.78}$$

While the first term corresponds to the ideal gas limit of non-interacting particles, the second term is the first in a series of corrections, in powers of the density, called the virial expansion; these corrections take systematically excluded volume effects into account. In the more general case of continuous potentials, the first correction to the ideal gas law is proportional to the second virial coefficient, which is easily derived from equation (2.76). In the low density limit, $y(r) = 1$, and a simple integration by parts of the integral leads to:

$$Z = 1 + B_2(T)\rho + \mathcal{O}(\rho^2) \tag{2.79}$$

$$B_2(T) = -2\pi \int_0^\infty [\exp(-\beta v(r)) - 1] r^2 \, dr \tag{2.80}$$

Note that, except in the athermal hard sphere case, $B_2(T)$ depends on temperature.

Scaled particle theory considers the probability of finding a cavity of radius between r and $r + dr$, within the hard sphere fluid, i.e. a spherical region void of the centre of any sphere. If the cavity were macroscopic, the probability would be given by:

$$p(r) \sim \exp(-\beta W(r)) \tag{2.81}$$

where $W(r)$ is the reversible work required to create a spherical cavity of radius r within the fluid; this reversible work is equal to the change in free energy of the fluid upon creation of the cavity, namely $W(r) = -P\Delta V + \gamma \Delta A$. The change in volume of the fluid is equal to minus the volume of the cavity, i.e. $\Delta V = -4\pi r^3/3$, while $\Delta A = 4\pi r^2$, so that:

$$W(r) = P\frac{4\pi r^3}{3} + \gamma 4\pi r^2 \tag{2.82}$$

This macroscopic result is assumed to hold even for a microscopic cavity, of radius comparable to molecular dimensions. The surface of such a cavity is highly curved, so that it is reasonable to assume that the surface tension must be corrected for curvature according to:

$$\gamma = \gamma_0(1 - \xi\sigma/r) \tag{2.83}$$

where γ_0 is the surface tension for a planar surface, and ξ is a dimensionless correction factor. The basic assumption of scaled particle theory is that equation (2.81), with $W(r)$ given by (2.82) and (2.83), remains valid for all cavities of radius $r > \sigma/2$. If $r < \sigma/2$, at most one sphere of diameter σ could fit into such a cavity. Hence $p(r)$ is simply the probability that no sphere is present in the cavity, i.e.

$$p(r) = 1 - \frac{4\pi}{3}\rho r^3 \qquad r < \sigma/2 \tag{2.84}$$

To determine the pressure of the hard sphere fluid, contact must be made with equation (2.77). The probability $dp = p(r + dr) - p(r)$ of finding a cavity with radius in the interval $[r, r + dr]$ may also be expressed by the product of the probability $p(r)$ of finding a cavity of radius larger than or equal to r, by the conditional probability of finding the centre of a sphere at a distance between r and $r + dr$ from the centre of the cavity, provided there is no other centre of a sphere inside the cavity. This conditional probability is expressed as $4\pi r^2 \rho G(r) \, dr$, which defines the function $G(r)$. Hence we have

$$dp = p(r)(4\pi r^2 \rho G(r)\,dr) \tag{2.85}$$

Now, since a sphere of diameter σ creates a cavity of radius σ around it, clearly $G(r = \sigma) = g(\sigma)$, so that the pressure (2.77) may be re-expressed as:

$$Z = 1 + \frac{2\pi\rho}{3}\sigma^3 G(\sigma) \tag{2.86}$$

Introducing the dimensionless quantities $\Gamma = \pi\gamma_0\sigma^2/6k_B T$ and $x = r/\sigma$ and combining equations (2.81) and (2.82) with (2.85), we find that:

$$G(x) = Z + \frac{2\Gamma}{\phi x} - \frac{\Gamma\xi}{\phi x^2} \tag{2.87}$$

Substituting this into the virial expression (2.86), we obtain the following relation for the reduced pressure:

$$Z = \frac{1 + 8\Gamma - 4\Gamma\xi}{1 - 4\phi} \tag{2.88}$$

This equation for Z still involves the two unknowns Γ and ξ. These are determined by matching the functional form (2.87) of G, to the form valid for $r < \sigma/2$, which follows from equations (2.84) and (2.85), namely:

$$G(x) = \frac{1}{1 - 8\phi x^3} \tag{2.89}$$

By matching the two forms of $G(x)$ and their derivatives $G'(x)$ at $x = 1/2$, we arrive at the following two relations:

$$Z + \frac{4\Gamma}{\phi} - \frac{4\Gamma\xi}{\phi} = \frac{1}{1 - \phi} \qquad 4\Gamma(2\xi - 1) = \frac{3\phi^2}{(1 - \phi)^2} \tag{2.90}$$

which yield Γ and ξ in terms of ϕ and Z. Upon substituting these values into (2.88), one arrives at the desired result for the equation of state of a hard sphere fluid:

$$Z = \frac{P}{\rho k_B T} = \frac{1 + \phi + \phi^2}{(1 - \phi)^3} \tag{2.91}$$

When compared to the results of computer simulations, this equation of state turns out to be remarkably accurate, up to packing fractions near freezing, where it overestimates the 'exact' equation of state by 7%. A semi-empirical correction to (2.91) leads to the very accurate Carnahan–Starling equation of state:

$$Z = \frac{1 + \phi + \phi^2 - \phi^3}{(1 - \phi)^3} \tag{2.92}$$

Expanding (2.91) or (2.92) in powers of ϕ, one recovers the coefficients of ϕ and ϕ^2 in the exact virial expansion, i.e. the exact second and third virial coefficients.

It is worth noting that the above scaled particle calculation yields a negative surface tension for the hard sphere fluid, namely:

$$\gamma = \frac{\pi \gamma_0 \sigma^2}{6 k_B T} = \frac{-3\phi^2(1+\phi)}{4(1-\phi)^3} \tag{2.93}$$

This is not surprising, since the hard sphere model lacks cohesive interactions. The liquid–vapour transition and hence the corresponding interface do not physically exist in such a fluid, since they are intimately related to the existence of attractive interactions (see part III).

Scaled particle theory is easily extended to treat mixtures of hard spheres of different diameters. It has also been generalized to non-spherical convex hard bodies, like rods, platelets and ellipsoids. A disadvantage of scaled particle theory is that it does not yield any structural information, like the pair distribution function.

> **Exercise:** Adapt the above calculation to the two-dimensional case, to derive the equation of state of a fluid of hard discs.

2.5 Application 2: particle insertion

The chemical potential μ_α is the work required to bring a single particle of species α from infinity, to be inserted into the bulk of a many-particle system. If the process is carried out at fixed total volume and temperature then, in the simplest case of a single chemical species, μ may be calculated by differentiating the Helmholtz free energy, according to equation (2.12):

$$\mu = \left(\frac{\partial F}{\partial N}\right)_{V,T} = F(N+1, V, T) - F(N, V, T) = -k_B T \ln\left(\frac{Q_{N+1}(V, T)}{Q_N(V, T)}\right) \tag{2.94}$$

Considering spherical particles for the sake of simplicity, and exploiting equation (2.38), the activity $z = \exp(\beta \mu)$ may be cast in the form:

$$z = (N+1)\Lambda^3 \frac{Z_N}{Z_{N+1}} \tag{2.95}$$

The ratio Z_{N+1}/Z_N of the configuration integrals has a physically appealing interpretation. The additional, $(N+1)$th particle may be regarded as a 'test particle' which explores the configuration space of the other N particles. The total potential energy V_{N+1} may be split into the interaction energy V_N of the initial particles, plus the energy of interaction W of the 'test particle' with the N other particles, so that

$$Z_{N+1} = \int d\mathbf{r}^N \int d\mathbf{r} \exp\left[-\beta\left(V_N(\{\mathbf{r}_i\}) + W(\mathbf{r}, \{\mathbf{r}_i\})\right)\right] \tag{2.96}$$

Due to translational invariance, the test particle may be placed at an arbitrary fixed point within the volume V occupied by the N particles, so that Z_{N+1} may be recast in the form:

$$Z_{N+1} = V \int d\mathbf{r}^N \exp(-\beta V_N) \exp(-\beta W) = V Z_N \langle \exp(-\beta W) \rangle_N \qquad (2.97)$$

where the statistical average is over the configuration space of the N-particle system, with the normalized Boltzmann weight $\exp(-\beta V_N)/Z_N$. Substitution into (2.95) leads to the desired expression for the activity or the chemical potential:

$$\frac{n\Lambda^3}{z} = \langle \exp(-\beta W) \rangle_N \implies \mu^{ex} = -k_B T \ln \langle \exp(-\beta W) \rangle_N \qquad (2.98)$$

where use was made of equation (2.41).

Exercise: By following similar steps, starting from Z_N rather than Z_{N+1}, show that

$$\frac{n\Lambda^3}{z} = \langle \exp(+\beta W) \rangle_{N+1} \implies \mu^{ex} = -k_B T \ln \langle \exp(\beta W) \rangle_{N+1} \qquad (2.99)$$

where the canonical average is now over an $N+1$ particle system, and W is the interaction energy of the $(N+1)$th particle with the others.

Equation (2.98) is the basis of the Widom particle insertion method for the direct calculation of chemical potentials in Monte-Carlo simulations. Due to translational invariance of the fluid, equation (2.98) is valid for any position of the test particle within the volume V, so that an equivalent expression involves averaging over all such positions, i.e.:

$$\mu^{ex} = -k_B T \ln \left[\frac{1}{V} \int_V d\mathbf{r} \langle \exp(-\beta W) \rangle_N \right] \qquad (2.100)$$

The integration is achieved by inserting the test particle at random positions, uniformly distributed throughout V, in a conventional Monte-Carlo simulation. For each such position, the interaction energy of the test particle with the N particles of the system is calculated, and accumulated to obtain an estimate of the integral in (2.100). Note that the test particle is a 'ghost' particle, which is removed once W has been calculated after a random insertion: it has no influence on the configurations of the other N particles, generated by the Metropolis algorithm mentioned in section (1.6). In the simple case of hard spheres, the Boltzmann factor in equation (2.100) is either 1, when the insertion is successful (i.e. the test particle overlaps with none of the N spheres within the volume V), or 0 when the insertion leads to overlap with at least one of the N spheres. μ^{ex} is then proportional to the logarithm of the 'success rate' of particle insertions. Clearly, at high packing fractions the available free volume is small compared to V, so that the success rate of particle insertion is very small, and hence the excess chemical potential will be large. Due to the small success rate, the Widom insertion method for the direct calculation of the chemical potential tends to become inefficient at high densities.

Particle insertions and removals are also central to the grand canonical Monte-Carlo method, which simulates open systems, allowed to exchange particles with a reservoir which fixes the chemical potentials of the various species.

2.6 Application 3: critical micellar concentration

Amphiphilic molecules were briefly described in section 1.2 as being relatively short chain molecules involving a hydrophilic head-group and a hydrophobic tail made up of one or more hydrocarbon chains. Such molecules have a very low solubility in either oil or water, but because of their amphiphilic nature, they readily adsorb at oil–water interfaces, where their head-groups will spontaneously go towards the aqueous side of the interface, while the tail chains will stick out into the oil phase, to optimize the hydrophilic and hydrophobic interactions. This adsorption leads to a strong reduction of the oil/water interfacial tension. At sufficiently low adsorption of the amphiphilic molecules, their mutual interactions may be neglected. For a surface density Γ of molecules adsorbed at the interface[2], the surface free energy may be cast in the simple form:

$$F^{(s)} = A\left\{\gamma_0 + k_B T\Gamma\left[\ln(\Gamma a) - 1\right] + \Gamma u\right\} \tag{2.101}$$

where A is the interfacial area, γ_0 is the interfacial tension of the bare oil–water interface (i.e. in the absence of amphiphilic molecules), a is an irrelevant area scale, and u is a characteristic energy per adsorbed molecule; the second term in braces is the ideal surface entropy. The resulting surface tension is:

$$\gamma = \left(\frac{\partial F^{(s)}}{\partial A}\right)_{T,N^{(s)}} = \gamma_0 - \Gamma k_B T \tag{2.102}$$

where we have taken into account that the derivative is taken at constant $N^{(s)} = \Gamma A$. Equation (2.102) shows a reduction in surface tension proportional to the adsorption of amphiphilic molecules. This surface activity has gained them the generic name of surfactants. The latter are ionic or non-ionic, depending on whether the head-group ionizes in water, or is merely highly polar. The name surfactant has been coined for synthetic amphiphiles, used as detergents. Natural amphiphiles, occurring in biological membranes, are called lipids, but we shall refer to natural or synthetic amphiphiles as surfactants throughout.

We now consider solutions of surfactant molecules in water. Due to their very low solubility, the surfactant will tend to phase separate when the concentration

---

[2] A precise thermodynamic definition of Γ will be given in section 6.1.

Figure 2.4. Schematic
representation of
surfactant molecules and
their packing. Cone
shaped molecules
($\xi = 1/3$) pack into
spherical micelles, while
cylinders ($\xi = 1$) tend to
form lamellar structures.

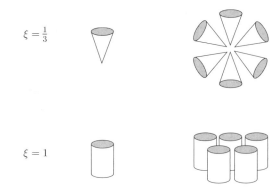

$$\xi = \frac{1}{3}$$

$$\xi = 1$$

increases. However, due to the amphiphilic nature of the surfactant, another scenario competes with complete phase separation, namely the spontaneous self-
assembly of surfactant molecules into supramolecular aggregates, called micelles. The size and shape of the micelles is determined by a balance between
the hydrophobicity of the hydrocarbon tails and the repulsion between the polar
or ionic head-groups. The overall shape of a surfactant molecule roughly resem-
bles a cone or a truncated cone, with a base area a largely determined by the
effective size of the head-group, and a height h related to the molecular weight,
or length of the hydrocarbon tail, as illustrated in figure 2.4. If ω denotes the
volume associated with a surfactant molecule, then the shape of the micelle is
largely determined by simple geometric considerations, and more specifically
by the aspect ratio $\xi = \omega/al$. If the surfactant tail consists of a single hydrocarbon chain, ξ is generally small ($\xi < 1/3$), favouring a large curvature of the
surface formed by the hydrophilic head-groups in contact with the water, and
hence spherical micelles. However, $\xi \simeq 1$ corresponds to cylindrical surfactant
molecules which pack easily into a planar bilayer, of zero curvature, as illustrated
in figure 2.4. If ξ is slightly less than unity, a supramolecular aggregate with a
finite curvature is favoured, leading to the bending of the planar bilayer into a
closed vesicle, thus avoiding the energetic penalty associated with edge effects,
where the hydrophobic tails would be exposed to water. Intermediate values of ξ
(say $\xi \simeq 1/2$) favour non-spherical, lamellar or cylindrical (rod-like) micelles.

The surfactant concentration c^* beyond which self-assembly into supramolecular aggregates sets in, is called the critical micellar concentration (cmc). It may
be estimated by expressing the 'chemical' equilibrium between non-associated
surfactant molecules (or monomers), and aggregates of two or more molecules.
This equilibrium is easily expressed in the dilute case, where interactions between different aggregates can be ignored, so that they form a polydisperse ideal
solution.

Consider a solution of N surfactant molecules in a volume V of water. These
molecules can form micelles of various sizes characterized by aggregation

numbers α (equal to the numbers of molecules in the corresponding micelles). Let N_α be the number of molecules belonging to micelles of aggregation number α (or α–micelles) and $n_\alpha = N_\alpha/\alpha$ the corresponding number of micelles. Clearly,

$$\sum_\alpha N_\alpha = N \qquad (2.103)$$

The situation where all $N_\alpha = 0$, except for $\alpha = N$, would correspond to a single, macroscopic aggregate, and hence to complete phase separation of surfactant and water. For any given distribution, the partition function for the non-interacting micelles reads, according to equation (2.42)

$$Q^{(\mathrm{id})} = \prod_{\alpha=1}^{N} \frac{1}{n_\alpha!} \left(\frac{V q_\alpha^{\mathrm{int}}}{\Lambda_\alpha^3} \right)^{n_\alpha} \qquad (2.104)$$

where q_α^{int} is the internal partition function of an α-micelle in an aqueous environment. Note that while the centre of a micelle can explore the whole available volume V of the solution, the positions of the surfactant molecules within the aggregate are restricted to the micellar volume. The resulting free energy is:

$$F = -k_B T \ln Q^{(\mathrm{id})} = \sum_{\alpha=1}^{N} \left[n_\alpha k_B T \left(\ln \left(\frac{n_\alpha \Lambda_\alpha^3}{V} \right) - 1 \right) + n_\alpha f_\alpha^{\mathrm{int}} \right] \qquad (2.105)$$

where $f_\alpha^{\mathrm{int}} = -k_B T \ln q_\alpha^{\mathrm{int}}$ is the internal free energy associated with an α-micelle. The chemical potential of a surfactant molecule belonging to an α-micelle is:

$$\mu_\alpha = \left(\frac{\partial F}{\partial N_\alpha} \right)_{T,V} = \frac{1}{\alpha} \left(\frac{\partial F}{\partial n_\alpha} \right)_{T,V} = \frac{k_B T}{\alpha} \ln \left(\frac{n_\alpha \Lambda_\alpha^3}{V} \right) + \epsilon_\alpha' \qquad (2.106)$$

where $\epsilon_\alpha' = f_\alpha^{\mathrm{int}}/\alpha$ is the internal free energy per surfactant molecule in a micelle of aggregation number α. In the special case of monomers ($\alpha = 1$), ϵ_1' is seen to reduce to the excess part of the chemical potential μ_1 of an isolated surfactant molecule in solution. If N_t denotes the total number of molecules (surfactant and water) in the solution, the chemical potentials (2.106) may be rewritten in the form

$$\mu_\alpha = \frac{k_B T}{\alpha} \ln \left(\frac{x_\alpha}{\alpha} \right) + \epsilon_\alpha \qquad (2.107)$$

where $\epsilon_\alpha = \epsilon_\alpha' - (k_B T/\alpha) \ln(v/\lambda_\alpha^3)$, v is the mean volume per molecule in the solution, and $x_\alpha = N_\alpha/N_t$ is the mole fraction of surfactant molecules that belong to α-micelles. Note that at the low surfactant concentrations ($x = N/N_t \ll 1$) considered here, v is practically equal to the volume per solvent molecule.

The condition of chemical equilibrium between micelles of different aggregation number requires all μ_α to be equal to the same value μ, which implies

$$x_\alpha = \frac{N_\alpha}{N_t} = \alpha \exp \left(\alpha(\mu - \epsilon_\alpha)/k_B T \right) \qquad (2.108)$$

Since $x_1 = \exp((\mu - \epsilon_1)/k_B T)$, the x_α may be written in a form reminiscent of the law of mass action

$$x_\alpha = \alpha x_1^\alpha \exp(\alpha(\epsilon_1 - \epsilon_\alpha)/k_B T) \tag{2.109}$$

Since $\sum_\alpha x_\alpha = x$, the mean aggregation number is simply

$$\bar{\alpha} = \frac{1}{x} \sum_\alpha x_\alpha \tag{2.110}$$

Exercise: Show that the free energy (2.105) may be cast in the equivalent form:

$$F = \sum_\alpha \left\{ n_\alpha k_B T \ln \frac{x_\alpha}{\alpha} + f_\alpha \right\} \tag{2.111}$$

Give the expression of f_α, and argue that it is independent of n_α. Show that the expression (2.108) for the mole fractions may be recovered by minimizing the free energy with respect to the x_α, subject to the constraint (2.103), using the method of Lagrange multipliers.

Since aggregation of surfactants into micelles is a means of reducing the hydrophobic interactions, the internal free energy per molecule ϵ_α is expected to decrease with increasing α, so that the Boltzmann factor in equation (2.109) favours aggregation, while the prefactor will favour monomers, since $x_1 < 1$. At low overall surfactant concentrations, μ will be much less than all ϵ_α, and all x_α will be negligibly small compared to x_1 so that $x \simeq x_1$. As the surfactant concentration increases, μ increases and becomes comparable to the lowest lying ϵ_α, so that the formation of finite aggregates becomes highly probable, while the fraction x_1 of monomers tends to saturate; the cmc is conventionally determined by expressing that the fraction of surfactant molecules within micelles of aggregation number $\alpha \geq 2$ equals the fraction of monomers,

$$\sum_{\alpha \geq 2} x_\alpha = x - x_1 = x_1 \tag{2.112}$$

During the micellization process, the chemical potential μ of surfactant molecules is pinned at its cmc value μ_c, since x_1 and hence μ_1 remain practically constant for $x > x_c$, a regime where any added surfactant will aggregate. The loss in translational entropy due to aggregation is compensated by a lowering of the total interaction free energy of molecules within micelles.

Broadly speaking, one may now distinguish between two scenarios, depending on how ϵ_α varies with α. If the geometry of the surfactant molecules favours the formation of spherical micelles, the energy ϵ_α will go through a sharp minimum for that aggregation number, say α^*, which allows the optimum curvature. In this case the largest contribution to the sum on the left-hand side of equation (2.112) will come from the term α^*, such that $\partial \epsilon_\alpha / \partial \alpha = 0$.

$$\sum_{\alpha \geq 2} x_\alpha \simeq x_{\alpha^*} = x_1 = x_c/2 \tag{2.113}$$

Substitution of (2.108) into (2.113) and elimination of the chemical potential $\mu = \mu_c$ at the cmc, leads to the following expression for the cmc:

$$x_c = 2(\alpha^*)^{\frac{1}{\alpha^*-1}} \exp\left(\beta \frac{(\epsilon_{\alpha^*} - \epsilon_1)\alpha^*}{\alpha^* - 1}\right) \simeq 2 \exp\left(\beta(\epsilon_{\alpha^*} - \epsilon_1)\right) \tag{2.114}$$

since $\alpha^* \gg 1$ (typically $\alpha^* = 10^2$). For most surfactants, $x_c \simeq 10^{-5}$–10^{-4}, and the spherical micelles which are formed are reasonably monodisperse, i.e. have the same aggregation number $\alpha = \alpha^*$, reflecting the sharp increase of ϵ_α for α slightly less or above α^*.

The formation of cylindrical rod-like micelles follows a quite different scenario, which is strongly influenced by end-effects. Contact of the hydrophobic tails with water at both ends of a cylindrical micelle is prevented by the formation of hemi-spherical caps, with a free energy cost δf, which is independent of the aggregation number α. The internal free energy per surfactant molecule of a cylindrical micelle is hence expected to be of the form:

$$\epsilon_\alpha = \epsilon_\infty + \delta f / \alpha \tag{2.115}$$

where ϵ_∞ would be the internal free energy per molecule of an infinite cylinder. For large values of α, ϵ_α varies slowly with aggregation number, so that one expects a rather broad (or polydisperse) distribution of micellar sizes beyond the cmc. Substitution of equation (2.115) into equation (2.109) yields

$$x_\alpha = \alpha \left[x_1 \exp(\beta \, \delta f)\right]^\alpha \exp(-\beta \, \delta f) \tag{2.116}$$

If this result is in turn substituted into the normalization condition $\sum_\alpha x_\alpha = x$, one arrives at the following relation between x_1 and x:

$$\sum_{\alpha=1}^{N} x_\alpha \simeq \exp(-\beta \, \delta f) \sum_{\alpha=1}^{\infty} \alpha \left[x_1 \exp(\beta \, \delta f)\right]^\alpha = \frac{x_1}{(1 - x_1 \exp(\beta \, \delta f))^2} = x \tag{2.117}$$

According to the definition (2.112), $x = 2x_1$ at the cmc, so that the critical concentration is $x_c = \exp(-\beta \delta f)$.

For arbitrary surfactant concentration, (2.117) admits the following physical root (ensuring that $x_1 < x$)

$$x_1 = \frac{x_c^2}{2x}(1 + 2x/x_c - \sqrt{1 + 4x/x_c}) \tag{2.118}$$

At very low surfactant concentration $x \ll x_c$, (2.118) yields $x_1 = x$, i.e. only monomers are present, as expected. However, well above the cmc, such that $x \gg x_c$, (2.118) reduces to $x_1 = x_c(1 - \sqrt{x_c/x})$ showing that the concentration of monomers remains practically pinned at its cmc value. The distribution of

molecules within cylindrical micelles of aggregation number α follows from
(2.116):

$$x_\alpha = \alpha x_c (1 - \sqrt{x_c/x})^\alpha \qquad (2.119)$$

The most probable value α^* of the aggregation number satisfies $dx_\alpha/d\alpha = 0$,
leading to $\alpha^* = \sqrt{x/x_c}$. The size of cylindrical micelles increases like the square
root of the surfactant concentration. For large α, the distribution (2.119) is well
approximated by an exponential, $x_\alpha \propto \alpha \exp(-\alpha/\alpha^*)$ confirming the high degree
of polydispersity of cylindrical micelles

Exercise: Using equation (2.116), calculate the mean coordination number $\bar{\alpha}$.
Discuss the result in the limits $x \ll x_c$ and $x \gg x_c$. Illustrate the high degree of
polydispersity in the latter limit, by calculating the relative root mean square
deviation $(\overline{\alpha^2} - \bar{\alpha}^2)/\bar{\alpha}$.

Hint: For $x \gg x_c$, α may be treated as a continuous variable.

The cmc may be determined experimentally by measuring colligative properties, like the osmotic pressure of the surfactant. The latter is observed to increase
linearly with surfactant concentration x up to $x = x_c$, beyond which it saturates,
signalling the aggregation of surfactant molecules into micelles, which make a
negligible contribution to the osmotic pressure.

Similarly the self-diffusion coefficient D of surfactant molecules is practically
constant as long as they remain unassociated; beyond the cmc, D is observed to
decrease because an increasing fraction of surfactant is 'bound' in micelles.
Finally, the surface tension γ of an aqueous surfactant solution drops rapidly
with increasing concentration, as indicated by equation (2.102); however, when
x exceeds x_c, γ saturates at its cmc value, since any additional surfactant will go
into the formation of micelles, and not to the solution–air interface.

Finally, let us mention that the aggregation process is not a static one, but results
from a dynamical equilibrium between all types of micelles. Surfactant molecules
are constantly being exchanged between different micelles, which can break and
reform at a rate that depends, for cylindrical micelles, on the temperature and of
the end energy δf. For this reason solutions of long cylindrical micelles are often
described as 'living polymer' solutions.

2.7 Application 4: depletion interactions and solvation forces

When considering suspensions of colloidal particles in a solvent, one is immediately faced with a system involving widely different length scales: the mesoscopic

scale of the colloidal particles with dimensions typically of the order of hundreds of nanometres or more, and the microscopic scale of the solvent molecules or dissolved ions, with diameters of the order of a fraction of a nanometre. In many practical situations, the system under consideration may also contain intermediate sized particles, like non-adsorbing polymer, or small colloidal particles or micelles, with characteristic sizes (e.g. the radius of gyration of polymer coils) of several nanometres.

Clearly, a statistical description of such highly asymmetric multi-component fluids, with a clear separation of length scales, requires a substantial amount of coarse-graining to be tractable. This is achieved by formally integrating (or 'tracing') out the degrees of freedom of the smaller particles. The resulting reduction of the total phase space of the initial multi-component system to the much smaller phase space of the large particles alone amounts to replacing the bare or direct interactions between the latter by effective interactions which involve the free energy (or the grand potential) of the small particles for any given configuration of the larger (colloidal) particles. Because these effective interactions derive from a free energy, rather than merely from a potential energy, they are generally state dependent, and have an entropic component. In many situations the latter dominates, and the effective forces are referred to as entropic forces.

We shall return to the formal reduction procedure in chapter 7, but the basic concept can be illustrated quite simply in the case of the so-called depletion forces between colloidal particles, as first investigated in the 1950s by S. Asakura and F. Oosawa. Two impenetrable bodies immersed in a solution of non-interacting (or weakly interacting) particles, like non-adsorbing polymer coils, are found to attract each other when their surfaces are closer than the characteristic size of the smaller particles. Consider first the case of two parallel plates (which may represent e.g. mineral platelets in a clay suspension) as illustrated in figure 2.5. The polymer coils are considered to be interpenetrable spheres of diameter σ (equal to twice their radius of gyration); their centres are excluded from slabs of width σ centred on the two plates. If the polymer coils are in equilibrium with a large reservoir, the work required to push the plates together to a separation d from an initially large separation (formally infinite), is given by the change in grand potential of the polymers

$$W(d) = \Omega(d) - \Omega(d \gg \sigma) = -\Pi[V'(d) - V'(d \gg \sigma)] \qquad (2.120)$$

where Π is the osmotic pressure of the polymer, and $V'(d)$ is the total volume available (or accessible) to the polymer when the two plates are a distance d apart. If V is the total volume of the system, then

$$V'(d) = V - 2V_{ex} + V_{ov}(d) \qquad (2.121)$$

The excluded volume V_{ex} around each of the two plates is equal to the product of their area A by σ. When the two plates are closer than the diameter of

A reminder of thermodynamics

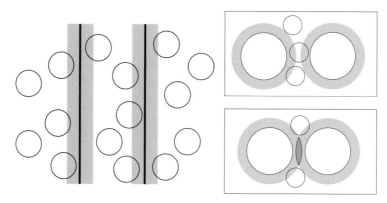

Figure 2.5. The origin of the depletion attraction between two planes (left) or two spheres (right), within the Asakura–Oosawa model. The centres of the spherical depletant particles (polymer coils or small colloidal particles) are excluded from the hatched areas around the planes or large spheres. The free volume, i.e. the volume (or area in this two-dimensional representation) accessible to the centres of the depletant particles, increases when the excluded volumes overlap, as in the lower right frame.

the polymer coils ($d < \sigma$), their exclusion zones overlap, and the corresponding overlap volume is obviously $V_{\text{ov}}(d) = A(\sigma - d)$; clearly $V_{\text{ov}}(d > \sigma) = 0$. Overlap of the excluded (or depletion) volumes associated with the two plates thus increases the volume accessible to the polymer, and hence its ideal entropy; the corresponding free energy, or grand potential, is consequently lowered when the two plates come together. Substituting these expressions into (2.120) one arrives at the desired result:

$$W(d) = -\Pi V_{\text{ov}}(d) = -\Pi A(\sigma - d)\theta(\sigma - d) \tag{2.122}$$

where θ is Heaviside's step function. $W(d)$ is hence always negative, corresponding to an effective attraction between the plates induced by the polymer; since the coils do not interact (ideal gas), $\Pi = \rho k_B T$, where ρ is the concentration (number density) of polymer chains. This shows that the effective interaction between the plates is state dependent, i.e. it depends on the temperature and polymer concentration. The physical interpretation of the result (2.122) is quite obvious: when the two plates are closer than σ, the polymer coils are 'squeezed out', so that they no longer balance the osmotic pressure exerted by the polymer on the outside of the plates.

The above calculation of the depletion potential is easily generalized to the physically more interesting case of two colloidal spheres, of radius R, immersed in a solution of N ideal (non-interacting) polymers (cf. figure 2.5), occupying a volume V; the centres of the ideal polymer coils are now excluded from a spherical zone of volume $V_{\text{ex}} = \pi(2R + \sigma)^3/6$ around each of the two spheres, while the overlap volume of the two exclusion zones is non-zero when the distance d between the surfaces of the colloid spheres is less than σ, i.e. when their centres are separated by $r < 2R + \sigma$; an elementary calculation yields:

$$V_{\text{ov}}(r) = \frac{\pi}{6}(2R + \sigma)^3 \left(1 - \frac{3r}{2(2R + \sigma)} + \frac{r^3}{2(2R + \sigma)^3} \right) \qquad 2R < r < 2R + \sigma$$

$$(2.123)$$

Gathering results, we find that:

$$V'(r) = V - \frac{\pi}{6}D^3 \left(1 + \frac{3r}{2D} - \frac{r^3}{2D^3} \right) \qquad 2R < r < D \qquad (2.124)$$

where $D = 2R + \sigma$. The total phase space volume of the polymer is $V'^N / \Lambda^{3N} N!$, reminiscent of the free volume result (1.16). The purely entropic free energy is accordingly

$$F = -TS = -k_B T \ln \left(\frac{V'^N}{\Lambda^{3N} N!} \right) = F_{\text{id}} - N k_B T \ln \left(\frac{V'(r)}{V} \right) \qquad (2.125)$$

where F_{id} is the usual ideal contribution to the free energy of the polymer, which is independent of r. The effective radial force induced between the two spheres is simply given by $-dF/dr = -dv_{\text{eff}}/dr$, where $v_{\text{eff}}(r)$ is the effective depletion potential, which differs from $F(r)$ only by an irrelevant constant, conventionally chosen such that $v_{\text{eff}}(r = D) = 0$. Substituting (2.124) into equation (2.125), linearizing the logarithm, which is justified since the macroscopic volume V is much larger than the exclusion volume $\pi D^3/6$, and imposing the above convention that v_{eff} vanish for $r = D$, one arrives at the desired result for the depletion potential between two spheres:

$$v_{\text{eff}}(r) = -\rho k_B T \frac{\pi D^3}{6} \left[1 - \frac{3r}{2D} + \frac{r^3}{2D^3} \right] \qquad \sigma < r < D \qquad (2.126)$$

The interaction is always attractive and, as in the case of two plates, it is proportional to the temperature and polymer concentration.

Exercise: By neglecting terms of quadratic and higher order in σ/R, show that the result (2.126) leads back, within an irrelevant constant, to the result (2.122) for the depletion potential between two plates, with $A = \pi R^2$, in the limit $R \gg \sigma$.

So far only isolated pairs of colloidal particles have been considered, corresponding to the infinite dilution limit of colloidal dispersions. For finite colloid concentrations, the basic relation (2.120) still holds, but the available (or free) volume of the polymer is now a more complicated function of the positions of all colloidal particles; in addition to the effective pair potential, there will be three and more-body terms corresponding to the overlap of three or more exclusion zones around neighbouring particles. However, the attractive effective

Figure 2.6. Measured depletion potential (in units of k_BT) between two silica spheres of diameter $\sigma = 1, 2\,\mu$m, induced by solutions of DNA molecules, versus separation r between the spheres, for increasing polymer concentrations (from top to bottom). The energy scales are indicated by the vertical bars in the top panel (corresponding to the dilute regime) and in the lower panel, where the concentrations are in the semi-dilute regime. After R. Verma, J.C. Crocker, T.C. Lubensky and A.G. Yodh, *Phys. Rev. Lett.* **81**, 4004 (1998).

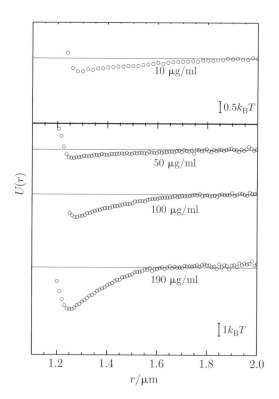

pair potential between spherical colloids makes the dominant contribution to the total effective interaction energy of the particles. $v_{\text{eff}}(r)$ may be measured using 'optical tweezers' and video microscopy techniques, and an example of experimental data is shown in figure 2.6. These data confirm the scaling of the amplitude of v_{eff} with polymer concentration ρ. Since the depth of the depletion attraction is of the order of k_BT, one might expect that it could induce a phase separation into dilute and concentrated dispersions, reminiscent of the gas–liquid (or condensation) transition of simple molecular liquids, driven by the van der Waals attraction between molecules. That this is indeed the case was shown both theoretically and experimentally, provided that the polymer-to-colloid size ratio $\xi = \sigma/2R$ is not too small, typically $\xi > 0.3$. Note that such phase separations lead to a partitioning of the polymer between the coexisting colloid-poor and colloid-rich phases; the most convenient variables to plot phase diagrams at a fixed temperature are the colloid packing fraction and the chemical potential μ_p of the polymer, which must be the same in the dilute and concentrated phases, rather than the polymer concentration, which is not the same in the two coexisting phases (polymer partitioning).

The depletant considered so far (e.g. the polymer) was assumed to be ideal, i.e. mutual interactions between depletant particles were neglected. This is an excellent approximation for polymers in θ-solvent (see section 4.4), but no longer holds when the depletant particles are, for instance, small colloidal particles with mutual excluded volume interactions. The simple free volume arguments presented earlier no longer apply, and correlation effects become important, particularly so at high depletant concentration. In particular, the smaller particles will tend to form layers near the surfaces of the larger colloids, and the interference between the layers around neighbouring colloidal particles leads to an oscillatory depletion force, which may be studied within the framework of the density functional formalism to be introduced in chapter 7.

Depletion interactions between colloidal particles induced by smaller, nanometric particles still take no account of the granularity of the molecular solvent. However, when the surfaces of the mesoscopic particles come within distances of the order of a few molecular diameters, the solvent granularity can no longer be neglected, and the solvent itself will induce effective interactions between the surfaces called solvation forces (or hydration forces in the important case where the solvent is water). Except for the change in scale, solvation forces are reminiscent of depletion forces induced by concentrated solutions of strongly interacting nanometric particles. Solvent molecules in the vicinity of a surface or a macromolecule (say of DNA) will order and adopt a layered structure which arises from a balance between their mutual interactions and the force field exerted by the surface. Thus a solvation zone is created around each mesoscopic particle or surface, where the structure and dynamics of the solvent differ significantly from the corresponding bulk properties. Solvation forces between macromolecules or colloidal particles arise when the solvation zones of neighbouring particles overlap, leading to a modification of the zones around individual particles. If this modification amounts to disrupting initially ordered solvation shells, the local solvent entropy is expected to increase, leading to a lowering of the free energy, and hence to an effective attraction between the surfaces. However, as the distance between the surfaces is varied, the layering of solvent molecules may be successively enhanced or reduced, resulting in an oscillatory solvation force, as measured with a surface force apparatus, capable of resolving distances of the order of 0.1 nm or less, and forces of the order of nanonewtons.

In the important special case where the solvent is water, the hydration forces depend very much on the hydrophilic or hydrophobic nature of the solvated surfaces, so that excluded volume considerations alone are not sufficient to explain solvation forces. The highly polar nature and hydrogen-bonding tendency of the water molecules must be included in a quantitative analysis of hydration forces.

Further reading

Thermodynamics is of course covered in a large number of textbooks. At the undergraduate level, and in relation to the present volume, let us mention P. Atkins, *Physical Chemistry*,

W.H. Freeman, 1997. At a more advanced level, the book by H.E. Callen, *Thermodynamics and an Introduction to Thermostatistics*, 2nd edn., Wiley, New York, 1985, gives a very rigorous discussion of thermodynamic principles. A convenient reference for thermodynamics of mixtures is E.A. Guggenheim, *Thermodynamics*, North Holland, Amsterdam, 1967.

An excellent introductory text to statistical mechanics is the book by D. Chandler, *Introduction to Modern Statistical Physics*, Oxford University Press, Oxford, 1987.

Several topics mentioned in our applications (in particular critical micellar concentration and depletion forces) are illustrated in the book by J. Israelachvili, *Intermolecular and Surface Forces*, Academic Press, New York, 1992.

3 Equilibrium fluctuations

3.1 Gaussian distribution of fluctuations

The equilibrium properties of macroscopic systems are stationary, i.e. they do not vary in time. Depending on the external or boundary conditions, some of the thermodynamic variables characterizing a macroscopic sample of matter are strictly constant or fixed, like the total energy if the sample is thermally insulated, while others may fluctuate slightly around their mean value, as is the case of the internal energy if the sample is in thermal equilibrium with a thermostat. However, if we rewrite equation (2.46) in the form

$$\frac{\sqrt{\langle H_N^2 \rangle - \langle H_N \rangle^2}}{\langle H_N \rangle} = \frac{\sqrt{C_v k_B T^2}}{U} \tag{3.1}$$

it is immediately clear that the relative fluctuation is of the order of $1/\sqrt{N}$, since both C_v and U are extensive variables; this is entirely negligible in a macroscopic sample containing a number of molecules comparable to Avogadro's number. In the thermodynamic limit, where $N \to \infty$, relative fluctuations vanish, and all thermodynamic properties are strictly constant.

The situation is, however, quite different if one examines mesoscopic subsystems of a macroscopic system in thermodynamic equilibrium. If the length scale of the subsystem is of the order of a micrometre, it will contain typically 10^{10}–10^{12} molecules in a condensed phase, so that relative fluctuations of order $1/\sqrt{N}$ are no longer totally negligible and have measurable consequences, like the Brownian motion of suspended colloidal particles or the scattering of visible light.

In this section we derive the probability density for the fluctuations of *local* thermodynamic variables, associated with mesoscopic subsystems of a macroscopic fluid in equilibrium, using essentially arguments borrowed from bulk thermodynamics.

The maximum entropy principle states that for a thermally insulated system of fixed total internal energy, the total entropy takes its maximum value when all parts of the system are in thermodynamic equilibrium. Due to the relentless thermal motion of the molecules, the local thermodynamic variables associated

with a mesoscopic subsystem may momentarily deviate from their overall equilibrium values; such a fluctuation will temporarily remove the system from overall
equilibrium, and will hence be characterized by a small reduction of the total
entropy of the system. According to Boltzmann's statistical definition of entropy,
and to the fundamental principle of equal probabilities in phase space (valid for
a closed, isolated system), the probability of the fluctuation under consideration
is given by Einstein's relation:

$$w \sim \exp(\Delta S_t / k_B) \tag{3.2}$$

Generally one is interested in the fluctuations of an arbitrary number of local
variables within all suitably defined subsystems of a macroscopic sample of
matter, and in the correlations between the fluctuations of variables associated
with different subsystems. Variables of interest may be the local density, the local
concentrations of various species within a mixture, the local charge density in
electrolytes, the local electric polarization in fluids of highly polar molecules
or particles, the local stress, or the local orientation (or director) of mesogenic
molecules in liquid crystals. Let ξ_k^j denote the jth variable associated with the
kth subsystem; in many applications it will be convenient to consider Fourier
components of spatially varying quantities, and k will then label such components.
For the sake of notational convenience, the indices k and j will be lumped into a
single index $i = (k, j)$ ranging from 1 to n. If $\bar{\xi}_i$ denotes the equilibrium value of
the ith variable, the quantities of main interest are the deviations from the mean
(or fluctuations):

$$x_i = \xi_i - \bar{\xi}_i \tag{3.3}$$

The entropy of the total system is a function of all these local fluctuations,

$$S_t = S_t(x_1, \ldots, x_n) \tag{3.4}$$

Since S_t reaches its maximum $S_t^{\max}$ at equilibrium ($x_i = 0$), the Taylor expansion
of (3.4) around $x_i = 0$ is a quadratic function of the variables x_i; to lowest non-
vanishing order:

$$\Delta S_t = S_t - S_t^{\max} = -\frac{1}{2} \sum_{i,j} g_{ij} x_i x_j \tag{3.5}$$

The matrix of coefficients g_{ij} is related to the second derivatives of S_t at its
maximum,

$$g_{ij} = -\left. \frac{\partial^2 S_t}{\partial x_i \partial x_j} \right|_{\{x_i = 0\}} \tag{3.6}$$

and is therefore symmetric and positive [1]. According to Einstein's general formula (3.2), the resulting probability density is of the multivariate Gaussian

---

[1] Positive, in reference to a matrix, means that the sum in equation (3.5) is always negative, for all
values of $\{x_i\}$. Diagonal elements of a positive matrix are always positive, while off-diagonal
elements may be negative.

form

$$w = a \exp\left[-\frac{1}{2k_B} \sum_{i,j} g_{ij} x_i x_j\right] \tag{3.7}$$

where the normalization constant a is determined by the condition

$$a \int_{-\infty}^{+\infty} dx_1 \ldots \int_{-\infty}^{+\infty} dx_n \exp\left[-\frac{1}{2k_B} \sum_{i,j} g_{ij} x_i x_j\right] = 1 \tag{3.8}$$

Note that, although the fluctuations $\{x_i\}$ are assumed to be small to justify truncation of the Taylor series (3.5) after quadratic order, the integrations are taken from $-\infty$ to $+\infty$, with negligible error in view of the rapid decay of the Gaussian function. The truncation can then be justified a posteriori by noting that typical values of x_i will be of order $N^{1/2}$, while the derivatives of S_t with respect to the extensive quantities x_i scale as N^{1-p} for a derivative of order p. Hence, the third-order term neglected in (3.5) is typically of order $N^{-1/2}$, while the second-order term is of order N^0. The integral (3.8) is easily calculated upon diagonalization of the argument of the exponential by a linear transformation, leading to the final result

$$w(x_1, \ldots, x_n) = \frac{(\det g)^{1/2}}{(2\pi k_B)^{n/2}} \exp\left[-\frac{1}{2k_B} \sum_{i,j} g_{ij} x_i x_j\right] \tag{3.9}$$

where $\det g$ denotes the determinant of the matrix g_{ij}.

The *thermodynamic forces* X_i conjugate to the fluctuations x_i are defined by

$$X_i = \frac{\partial \Delta S}{\partial x_i} = -\sum_k g_{ik} x_k \tag{3.10}$$

By analogy with Hooke's law for harmonic springs, the X_i act as 'restoring forces', tending to return the system to the thermodynamic equilibrium state of maximum entropy. In fact, as will be shown in part IV of this book, the thermodynamic forces control the entropy production during the relaxation (or regression) of spontaneous thermal fluctuations.

Statistical averages of fluctuating variables, weighted with the probability density (3.9), are easily evaluated by momentarily considering a modified distribution that produces non-zero $\langle x_i \rangle$, i.e. using the identity

$$a \int_{-\infty}^{+\infty} dx_1 \ldots \int_{-\infty}^{+\infty} dx_n x_i \exp\left[-\frac{1}{2k_B} \sum_{i,j} g_{ij}(x_i - x_i^0)(x_j - x_j^0)\right] = x_i^0 \tag{3.11}$$

valid for any x_i^0. Differentiating both sides of this equation with respect to x_i^0, and then setting all the x_i^0 equal to zero, one arrives at the desired result

$$\langle x_i X_j \rangle = -\left\langle \sum_k g_{jk} x_i x_k \right\rangle = -k_B \delta_{ij} \tag{3.12}$$

The n^2 relations (3.12) may be rewritten in matrix form, by defining the n-vectors $\mathbf{x}$ and $\mathbf{X}$ (with the components $\{x_i\}$ and $\{X_i\}$) and the $n \times n$ matrix g of coefficients g_{ij}. Equations (3.12) may then be cast in the form

$$\langle \mathbf{xX} \rangle = -k_{\mathrm{B}}\mathbf{I} \qquad \langle \mathbf{xx} \rangle = k_{\mathrm{B}}\mathbf{g}^{-1} \qquad \langle \mathbf{XX} \rangle = k_{\mathrm{B}}\mathbf{g} \tag{3.13}$$

where $\mathbf{I}$ denotes the unit matrix. Two fluctuations x_i and x_j will be statistically independent, or uncorrelated, if the (i, j) component of the inverse matrix $\mathbf{g}^{-1}$ is zero. Note that according to the first identity in (3.13), any fluctuation is uncorrelated with the $n - 1$ thermodynamic forces conjugate to all other fluctuations.

The continuum limit corresponds to the situation where the spatial extension of all subsystems, viewed on the macroscopic scale, is taken to zero, such that the discrete index labelling each subsystem becomes continuous. The variables x_i go over into one (or several) fields $x(\mathbf{r})$ associated with the local physical quantities (density, concentration, polarization, etc.) of interest, and the total entropy S_{t} becomes a functional of the various fields $x(\mathbf{r})$. Summations over i are to be replaced by integrals over the total volume V of the system. The correlation matrices (3.13) go over into correlation functions depending on the two continuous 'indices' $\mathbf{r}$ and $\mathbf{r}'$ defining the positions of infinitesimal subsystems within the sample. Such correlation functions, associated with local densities, will be the main object of section 3.4.

3.2 Density fluctuations in a one-component system

As a first example, let us consider a subsystem defined by a fixed volume V included in a much larger reservoir, in a fluid of average number density ρ. In a one-component fluid, this open subsystem is characterized by its energy U and number of particles N, with average values $\langle U \rangle$ and $\langle N \rangle = \rho V$. These two quantities play, in this particular case, the role of the x_i defined in section 3.1. The increase in total entropy associated with a fluctuation ΔN, ΔU of the subsystem can be obtained from

$$\Delta S_{\mathrm{t}} = \Delta S_{\mathrm{r}}(U_{\mathrm{r}} - \Delta U, N_{\mathrm{r}} - \Delta N) + \Delta S(\langle U \rangle + \Delta U, \langle N \rangle + \Delta N) \tag{3.14}$$

The index r refers to the reservoir with which the system is exchanging energy and particles, while the quantities S, U and N are relative to the subsystem under consideration. The entropy change for the reservoir is simply

$$\Delta S_{\mathrm{r}}(U_{\mathrm{r}} - \Delta U, N_{\mathrm{r}} - \Delta N) = -\frac{1}{T_{\mathrm{r}}}\Delta U + \frac{\mu_{\mathrm{r}}}{T_{\mathrm{r}}}\Delta N \tag{3.15}$$

For a macroscopic reservoir, higher order terms will involve the ratio of the subsystem volume to that of the reservoir, and can be neglected. The entropy variation of the subsystem, however, must be expanded to second order in the fluctuations, which gives

$$\Delta S(\langle U \rangle + \Delta U, \langle N \rangle + \Delta N) = \frac{1}{T_r} \Delta U - \frac{\mu_r}{T_r} \Delta N + \frac{1}{2} \Delta \left(\frac{1}{T} \right) \Delta U + \frac{1}{2} \Delta \left(\frac{-\mu}{T} \right) \Delta N \tag{3.16}$$

In deriving (3.16), we have used the property that for $\Delta U = 0$ and $\Delta N = 0$, the temperature and chemical potential of the subsystem are identical to those of the reservoir. The last two terms in (3.16) are a convenient expression of the second-order term in the expansion of ΔS in powers of ΔU and ΔN. $\Delta \left(\frac{1}{T} \right)$, for example, can be written as

$$\Delta \left(\frac{1}{T} \right) = \left(\frac{\partial S}{\partial U} \right)_{\langle U \rangle + \Delta U, \langle N \rangle + \Delta N} - \left(\frac{\partial S}{\partial U} \right)_{\langle U \rangle, \langle N \rangle} = \frac{\partial^2 S}{\partial^2 U} \Delta U + \frac{\partial^2 S}{\partial U \partial N} \Delta N \tag{3.17}$$

In the language of section 3.1, $-\mu/T$ and $1/T$ are the thermodynamic forces (or affinities) conjugate to N and U, respectively.

Using $\Delta U = T \Delta S + \mu \Delta N$, and expanding ΔS as $\Delta S = C_V \Delta T / T + (\partial S / \partial N)_T \Delta N$, and finally using the thermodynamic identity $(\partial S / \partial N)_{T,V} = -(\partial \mu / \partial T)_{N,V}$ (see equation (2.13)) the total entropy change associated with a fluctuation can be written as

$$\Delta S_t = -\frac{C_V}{2T^2} (\Delta T)^2 - \frac{1}{2k_B T} \left(\frac{\partial \mu}{\partial N} \right)_{T,V} (\Delta N)^2 \tag{3.18}$$

This is of the general form (3.5), with a diagonal g matrix. Number and temperature fluctuations are therefore uncorrelated. The mean squared amplitude of the temperature fluctuations is

$$\langle (\Delta T)^2 \rangle = \frac{k_B T^2}{C_V} \tag{3.19}$$

The physical significance of the fluctuations in an intensive parameter such as T is the following. The quantities that are *actually* fluctuating are extensive quantities, such as N or U. The temperature fluctuations are 'driven' by the fluctuations in these extensive quantities, i.e. the temperature of the subsystem in a configuration that deviates from equilibrium by an amount $\Delta N, \Delta U$ is, by definition, the temperature of an equilibrium system containing $\langle N \rangle + \Delta N$ particles and an energy $\langle U \rangle + \Delta U$.

The mean squared average of the number fluctuations is

$$\langle (\Delta N)^2 \rangle = \frac{k_B T}{(\partial \mu / \partial N)_{T,V}} \tag{3.20}$$

The right-hand side in (3.20) can be transformed by using the intensive character of μ. μ is a function of N/V and T, so that

$$\left(\frac{\partial \mu}{\partial N} \right)_{T,V} = -\frac{V}{N} \left(\frac{\partial \mu}{\partial V} \right)_{T,N} \tag{3.21}$$

The Maxwell relation $(\partial \mu / \partial V)_{T,N} = -(\partial P / \partial N)_{T,V}$ is then used to obtain finally

$$\left(\frac{\partial \mu}{\partial N}\right)_{T,V} = \frac{V}{N}\left(\frac{\partial P}{\partial N}\right)_{T,V} = -\left(\frac{V}{N}\right)^2 \left(\frac{\partial P}{\partial V}\right)_{T,N} \tag{3.22}$$

The fluctuations in particle number, or density, are therefore related to the isothermal compressibility χ_T of the fluid (see equation (2.28)), as was already established in section 2.2 (see equation (2.52)):

$$\langle (\Delta N)^2 \rangle = \frac{N^2 k_B T}{V^2 (-\partial P / \partial V)_{T,N}} = k_B T \rho \chi_T N \tag{3.23}$$

Equations (3.19) and (3.23) constitute first examples of a relationship between quantities that characterize the response to an external change, such as C_V or χ_T, and fluctuations. This connection will be studied in more detail in section 3.6

3.3 Concentration fluctuations in a mixture

As a further application of the thermodynamic fluctuation theory presented in section 3.1, consider now the case where the mesoscopic subsystem under consideration is again an open system of given volume V, the fluid being now a mixture of ν species. The fluctuating quantities will again be the number of particles of each species and the energy in the volume V. Following very closely the lines of section 3.2, one arrives at the following expression for the statistical weight of a fluctuation $\Delta T, \{\Delta N_\alpha\}$:

$$w(\Delta T, \{\Delta N_\alpha\}) \sim \exp\left(-\frac{C_v}{2k_B T^2}(\Delta T)^2 - \frac{1}{2k_B T}\sum_\alpha \sum_\beta \left(\frac{\partial \mu_\alpha}{\partial N_\beta}\right)_{V,T} \Delta N_\alpha \Delta N_\beta\right) \tag{3.24}$$

This is of the general form (3.7), with a $(\nu + 1) \times (\nu + 1)$ matrix of coefficients g_{ij}

$$g = \begin{pmatrix} C_v/T^2 & 0 \\ 0 & g' \end{pmatrix}$$

where g' is now a $\nu \times \nu$ matrix with elements $(\frac{\partial \mu_\alpha/T}{\partial N_\beta})_{V,T}$. A direct application of equation (3.13) shows that temperature and density fluctuations are uncorrelated, and leads back to the result (3.19) for the temperature fluctuation. Moreover

$$\langle \Delta N_\alpha \Delta N_\beta \rangle = (g')^{-1}_{\alpha\beta} = D_{\alpha\beta}/\det g' \tag{3.25}$$

where $D_{\alpha\beta}$ is the cofactor of $g'_{\alpha\beta}$ in the determinant of g'. Switching from the variables N_α to the conjugate variables μ_α, equation (3.25) leads back directly to the result (2.51) obtained within the statistical mechanics framework of the grand canonical ensemble, appropriate for open systems, namely

$$\langle \Delta N_\alpha \Delta N_\beta \rangle = G_{\alpha\beta} = \frac{\partial \langle N_\alpha \rangle}{\partial \mu_\beta / k_B T} \tag{3.26}$$

where the matrix G is the inverse of g'. The multicomponent generalization of equation (3.23) relating the isothermal compressibility to the density fluctuations is obtained from the extensivity property of the free energy F in the form

$$F(V, \{N_\alpha\}, T) = Vf(\{\rho_\alpha\}, T) \tag{3.27}$$

where f is the free energy per unit volume, which can only depend on the intensive quantities $\rho_\alpha = N_\alpha/V$. This implies that:

$$\frac{1}{k_B T \chi_T} = \frac{1}{k_B T} \sum_\alpha \sum_\beta \rho_\alpha \rho_\beta \left(\frac{\partial^2 f}{\partial \rho_\alpha \partial \rho_\beta}\right)_T = \frac{1}{k_B T} \sum_\alpha \sum_\beta \rho_\alpha \rho_\beta \left(\frac{\partial \mu_\alpha}{\partial \rho_\beta}\right)_T \tag{3.28}$$

or, in terms of the fluctuation matrix G defined in equation (3.26),

$$\frac{1}{k_B T \chi_T} = V \sum_\alpha \sum_\beta (G^{-1})_{\alpha\beta} \tag{3.29}$$

3.4 Local order and pair structure

The fluctuation relations established so far in this chapter are valid for mesoscopic subsystems, which are small compared to the total (macroscopic) system, yet large enough for thermodynamics to apply, as embodied in the assumption of local thermodynamic equilibrium. In the present section these considerations are generalized to the microscopic level. To account for the molecular graininess of a fluid, we consider microscopic subsystems by dividing up the total volume of the macroscopic system into a three-dimensional grid of small cells of volume of the order of the molecular volume σ^3 (for simplicity, the discussion will first be restricted to quasi-spherical molecules). This is the essence of a simple lattice gas model of fluids, but the subsequent considerations will apply to continuous fluids. Instead of considering the fluctuations of the molecular density in a mesoscopic subsystem, we focus on the microscopic density:

$$\hat{\rho}(\mathbf{r}) = \sum_{i=1}^N \delta(\mathbf{r} - \mathbf{r}_i) \tag{3.30}$$

where $\mathbf{r}_i$ denotes the position of the centre of the ith molecule. Integration of $\hat{\rho}(\mathbf{r})$ over the total volume V of the sample yields the total number of molecules, but the integral of the microscopic density over any cell of volume σ^3 within the total available volume V can only take the values 0 or 1, reflecting the excluded volume constraint. The statistical average of $\hat{\rho}(\mathbf{r})$ is the local density:

$$\rho(\mathbf{r}) = \langle \hat{\rho}(\mathbf{r}) \rangle \tag{3.31}$$

which tells us how many molecules will be found on average per unit volume around any position $\mathbf{r}$ within the fluid. If the latter is homogeneous, as is generally the case in the bulk (in the absence of an external field), translational invariance implies that $\rho(\mathbf{r})$ reduces to a constant, equal to the macroscopic density $\rho = N/V$.

The probabilities of finding molecules within neighbouring volume elements (or cells) are highly correlated in dense fluids, because of obvious packing considerations (cf. figure 1.2). To characterize these correlations, it appears quite natural to extend the macroscopic density fluctuation formula (3.20) to the microscopic scale.

This is achieved by considering the correlation between the fluctuations of the microscopic density (3.30) at two different points in space, say $\mathbf{r}$ and $\mathbf{r}'$; in practice one is interested in the fluctuations of $\hat{\rho}(\mathbf{r})$ away from its mean $\rho(\mathbf{r})$. For a homogeneous fluid, the relevant two-point correlation function is hence:

$$G(\mathbf{r}', \mathbf{r}'') = \langle [\hat{\rho}(\mathbf{r}') - \rho][\hat{\rho}(\mathbf{r}'') - \rho] \rangle = \langle \hat{\rho}(\mathbf{r}')\hat{\rho}(\mathbf{r}'') \rangle - \rho^2 \qquad (3.32)$$

Due to translational invariance, $G(\mathbf{r}, \mathbf{r}') = G(\mathbf{r} - \mathbf{r}')$ is a function of the difference $\mathbf{r} - \mathbf{r}'$ only. Substitution of (3.30) into (3.32), and separation of the $i = j$ and $i \neq j$ contributions in the resulting double sum, leads to

$$G(\mathbf{r} - \mathbf{r}') = \rho \delta(\mathbf{r} - \mathbf{r}') + \rho^2(g(\mathbf{r} - \mathbf{r}') - 1) \qquad (3.33)$$

The non-trivial factor g is the pair distribution function already introduced in section 1.1. For an isotropic fluid, rotational invariance implies that $g(\mathbf{r}' - \mathbf{r}'') = g(r)$, where $r = |\mathbf{r}' - \mathbf{r}''|$ is the distance between locations $\mathbf{r}'$ and $\mathbf{r}''$. According to the earlier definition of $g(r)$ in section 1.1, the probability of finding the centre of a molecule at a distance r, within dr, from another molecule is:

$$P(r)dr = \rho g(r) \times (4\pi r^2 dr) \qquad (3.34)$$

and it is this interpretation which allows $g(r)$ in equation (3.33) to be identified with the pair distribution of section 1.1.

Contrarily to crystalline solids, characterized by long-range positional order, fluids only exhibit short-range order which extends over a few molecular diameters at most, as illustrated in figure 1.3. In other words positional correlations decay on molecular scales, so that

$$G(\mathbf{r}' - \mathbf{r}'') \to 0 \qquad |\mathbf{r}' - \mathbf{r}''| \gg \sigma \qquad (3.35)$$

which is equivalent to $\lim_{r \to \infty} g(r) = 1$; it is sometimes convenient to introduce the pair correlation function $h(r) = g(r) - 1$, which vanishes for $r \gg \sigma$.

However, as already emphasized in section 1.1, the strong short-range repulsion between molecules, which prevents them from overlapping, is reflected in the excluded volume condition:

$$g(r) = 0 \qquad r \leq \sigma \qquad (3.36)$$

which holds independently of density.

It is interesting to establish the link between the macroscopic fluctuation formula (3.20), and the correlation function $G(\mathbf{r}', \mathbf{r}'')$. From the very definition (3.30) of the microscopic density, it is clear that:

$$\langle (N - \langle N \rangle)^2 \rangle = \left\langle \int_V [\hat{\rho}(\mathbf{r}') - \rho] d\mathbf{r}' \int_V [\hat{\rho}(\mathbf{r}'') - \rho] d\mathbf{r}'' \right\rangle$$

$$= \int_V d\mathbf{r}' \int_V d\mathbf{r}'' G(\mathbf{r}', \mathbf{r}'') \simeq V \int G(\mathbf{r}) d\mathbf{r} \qquad (3.37)$$

where the statistical average is taken over a grand canonical ensemble. The last equality in equation (3.37) is only approximate, in the sense that it neglects a term proportional to the surface of the subvolume V under consideration, times the range of the correlation function [2]. If the volume is macroscopic, the neglected contribution is indeed much smaller than the volume term, and the approximation is perfectly justified. From this argument it appears that a 'macroscopic' subvolume can be defined as a subvolume whose lateral dimensions are much larger than the range of the pair correlations.

Inserting equation (3.32) into (3.37), and remembering (3.20), one arrives at a key result, linking pair correlations on the molecular scale to macroscopic thermodynamics, and more specifically to the isothermal compressibility:

$$1 + \rho \int [g(r) - 1] d\mathbf{r} = \rho k_B T \chi_T \qquad (3.38)$$

Since the compressibility diverges at the critical point of the gas–liquid phase transition, the range of the pair correlation function $h(r) = g(r) - 1$ is expected to increase dramatically near the critical point, so as to yield a divergent integral. We shall return to this point when critical fluctuations will be discussed in more detail, but this observation prompted Ornstein and Zernike (OZ) to introduce an auxiliary function called the direct correlation function, related to $h(r)$ by the following convolution product:

$$h(r) = c(r) + \rho \int d\mathbf{r}' c(r') h(|\mathbf{r} - \mathbf{r}'|) \qquad (3.39)$$

The physical interpretation of this definition is that the correlation between the positions of two molecules is the sum of their direct correlation, due to their mutual interaction, and to the indirect correlations mediated by other neighbouring molecules; these indirect correlations, involving one or several 'intermediate' molecules, are embodied in the convolution term in (3.39), which can be rewritten formally as

$$h = c + \rho c * c + \rho^2 c * c * c + \dots \qquad (3.40)$$

---

[2] This can be seen in the following manner. The inner integral $\int_V G(\mathbf{r}' - \mathbf{r}'') d\mathbf{r}''$ is independent of the variable $\mathbf{r}'$, except when the point $\mathbf{r}'$ is so close to the boundary of V that the integration over $\mathbf{r}''$ cannot be replaced by an integration over all space.

At very low density the influence of the 'intermediate' molecules becomes negligible, so that $h(r) \simeq c(r)$, as implied by (3.39) in the limit $\rho \to 0$. In that limit $h(r)$, reduces to $\exp(-\beta v(r)) - 1$, and the key assumption is that the range of $c(r)$ remains the same, i.e. equal to the range of $-\beta v(r)$ at all densities, while the range of the total correlation function $h(r)$ may increase dramatically near the critical point. Approximate theories of pair correlations in dense fluids are always implicitly or explicitly based on the assumption that

$$c(r) = -\beta v(r) \qquad r \gg \sigma \qquad (3.41)$$

independently of density. In fact a very successful theory of pair correlations in fluids of hard spheres of diameter σ, due to Percus and Yevick, is based on the prescription:

$$h(r) = -1 \qquad r < \sigma \qquad (3.42)$$

$$c(r) = 0 \qquad r > \sigma \qquad (3.43)$$

Equation (3.42) expresses the requirement of non-penetrability of hard spheres, while (3.43) assumes that the conjectured asymptotic form holds for all distances $r \geq \sigma$, where hard spheres no longer interact. These 'closure' relations, together with the OZ relation (3.39) form a closed set of equations for the unknown functions $h(r)$ and $c(r)$, which can be solved analytically for $c(r)$ in the range $r < \sigma$. The resulting pair distribution function $g(r) = 1 + h(r)$ agrees well with computer simulation data for all but the highest fluid densities.

It proves convenient to introduce the Fourier ρ_k transform of the microscopic density (3.30), namely

$$\rho_k = \int \hat{\rho}(\mathbf{r}) \exp(i\mathbf{k} \cdot \mathbf{r}) d\mathbf{r} = \sum_{i=1}^{N} \exp(i\mathbf{k} \cdot \mathbf{r}_i) \qquad (3.44)$$

Since ρ_k depends on the positions of all the molecules, it is often referred to as a *collective coordinate*. Translational invariance implies that for a homogeneous fluid:

$$\langle \rho_k \rangle = N \delta_{k,0} \qquad (3.45)$$

The correlation function of ρ_k with its complex conjugate, $\rho_k^* = \rho_{-k}$ (i.e. the mean square of the density fluctuation with wavevector $\mathbf{k}$), is a real function of the wavevector, called the static structure factor:

$$S(\mathbf{k}) = \frac{1}{N} \langle \rho_k \rho_{-k} \rangle \qquad (3.46)$$

Substituting the definition (3.44) into (3.46), it is easily verified that $S(k)$ is the Fourier transform of the pair correlation function $h(r) = g(r) - 1$, except for unimportant constants. Indeed, remembering equation (3.32):

$$S(\mathbf{k}) = \frac{1}{N} \int d\mathbf{r}' \exp(i\mathbf{k} \cdot \mathbf{r}') \int d\mathbf{r}'' \exp(i\mathbf{k} \cdot \mathbf{r}'') \langle \hat{\rho}(\mathbf{r}') \hat{\rho}(\mathbf{r}'') \rangle$$

$$= \frac{V}{N} \int \exp(i\mathbf{k} \cdot \mathbf{r}) G(\mathbf{r}) d\mathbf{r} + \rho \delta_{\mathbf{k},0} \tag{3.47}$$

$$= 1 + \rho \int h(\mathbf{r}) \exp(i\mathbf{k} \cdot \mathbf{r}) d\mathbf{r}$$

It will be shown in section 3.8 that the structure factor is directly measurable by diffraction of X-rays, neutrons or visible light (in the case of colloidal dispersions). From equation (3.47) we may then conclude that an experimental determination of the pair distribution function follows from inverse Fourier transformation of the measured structure factor.

The method used to derive equation (3.47) can also be used to show that for a wavevector $\mathbf{k}' \neq \mathbf{k}$,

$$\langle \rho_{\mathbf{k}} \rho_{\mathbf{k}'} \rangle = 0 \qquad \text{for} \qquad \mathbf{k} \neq \mathbf{k}' \tag{3.48}$$

which is simply a consequence of translational invariance, $G(\mathbf{r}, \mathbf{r}') = G(\mathbf{r} - \mathbf{r}')$.

In the absence of spatial correlations, as in an ideal gas, $h(r) = 0$, and $S(\mathbf{k}) = 1$ for all $\mathbf{k} \neq 0$. In a dense fluid, $h(r)$ is highly structured, and so is $S(k)$, examples of which are shown in figure 3.1. The main peak occurs at $k \simeq 2\pi/d$, where d is the mean nearest-neighbour distance, comparable to the molecular diameter

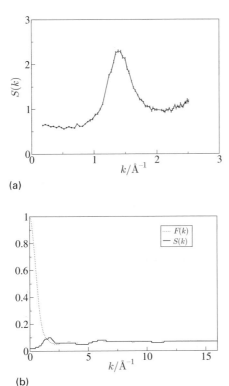

(a)

(b)

Figure 3.1. (a) Structure factor of a polymer melt (polybutadiene) at $T = 293$ K and $P = 1$ atm. The peak reflects the short-range order at the monomer scale. (b) Structure factor and molecular form factor in a simple organic liquid, toluene. (Courtesy of C. Alba, Orsay.)

σ. The main peak reflects the short-range order in the fluid and may be regarded as a remnant of the much sharper first Bragg peak of crystalline solids, which characterizes the periodic long-range order of the crystal. If the OZ relation (3.39) is substituted in the expression (3.47) for $S(\mathbf{k})$, it immediately follows that for $\mathbf{k} \neq 0$,

$$S(\mathbf{k}) = 1 + \rho h(\mathbf{k}) = \frac{1}{1 - \rho c(\mathbf{k})} \tag{3.49}$$

where $h(\mathbf{k})$ and $c(\mathbf{k})$ are the Fourier transforms of the total and direct correlation functions. Equation (3.49) shows that $\rho c(\mathbf{k}) = 1 - 1/S(\mathbf{k})$ is directly related to the inverse of the structure factor.

So far, we have only considered correlations between the centres of spherical molecules or colloidal particles. Most molecules or macromolecular aggregates are, of course, non-spherical (or anisometric), or even if they are quasi-spherical, they may carry large electric multipole moments, or interact via highly anisotropic forces. Good examples are water molecules, which form strong hydrogen bonds with their neighbours, elongated or flat mesogenic molecules assembling into liquid crystals, or colloidal particles with large embedded magnetic moments, capable of forming ferrofluids. In dense fluids or suspensions of such anisometric particles one may expect orientational short-range order, i.e. neighbouring particles will have preferential relative orientations. In liquid crystals made up of sufficiently elongated (or flat) molecules, the orientational order may extend over macroscopic distances. Orientational correlations between electric dipoles also control the dielectric behaviour of polar liquids, as will be shown in section 3.7. A detailed description of positional and orientational correlations between rigid molecules requires a molecular pair distribution function $g(\mathbf{R}, \Omega, \Omega')$ which depends on the vector $\mathbf{R}$ joining the centres of mass of two molecules and on the vectors of Euler angles, Ω and Ω', which define the orientations of the two molecules in a laboratory-fixed frame. Needless to say, this is a very complicated function, and orientational correlations are more usefully characterized by appropriate moments of the complete molecular pair distribution function. In the case of uniaxial molecules (e.g. rod-like or disc-like particles), one may define the following set of orientational correlation functions:

$$g_\ell(R) = \frac{1}{(4\pi)^2} \int \int g(\mathbf{R}, \Omega, \Omega') P_\ell(\mathbf{u} \cdot \mathbf{u}') d\Omega d\Omega' \tag{3.50}$$

where $\mathbf{u}$ and $\mathbf{u}'$ are unit vectors along the axes of the two molecules and $P_\ell(x)$ is the ℓth order Legendre polynomial. In particular $g_0(R)$ is just the pair distribution function of the centres of mass of the molecules. Since orientational correlations usually extend only over a few molecular diameters, the correlation functions $g_\ell(r)$ ($\ell \neq 0$) decay to zero over microscopic scales. One may then define correlation factors G_ℓ by integrating over all intermolecular distances:

$$G_\ell = \rho \int g_\ell(R) d\mathbf{R} \qquad (3.51)$$

Long-range orientational order in the nematic phase of a liquid crystal means that the range of $g_2(R)$ becomes macroscopic, and the corresponding correlation factor G_2 is of order N.

Molecular pair distribution functions are useful only for rigid particles, where the intramolecular structure is fixed. This is no longer true of highly flexible macromolecules, like polymers, which can exhibit very large numbers of different conformations.

In that case it makes more sense to consider monomer–monomer pair distribution functions inside the same polymer coil, or between two different coils. In principle this requires a very large number of different pair distribution functions (strictly speaking of order L^2), but for very long polymer chains, most monomers are equivalent (except for end effects), so that a full description of the pair structure requires only a small number of monomer–monomer correlation functions.

3.5 Link with thermodynamics

In the previous section it was already shown that there exists a rigorous link between the pair distribution function $g(r)$ and macroscopic thermodynamics, namely via the compressibility relation (3.38). If this relation is compared to equation (3.47) it is immediately clear that

$$\lim_{k \to 0} S(k) = \rho k_B T \chi_T \qquad (3.52)$$

showing that a measurement of the long wavelength limit of the intensity of scattered radiation, which will be shown in section 3.8 to be proportional to $S(k)$, gives direct access to a macroscopic thermodynamic property. In the case of mixtures one may define partial structure factors:

$$S_{\alpha\beta}(k) = \frac{1}{\sqrt{N_\alpha N_\beta}} \langle \rho_\mathbf{k}^\alpha \rho_{-\mathbf{k}}^\beta \rangle \qquad 1 \leq \alpha, \beta \leq \nu \qquad (3.53)$$

where $\rho_\mathbf{k}^\alpha$ is the collective coordinate (3.44) for particles of species α. The long wavelength limits of these partial structure factors are then directly related to the macroscopic fluctuation formulae (3.26) for mixtures. It is important to note that the compressibility equation (3.52) does not explicitly involve the intermolecular forces, although the latter determine, of course, the pair structure. Other thermodynamic properties can also be calculated from the pair distribution function if the particles (now assumed to be spherical for the sake of simplicity) interact via pair interactions only, i.e. provided the total interaction energy between particles is of the form (1.47), which may be rewritten, using the definition (3.30),

$$V_N(\{\mathbf{r}_i\}) = \frac{1}{2} \int \int \hat{\rho}(\mathbf{r})v(\mathbf{r} - \mathbf{r}')\hat{\rho}(\mathbf{r}')\mathrm{d}\mathbf{r}\mathrm{d}\mathbf{r}' \tag{3.54}$$

where it is understood that the infinite interaction energy of a particle with itself (corresponding to the $i = j$ self terms), is omitted. Taking the statistical average of both sides of (3.54), and remembering equations (3.32) and (3.33), one arrives at the following expression for the excess (non-ideal) internal energy:

$$
\begin{aligned}
U^{\mathrm{ex}} = \langle V_N \rangle &= \frac{\rho^2}{2} \int \mathrm{d}\mathbf{r} \int \mathrm{d}\mathbf{r}' v(\mathbf{r} - \mathbf{r}')g(\mathbf{r} - \mathbf{r}') \\
&= N\frac{\rho}{2} \int v(\mathbf{r})g(\mathbf{r})\mathrm{d}\mathbf{r} = N2\pi\rho \int_0^\infty v(r)g(r)r^2\mathrm{d}r
\end{aligned}
\tag{3.55}
$$

where the last equality holds for an isotropic fluid. The result (3.55) could have been easily guessed from the physical interpretation of $g(r)$.

Similarly, for pair-wise additive interactions, the pressure may be calculated from a knowledge of $g(r)$ and $v(r)$ via equation (2.73).

Both U^{ex} and P^{ex}, which are first derivatives of the free energy, are thus expressible in terms of $g(r)$ and $v(r)$ alone, but this is not the case of the free energy itself, or alternatively of the entropy, which require, in principle, a knowledge of distribution functions of all orders (i.e. involving not only two, but n particles, with $n \le N$), for any given thermodynamic state.

Alternatively, the excess free energy may be calculated by gradually 'turning on' the pair interactions between the molecules. This is formally achieved by multiplying the total interaction energy V_N by a coupling constant λ which is varied continuously from $\lambda = 0$ (ideal gas of non-interacting molecules) to $\lambda = 1$ (system of fully interacting molecules). For any intermediate value of λ, the configuration integral (2.38) is:

$$Z_N(\lambda) = \int \exp(-\beta\lambda V_N)\mathrm{d}\mathbf{r}_1 \ldots \mathrm{d}\mathbf{r}_N \tag{3.56}$$

and the excess Helmholtz free energy is $F^{\mathrm{ex}}(\lambda) = -k_\mathrm{B}T \ln Z_N(\lambda)/V^N$. Differentiation with respect to λ yields:

$$\frac{\partial F^{\mathrm{ex}}(\lambda)}{\partial \lambda} = \frac{1}{Z_N(\lambda)} \int \exp(-\beta\lambda V_N)V_N\mathrm{d}\mathbf{r}_1 \ldots \mathrm{d}\mathbf{r}_N = \langle V_N \rangle_\lambda \tag{3.57}$$

Integration of both sides of (3.57) leads to the following coupling constant integration expression for the free energy of the fully interacting system ($\lambda = 1$):

$$F^{\mathrm{ex}} = \int_0^1 \langle V_N \rangle_\lambda \mathrm{d}\lambda = \frac{N\rho}{2} \int_0^1 \mathrm{d}\lambda \int g_\lambda(r)v(r)\mathrm{d}r \tag{3.58}$$

since $F^{\mathrm{ex}}(\lambda = 0) = 0$. Since λ scales the inverse temperature in equation (3.56), it is clear that the expression (3.58) is equivalent to thermodynamic integration of the standard relation $U^{\mathrm{ex}} = \partial(\beta F^{\mathrm{ex}})/\partial\beta$, which follows directly from equations (2.37) and (2.39). Calculation of F^{ex} thus requires a knowledge of the pair distribution function $g_\lambda(r)$ over a continuous range of intermolecular couplings or temperatures.

A much more practical, albeit approximate, route to the free energy is provided by thermodynamic perturbation theory and the Gibbs–Bogoliubov inequality. As pointed out in section (1.2), the pair interaction between two molecules or colloidal particles naturally separates into a strong, steeply repulsive, short-range part, arising from excluded volume considerations, and a weaker, long-range, mostly attractive component, which includes van der Waals dispersion forces, Coulombic interactions, or effective entropic forces, arising e.g. from depletion, hydration or hydrophobicity. The total interaction energy of N particles accordingly splits into a dominant short-range part, $V_N^{(0)}$, and a weaker perturbation W_N which is more long range in character:

$$V_N(\{\mathbf{r}_i\}) = V_N^{(0)}(\{\mathbf{r}_i\}) + W_N(\{\mathbf{r}_i\}) \tag{3.59}$$

The auxiliary, hypothetical system of particles interacting only via the short-range part $V_N^{(0)}$, is called the reference system. Since $V_N^{(0)}$ accounts for excluded volume effects, the short-range local structure of the fully interacting system is expected to be close to that of the reference system, at least for dense fluids, where molecules are closely packed.

The configuration integral of the full system is easily related to that of the reference system by noting that:

$$Z_N = Z_N^{(0)} \int \frac{\exp\left(-\beta V_N^{(0)}\right) \exp(-\beta W_N) \, d\mathbf{r}_1 \ldots d\mathbf{r}_N}{Z_N^{(0)}} = Z_N^{(0)} \langle \exp(-\beta W_N) \rangle^{(0)} \tag{3.60}$$

The convexity of the exponential function immediately implies that $\langle \exp(-\beta W_N) \rangle^{(0)} \geq \exp(-\beta \langle W_N \rangle^{(0)})$, so that

$$Z_N \geq Z_N^{(0)} \exp(-\beta \langle W_N \rangle^{(0)}) \tag{3.61}$$

Taking logarithms of both sides of this inequality, and adding the trivial ideal contribution, one arrives at the Gibbs–Bogoliubov inequality:

$$F \leq F^{(0)} + \langle W_N \rangle^{(0)}$$
$$\leq F^{(0)} + \frac{N}{2} \rho \int g^{(0)}(\mathbf{r}) w(\mathbf{r}) d\mathbf{r} \tag{3.62}$$

where the second line holds if the perturbation W_N is pair-wise additive; $g^{(0)}$ is the pair distribution function of the reference system, while $F^{(0)}$ is its free energy. Particularly for dense fluids, a good approximation to the free energy of the system of interest is obtained by replacing the inequality (3.62) by an equality. More accurate estimates of the free energy may be obtained by taking full advantage of the variational aspect of the inequality (3.62): the separation of V_N into $V_N^{(0)}$ and W_N is in general not unique, and may be made dependent on one (or several) parameters, say ξ; both terms on the right-hand side of equation (3.62) will then be functions of ξ, and the best estimate of the free energy is obtained for that value of ξ which minimizes the sum of these two terms. For spherical particles with strong short-range repulsive forces, the hard sphere fluid provides a convenient reference system, with the hard sphere diameter σ as variational parameter.

3.6 Static linear response

From equation (3.52), it follows that the small wavevector limit of the structure factor, which is related to number fluctuations in a finite volume subsystem (see equations (3.37) and (3.52)), is also related to the compressibility, which characterizes the response of the volume to a change in external pressure. In this section, we show that the structure factor at finite wavevector also characterizes the linear response of a fluid to a weak external field which couples to its atoms. Let us assume the external potential to be weak, and periodically modulated, with a characteristic wavevector $\mathbf{k}$, i.e.:

$$\phi(\mathbf{r}) = \frac{1}{V} \delta\phi_{\mathbf{k}} \exp(-i\mathbf{k} \cdot \mathbf{r}) \tag{3.63}$$

where $\delta\phi_{\mathbf{k}}$ is an amplitude which is small compared to $k_{\mathrm{B}}T$. The total potential energy of the N atoms in the external field is:

$$\Phi_N = \int_V \phi(\mathbf{r})\hat{\rho}(\mathbf{r})d\mathbf{r} = \frac{1}{V} \delta\phi_{\mathbf{k}}\rho_{-\mathbf{k}} \tag{3.64}$$

In the absence of external potential, the fluid is homogeneous, i.e. its local density (3.31) is constant, equal to $\rho = N/V$. The external potential (3.63) breaks the translational invariance and induces a modulation of the local density $\rho(\mathbf{r})$, which may be characterized by a non-vanishing value of the thermal average of the Fourier component $\rho_{\mathbf{k}}$ which couples to $\delta\phi_{\mathbf{k}}$. To compute $\langle\rho_{\mathbf{k}}\rangle$, one proceeds through a Taylor expansion of the Boltzmann factor $\exp(-\beta(V_N + \Phi_N))$ in the formal expression

$$\langle\rho_{\mathbf{k}}\rangle = \frac{\int \exp(-\beta(V_N + \Phi_N)) \rho_{\mathbf{k}}d\mathbf{r}_1 \ldots d\mathbf{r}_N}{\int \exp(-\beta(V_N + \Phi_N)) d\mathbf{r}_1 \ldots d\mathbf{r}_N}$$

with respect to the small quantity $\beta\Phi_N$. Limiting the expansion to first order in $\beta\delta\phi_{\mathbf{k}}$ (the *linear response* limit) one finds after some simple algebra

$$\langle\rho_{\mathbf{k}}\rangle = \frac{\int \exp(-\beta V_N)(1 - \beta\delta\phi_{\mathbf{k}}\rho_{-\mathbf{k}}/V)\rho_{\mathbf{k}}d\mathbf{r}_1 \ldots d\mathbf{r}_N}{\int \exp(-\beta V_N) d\mathbf{r}_1 \ldots d\mathbf{r}_N} + \mathcal{O}((\beta\delta\phi_{\mathbf{k}})^2)$$

$$= -\frac{\beta}{V} \delta\phi_{\mathbf{k}} \langle\rho_{-\mathbf{k}}\rho_{\mathbf{k}}\rangle^{(0)} + \mathcal{O}((\beta\delta\phi_{\mathbf{k}})^2) \tag{3.65}$$

$$= -\beta\rho\delta\phi_{\mathbf{k}} S(\mathbf{k})$$

where the superscript $^{(0)}$ refers to the unperturbed, homogeneous fluid, for which $\langle\rho_{\mathbf{k}}\rangle^{(0)}$ vanishes according to equation (3.45). The induced density modulation is proportional to the amplitude of the external potential, and the corresponding *susceptibility* $\chi(\mathbf{k})$ is proportional to the static structure factor of the unperturbed fluid:

$$\chi(\mathbf{k}) = \frac{\delta\langle\rho_{\mathbf{k}}\rangle}{\delta\phi_{\mathbf{k}}} = -\frac{\rho S(\mathbf{k})}{k_B T} \tag{3.66}$$

which is a particular manifestation of a more general result of linear response theory. This linear response result (3.65) may be regarded as a generalization of the fluctuation result (3.52) to finite wavelength, provided $\lim_{\mathbf{k}\to 0}\phi_{\mathbf{k}}$ is identified with $-PV$.

Note that we have limited ourselves to the calculation of the response function for the wavevector $\mathbf{k}$ that characterizes the potential (3.63). It is easily shown that the response associated with a wavevector $\mathbf{k}'$ is proportional to $\langle\rho_{\mathbf{k}}\rho_{\mathbf{k}'}\rangle^{(0)}$, and therefore vanishes for $\mathbf{k} \neq \mathbf{k}'$ (see equation (3.48)). That the system responds, at linear order in the perturbation, only at the wavevector of the external potential, is a consequence of the invariance of the system under translations.

Finally, the response to *any* external potential can, at the same order, be calculated by superposing the responses to each Fourier component of the potential, computed using equation (3.6). Note that in the latter formula $\delta\phi_{\mathbf{k}}$ is actually the Fourier component of the potential, $\int d\mathbf{r}\phi(\mathbf{r})\exp(i\mathbf{k}\cdot\mathbf{r})$.

3.7 Application 1: dipole moment fluctuations and dielectric response

Non-conducting dielectric materials become polarized under the action of an external electric field $\mathbf{E}_0$; they acquire a polarization $\mathbf{P}$ materialized by surface charges of opposite sign on both sides of the dielectric sample, such as to oppose the applied field. The macroscopic field inside the dielectric is $\mathbf{E} = \mathbf{E}_0 - \nu\mathbf{P}$, where ν is a geometric factor which depends on the shape of the dielectric sample ($\nu = 4\pi$ for a slab perpendicular to $\mathbf{E}_0$; $\nu = 4\pi/3$ for a sphere). Except for very strong fields (which could lead to dielectric breakdown), $\mathbf{P}$ is linearly related to $\mathbf{E}$ by:

$$\mathbf{P}(\mathbf{r}) = \epsilon_0 \int_V \chi(\mathbf{r} - \mathbf{r}')\mathbf{E}(\mathbf{r}')d\mathbf{r}' \tag{3.67}$$

where χ is the dielectric susceptibility (or response function), which is a material property of the dielectric, assumed here to be isotropic, so that $\chi(\mathbf{r})$ is a scalar function. The corresponding relation between Fourier components is $\mathbf{P}(\mathbf{k}) = \chi(\mathbf{k})\mathbf{E}(\mathbf{k})$. In the limit of a uniform applied field $\mathbf{E}_0$, this reduces to the familiar constitutive relation $\mathbf{P} = \mathbf{P}(\mathbf{k} = 0) = \epsilon_0\chi\mathbf{E} = \mathbf{D} - \epsilon_0\mathbf{E}$, where $\mathbf{D}$ is the electric displacement vector $\mathbf{D} = \epsilon_0\epsilon\mathbf{E}$; the susceptibility and dielectric permittivity are hence related by $\chi = \epsilon - 1$. The objective of this section is to relate ϵ to fluctuations of molecular variables.

In the absence of free charges, the macroscopic polarization $\mathbf{P}$ can only result from a partial alignment of permanent and induced molecular dipoles under the action of the applied field. The electric dipole moment of molecule i is:

$$\mathbf{m}_i = \boldsymbol{\mu}_i + \alpha_i \mathbf{E}(\mathbf{r}) \tag{3.68}$$

where $\boldsymbol{\mu}_i$ is the permanent dipole (which vanishes only for centrosymmetric molecules), α_i is the molecular polarizability tensor (which reduces to a scalar for quasi-spherical molecules), and $\mathbf{E}(\mathbf{r})$ is the local electric field felt by the molecule. The microscopic polarization may be defined as:

$$\mathcal{P}(\mathbf{r}) = \sum_i \mathbf{m}_i \delta(\mathbf{r} - \mathbf{r}_i) \tag{3.69}$$

and the macroscopic polarization is the statistical average $\mathbf{P}(\mathbf{r}) = \langle \mathcal{P}(\mathbf{r}) \rangle$. The total dipole moment of the sample

$$\mathbf{M} = \sum_i \mathbf{m}_i = \int_V \mathcal{P}(\mathbf{r}) d\mathbf{r} \tag{3.70}$$

couples to the externally applied field.

To relate the permittivity ϵ to fluctuations of $\mathbf{M}$ we proceed in two steps. First the mean value of the total dipole moment induced by a uniform external field, $\langle \mathbf{M} \rangle_{\mathbf{E}_0}$, is calculated by linear response. Next, macroscopic electrostatics is used to relate the corresponding polarization $\mathbf{P} = \langle \mathbf{M} \rangle_{\mathbf{E}_0} / V$ to the external field $\mathbf{E}_0$, and elimination of the latter leads then to the required fluctuation formula.

If V_N is the total potential energy of the molecules in the absence of $\mathbf{E}_0$, and the coupling energy of $\mathbf{M}$ to the latter is $-\mathbf{M} \cdot \mathbf{E}_0$, the average total dipole moment is given by

$$\langle \mathbf{M} \rangle_{\mathbf{E}_0} = \frac{1}{Z_N} \int_{\Gamma_N} \mathbf{M} \exp\left(-\beta \left(V_N - \mathbf{M} \cdot \mathbf{E}_0\right)\right) d\Gamma_N \tag{3.71}$$

where the configuration space integral is over the centre-of-mass and orientational coordinates of the N molecules, and Z_N is the corresponding configuration integral. $\langle \mathbf{M} \rangle_{\mathbf{E}_0}$ is obtained to first order in the applied field by linearizing the Boltzmann factor with respect to the coupling $-\beta \mathbf{M} \cdot \mathbf{E}_0$. This leads directly to the desired linear response result

$$\langle \mathbf{M} \rangle_{\mathbf{E}_0} = \frac{1}{3 k_{\mathrm{B}} T} \langle \mathbf{M}^2 \rangle \mathbf{E}_0 \tag{3.72}$$

where the right-hand side is proportional to the fluctuation of the total dipole moment of the unperturbed ($\mathbf{E}_0 = 0$) system for which $\langle \mathbf{M} \rangle = 0$.

The next step is to relate $\mathbf{P}(\mathbf{r})$ to $\mathbf{E}_0(\mathbf{r})$ in the limit where $\mathbf{E}_0(\mathbf{r})$ becomes uniform. According to elementary electrostatics, the local electric field at any point inside the dielectric is

$$\mathbf{E}(\mathbf{r}) = \mathbf{E}_0(\mathbf{r}) + \int_V \mathbf{T}(\mathbf{r} - \mathbf{r}') \cdot \mathbf{P}(\mathbf{r}') d\mathbf{r}' \tag{3.73}$$

where $\mathbf{T}$ is the dipole tensor defined in equation (1.11). Some care must be taken in evaluating the convolution integral in equation (3.73), since $\mathbf{T}(\mathbf{r})$ is singular in the limit $|\mathbf{r}| \to 0$. The singularity is isolated by separating the convolution integration in equation (3.73) into a small spherical domain $|\mathbf{r} - \mathbf{r}'| < s$, and the remainder, where $\mathbf{T}$ is a regular function of its argument, and taking the limit $s \to 0$. The first integral, containing the singularity, yields $-\mathbf{P}(\mathbf{r})/3\epsilon_0$ so that

$$\mathbf{E}(\mathbf{r}) = \mathbf{E}_0(\mathbf{r}) - \frac{1}{3\epsilon_0}\mathbf{P}(\mathbf{r}) + \int_{V'} \mathbf{T}(\mathbf{r} - \mathbf{r}') \cdot \mathbf{P}(\mathbf{r}')d\mathbf{r}' \tag{3.74}$$

where the convolution is now over the domain from which the singularity has been excluded. In k-space:

$$\mathbf{E}(\mathbf{k}) = \mathbf{E}_0(\mathbf{k}) - \frac{1}{3\epsilon_0}\mathbf{P}(\mathbf{k}) + \mathbf{T}_{\text{reg}}(\mathbf{k}) \cdot \mathbf{P}(\mathbf{k}) \tag{3.75}$$

where $\mathbf{T}_{\text{reg}}(\mathbf{k})$ denotes the Fourier transform of $\mathbf{T}$ from which the singularity has been substracted, as explained above. Equation (3.75) is now substituted into the Fourier space version of the constitutive relation (3.67) with the result:

$$\left([1 + \chi(\mathbf{k})/3]\mathbf{I} - \epsilon_0\chi(\mathbf{k})\mathbf{T}_{\text{reg}}(\mathbf{k})\right) \cdot \mathbf{P}(\mathbf{k}) = \epsilon_0\chi(\mathbf{k})\mathbf{E}_0(\mathbf{k}) \tag{3.76}$$

In the special case of a uniform applied field $\mathbf{E}_0$, only the $\mathbf{k} = 0$ component is to be considered; $\mathbf{E}_0(\mathbf{k} = 0) = V\mathbf{E}_0$. According to our previous definitions, $\chi(\mathbf{k} = 0) = \chi$, $\mathbf{P}(\mathbf{k} = 0) = V\mathbf{P} \equiv \langle\mathbf{M}\rangle_{\mathbf{E}_0}$ while the zero wavevector component of the regular part of the dipolar tensor is easily seen to vanish, $\mathbf{T}_{\text{reg}}(\mathbf{k} = 0) = 0$. This leads immediately to the required result:

$$\frac{1}{V}\langle\mathbf{M}\rangle_{\mathbf{E}_0} = \frac{\epsilon_0\chi}{1 + \chi/3}\mathbf{E}_0 = 3\epsilon_0\frac{\epsilon - 1}{\epsilon + 2}\mathbf{E}_0 \tag{3.77}$$

Identification with the linear response result leads to the generalized Clausius–Mossotti relation

$$\frac{\epsilon - 1}{\epsilon + 2} = \frac{1}{9\epsilon_0 k_B T}\frac{\langle\mathbf{M}^2\rangle}{V} \tag{3.78}$$

This result is valid for a finite dielectric sample of volume V placed in empty space.

To take into account the long range of the dipolar interactions between molecules in explicit statistical mechanics calculations, it is convenient to consider the dipole moment fluctuations of a mesoscopic sample of dielectric assumed to be carved out of an infinite continuum of the same material. In practice, one considers N polar molecules in a spherical cavity of radius a surrounded by a dielectric continuum of permittivity ϵ. The electric field far from the cavity is $\mathbf{E}$, and a standard electrostatic calculation shows that inside the cavity, the corresponding cavity field, acting on the total dipole moment $\mathbf{M}$, is:

$$\mathbf{G} = \frac{3\epsilon}{2\epsilon + 1}\mathbf{E} \tag{3.79}$$

The linear response calculation leading from (3.71) to (3.72) carries through, provided $\mathbf{E}_0$ is replaced by $\mathbf{G}$ in the Boltzmann factor; this leads to:

$$\langle \mathbf{M} \rangle_{\mathbf{E}} = \frac{1}{k_{\mathrm{B}} T} \frac{\epsilon}{2\epsilon + 1} \langle M^2 \rangle \mathbf{E} \tag{3.80}$$

If the spherical sample of volume V is sufficiently large, its polarization $\mathbf{P} = \langle \mathbf{M} \rangle_{\mathbf{E}} / V$ is identical to the overall polarization $\mathbf{P} = \epsilon_0 \chi \mathbf{E}$ of the dielectric medium with which it is in equilibrium, so that $\langle \mathbf{M} \rangle_{\mathbf{E}} = \epsilon_0 \chi V \mathbf{E}$; substituting this on the left-hand side of (3.80) we arrive at the Kirkwood relation between the permittivity ϵ and the fluctuation of the total dipole moment of a spherical sample surrounded by a dielectric medium of identical permittivity:

$$\frac{(\epsilon - 1)(2\epsilon + 1)}{\epsilon} = \frac{1}{\epsilon_0 k_{\mathrm{B}} T} \frac{\langle M^2 \rangle}{V} \tag{3.81}$$

The difference between the two relations (3.78) and (3.81) of the permittivity (a macroscopic material property) clearly shows that the fluctuations of $\mathbf{M}$ depend strongly on the boundary conditions of the dielectric sample under investigation. This is a consequence of the $1/r^3$ range of the dipolar interactions, which imply that surface contributions are never negligible compared to bulk contributions.

In the special case of rigid (non-polarizable) molecules carrying dipole moments $\boldsymbol{\mu}_i = \mu \hat{\boldsymbol{\mu}}_i$ the Kirkwood formula may be rewritten in the form:

$$\frac{(\epsilon - 1)(2\epsilon + 1)}{\epsilon} = \frac{\rho \mu^2}{\epsilon_0 k_{\mathrm{B}} T} g_K \tag{3.82}$$

where the Kirkwood g-factor is

$$g_K = 1 + \sum_{j \neq 1} \langle \hat{\boldsymbol{\mu}}_1 \cdot \hat{\boldsymbol{\mu}}_j \rangle = 1 + \sum_{j \neq 1} \langle \cos \theta_j \rangle \tag{3.83}$$

θ_j is the angle between dipole j and an axis taken along the singled-out dipole $\boldsymbol{\mu}_1$. Although dipolar interactions are long ranged, the orientational correlations are short ranged in the absence of an applied field, so that the sum in equation (3.83) may be restricted to near neighbours. In the absence of any orientational correlations, $g_K = 1$ and equation (3.82) reduces to Onsager's mean field result.

The case of polarizable molecules ($\alpha \neq 0$) is significantly more difficult, because of the intrinsic many-body nature of the problem: the instantaneous dipole moment of a molecule now depends on the local electric field at the position of the molecule, and this local field is determined by the instantaneous dipole moments of all surrounding molecules, which are themselves polarizable; in other words the instantaneous dipole moments of neighbouring molecules must be determined self-consistently, since the value of the local electric field is a function of the surrounding dipole moments which, in turn, depend on the local field.

Moreover the linear response of a dielectric medium depends on the frequency ω of the applied electric field if the latter varies periodically with time.

For a homogeneous macroscopic medium, the dielectric response is then charac-terized by a complex, frequency-dependent permittivity $\epsilon(\omega) = \epsilon'(\omega) + i\epsilon''(\omega)$ (see also section 10.3). At very high frequencies, the rotational inertia of the polar molecules will prevent them from reorienting in response to the rapidly varying field; however, the induced polarization is practically instantaneous, because of the negligible inertia of the valence electrons which can follow the variation of a rapidly oscillating field, up to optical frequencies. The rapid dielectric response due to electron polarization is thus characterized by the (real) high frequency permittivity $\epsilon_\infty = \lim_{\omega\to\infty} \epsilon(\omega)$, while molecular reorientations also contribute to the low frequency (or static) permittivity $\epsilon_s \equiv \epsilon = \lim_{\omega\to 0} \epsilon(\omega)$.

Since two mechanisms contribute to ϵ_s, while only electronic polarizability determines ϵ_∞, it is clear that $\epsilon_\infty < \epsilon_s$. For non-polar molecules ($\mu = 0$), $\epsilon_s = \epsilon_\infty$. Moreover, ϵ_∞ directly determines the macroscopic optical properties of the medium through the relation $\epsilon_\infty = n^2$, where n is the index of refraction.

To estimate the effect of molecular polarizability on ϵ, consider Onsager's mean field model, whereby each molecule is assumed to be placed at the centre of a microscopic cavity, of volume $v = 4\pi a^3/3 = 1/\rho$ (equal to the volume per molecule). The medium outside the cavity is treated as a dielectric continuum of permittivity ϵ (to be determined). If $\mathbf{E}$ is the electric field in the medium, far from the cavity, the electric field inside the cavity, in the absence of the molecule, would be just the cavity field $\mathbf{G}$ calculated from macroscopic electrostatics as given by (3.79). The dipole moment $\mathbf{m}$ of the molecule will polarize the medium outside the cavity, and this polarization will in turn create a reaction field $\mathbf{R}$ inside the cavity; again according to macroscopic electrostatics:

$$\mathbf{R} = \xi\mathbf{m} = \frac{1}{4\pi\epsilon_0}\frac{2(\epsilon-1)}{2\epsilon+1}\frac{\mathbf{m}}{a^3} \tag{3.84}$$

Setting $\gamma = 3\epsilon/(2\epsilon+1)$, the local electric field acting on the molecule is, accord-ing to the superposition principle, $\mathbf{E}_{loc} = \mathbf{G} + \mathbf{R} = \gamma\mathbf{E} + \xi\mathbf{m}$. The total dipole moment is:

$$\mathbf{m} = \boldsymbol{\mu} + \alpha\mathbf{E}_{loc} = \frac{1}{1-\alpha\xi}[\boldsymbol{\mu} + \alpha\gamma\mathbf{E}] \tag{3.85}$$

while the resulting local field is:

$$\mathbf{E}_{loc} = \frac{1}{1-\alpha\xi}[\gamma\mathbf{E} + \xi\boldsymbol{\mu}] \tag{3.86}$$

Calculation of the average dipole moment $\langle\mathbf{m}\rangle$ thus only requires a calculation of $\langle\boldsymbol{\mu}\rangle$. If θ is the polar angle between $\mathbf{E}$ and $\boldsymbol{\mu}$, the coupling of $\mathbf{E}_{loc}$ to $\boldsymbol{\mu}$ is:

$$-\boldsymbol{\mu}\cdot\mathbf{E}_{loc} = -\frac{1}{1-\alpha\xi}[\gamma E\mu\cos\theta + \xi\mu^2] \tag{3.87}$$

so that:

$$\langle \cos \theta \rangle = \frac{\int_0^\pi \cos \theta \, \exp(\beta \boldsymbol{\mu} \cdot \mathbf{E}_{\mathrm{loc}}) \sin \theta \, d\theta}{\int_0^\pi \exp(\beta \boldsymbol{\mu} \cdot \mathbf{E}_{\mathrm{loc}}) \sin \theta \, d\theta} = \gamma \frac{\beta \mu E}{3(1 - \alpha \xi)} \tag{3.88}$$

By symmetry, $\langle \boldsymbol{\mu} \rangle$ can only be parallel to $\mathbf{E}$, so that substitution of (3.88) into (3.85) yields the required relation between the polarization and the field $\mathbf{E}$ inside the dielectric:

$$\mathbf{P} = \frac{\langle \mathbf{M} \rangle_{\mathbf{E}}}{V} = \frac{N \langle \mathbf{m} \rangle_{\mathbf{E}}}{V} = \frac{\gamma \rho}{1 - \alpha \xi} \left[\frac{\beta \mu^2}{3(1 - \alpha \xi)} + \alpha \right] \mathbf{E} \tag{3.89}$$

Remembering now that $\mathbf{P} = \epsilon_0 (\epsilon - 1) \mathbf{E}$, one may conclude from equation (3.89) that within Onsager's mean field theory:

$$\epsilon - 1 = \frac{\gamma \rho}{\epsilon_0 (1 - \alpha \xi)} \left[\frac{\beta \mu^2}{3(1 - \alpha \xi)} + \alpha \right] \tag{3.90}$$

Consider first the case on non-polar molecules, $\mu = 0$. A little algebra leads then to the Clausius–Mossotti relation:

$$\frac{\epsilon - 1}{\epsilon + 2} = \frac{n^2 - 1}{n^2 + 2} = \frac{\alpha \rho}{3 \epsilon_0} \tag{3.91}$$

which is reasonably well verified experimentally for fluids of non-polar molecules.

The case of polar, non-polarizable molecules ($\boldsymbol{\mu} \neq 0$, $\alpha = 0$) leads back to Kirkwood's result (3.82) with $g_K = 1$. Finally, in the most general case of polarizable molecules, equation (3.90), combined with (3.91), which remains valid for $\epsilon_\infty = n^2$, yields Onsager's relation for $\epsilon \equiv \epsilon_s$:

$$\frac{(\epsilon - n^2)(2\epsilon + n^2)}{\epsilon} = \left[\frac{n^2 + 2}{3} \right]^2 \frac{\rho \mu^2}{\epsilon_0 k_B T} \tag{3.92}$$

Kirkwood's generalization of the result (3.82), valid for rigid ($\alpha = 0$) molecules, to the polarizable case leads to Onsager's relation, modified so as to account for orientational correlations by multiplying the right-hand side by the Kirkwood factor defined in equation (3.83). It is worth noting that g_K is particularly large in the case of water, because of the very strong orientational correlations between neighbouring water molecules due to the hydrogen bonds; this results in the particularly large value of the permittivity, $\epsilon = 78$, under normal conditions (room temperature and atmospheric pressure).

3.8 Application 2: determination of the structure from diffraction experiments

In this section it will be shown how microscopic density fluctuations, as embodied in the static structure factor $S(\mathbf{k})$, defined in equation (3.46), can indeed be probed

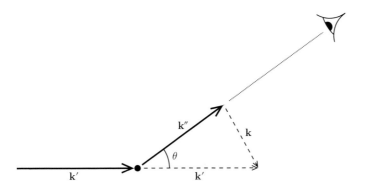

Figure 3.2. Geometry of a typical scattering experiment; $\mathbf{k}'$ and $\mathbf{k}''$ are the wavevectors of the incoming (incident) and outgoing (scattered) radiation; $\mathbf{k} = \mathbf{k}' - \mathbf{k}''$ is the scattering wavevector, and θ is the scattering angle; in the case of elastic scattering, $|\mathbf{k}'| = |\mathbf{k}''|$ and $k = 2k' \sin(\theta/2)$.

experimentally by elastic scattering (or diffraction) of radiation. The wavelengths λ of X-rays, or thermal [3] neutrons from a reactor, are typically of the order of a few angstroms, and hence capable of resolving structural details on a molecular scale, when they are scattered from samples of fluid or solid material. The wavelength of visible light is better suited for the mesoscale structures of supramolecular aggregates or colloids. In the case of regular arrays of particles, exhibiting long-range order, as is the case for crystalline solids, or lamellar (smectic) phases, the radiation is scattered in well defined directions obeying the Bragg diffraction conditions. In the case of disordered systems, like fluids, the scattering is less selective, and the elastic scattering cross-section is directly proportional to $S(\mathbf{k})$, which is a continuously varying function of wavevector $\mathbf{k}$, rather than a succession of Bragg peaks.

The scattering geometry is depicted schematically in figure 3.2. Let $\hbar \mathbf{k}'$ and $\hbar \mathbf{k}''$ denote the momenta of an incoming and scattered particle (photon or neutron); the particle is scattered by the nucleus (in the case of neutrons) or by one of the electrons (for photons) of an atom in the sample. Neglecting multiple scattering of the same incoming particle (a reasonable assumption in the case of neutrons and high energy photons), the momentum transferred from the radiation particle to the sample is:

$$\hbar \mathbf{k} = \hbar \mathbf{k}' - \hbar \mathbf{k}'' \tag{3.93}$$

For elastic scattering $|\mathbf{k}'| = |\mathbf{k}''|$, so that $\mathbf{k}$ is related to the scattering angle θ by:

$$|\mathbf{k}| = 2|\mathbf{k}'| \sin(\theta/2) \tag{3.94}$$

Let $|\mathbf{k}'\rangle \sim \exp(i\mathbf{k}' \cdot \mathbf{r})$ and $|\mathbf{k}''\rangle \sim \exp(i\mathbf{k}'' \cdot \mathbf{r})$ denote the plane wave states of the incident and scattered radiation, and let $\Phi(\mathbf{r})$ denote the interaction potential felt by the neutron or photon in the scattering medium; $\Phi(\mathbf{r})$ may be written as the

---

[3] 'Thermal' neutrons have a kinetic energy of the order of $k_B T$ with $T \simeq 300$ K.

sum over all atoms, assumed here to be identical for the sake of simplicity:

$$\Phi(\mathbf{r}) = \sum_{i=1}^{N} \phi(\mathbf{r} - \mathbf{r}_i) \tag{3.95}$$

If the coupling between the radiation and the atoms is weak, the rate of transitions from $|\mathbf{k}'\rangle$ to $|\mathbf{k}''\rangle$ is given by Fermi's golden rule, i.e. the differential cross-section per unit solid angle Ω is:

$$\frac{d\sigma}{d\Omega} \sim |\langle \mathbf{k}'|\Phi|\mathbf{k}''\rangle|^2 \tag{3.96}$$

Substituting (3.95) into (3.96), the matrix element is easily calculated to be:

$$\langle \mathbf{k}'|\Phi|\mathbf{k}''\rangle = \phi(\mathbf{k})\rho_{-\mathbf{k}} \tag{3.97}$$

where $\phi(\mathbf{k})$ is the Fourier transform of $\phi(\mathbf{r})$, while $\rho_{-\mathbf{k}}$ is a Fourier component (3.44) of the microscopic density of atoms corresponding to the scattering vector (3.93). The result (3.97) is valid for a given configuration of the N atoms in the sample. However, in a scattering experiment, measurements will extend over a macroscopic time, so that the resulting cross-section involves a statistical average over atomic configurations:

$$\frac{d\sigma}{d\Omega} \sim |\phi(\mathbf{k})|^2 \langle \rho_{\mathbf{k}}\rho_{-\mathbf{k}}\rangle \sim N|\phi(\mathbf{k})|^2 S(\mathbf{k}) \tag{3.98}$$

A measurement of the scattering cross-section $\frac{d\sigma}{d\Omega}$ as a function of scattering angle θ hence yields the static structure factor $S(\mathbf{k})$. In the case of neutron diffraction, $\phi(\mathbf{k})$ is essentially independent of $\mathbf{k}$[4] and proportional to a scattering length, which depends on the atomic nuclear species. In the case of diffraction of electromagnetic radiation, photons are scattered by the atomic electrons and equation (3.98) must be generalized to account for the internal electronic structure of the atoms. This introduces an atomic form factor $P(\mathbf{k})$, which follows from a quantum-mechanical calculation of the electronic structure of an isolated atom, such that:

$$\frac{d\sigma}{d\Omega} \sim N P(\mathbf{k}) S(\mathbf{k}) \tag{3.99}$$

At sufficiently low density, where $S(\mathbf{k}) = 1$, the scattering cross-section would correspond to that of N independent atoms. At fluid densities, the atomic form factor $P(\mathbf{k})$ is essentially unchanged, since electronic binding energies greatly exceed thermal energies, so that the electronic structure of atoms is not affected by neighbouring atoms. In practice, this form factor is closely related to the Fourier transform of the electronic density measured around a given atomic nucleus. This introduces an intrinsic upper cut-off, k_{max}, of the order of the inverse atomic size, and above which the scattering of X-rays becomes weak. Note also that

---

[4] See also section 11.6.

light elements such as hydrogen scatter X-rays only very weakly, because of their small number of electrons.

3.9 Application 3: form factors of complex objects or molecules

Radiation scattering processes considered so far in section 3.8 involved systems of point scatterers, as in simple atomic fluids. We now generalize these considerations to the case of scattering by composite particles, like molecules, macromolecules or colloidal dispersions, each of which contains several up to a large number of atoms. Without going into the details of scattering theory, and restricting the discussion to single scattering, within the first Born approximation, the amplitude of the radiation (e.g. the amplitude of the electric field of an electromagnetic wave) scattered by particle i is generally proportional to:

$$B_i(\mathbf{k}) = \int B_i(\mathbf{r}) \exp(i\mathbf{k} \cdot \mathbf{r}) \, d\mathbf{r} \qquad (3.100)$$

where $\mathbf{k}$ is the scattering wavevector defined in equation (3.93). $B_i(\mathbf{r})$ characterizes the distribution of scattering centres within the particle. In the case of thermal neutrons, which resolve atomic detail, $B_i(\mathbf{r})$ would be the sum over the M discrete scattering centres (nuclei) within the particle.

$$B_i(\mathbf{r}) = \sum_{\nu=1}^{M} b_\nu(\mathbf{r} - \mathbf{r}_{i\nu}) \qquad (3.101)$$

In the case of scattering of light by mesoscopic colloidal particles, $B_i(\mathbf{r})$ would just reflect the variation of the local index of refraction. If n_0 denotes the index of the suspending fluid, and $n_i(\mathbf{r})$ the (possibly variable) index within the particle regarded as a continuous medium, then:

$$B_i(\mathbf{r}) = n_i(\mathbf{r}) - n_0 \qquad (3.102)$$

The total amplitude scattered by the assembly of N particles, for any given configuration $\mathbf{r}_i$ of their centres, is then proportional to:

$$F(\mathbf{k}) = \sum_{i=1}^{N} B_i(\mathbf{k}) \exp(i\mathbf{k} \cdot \mathbf{r}_i) \qquad (3.103)$$

while the $\mathbf{k}$-dependent intensity of the scattered radiation finally follows from a statistical (ensemble) average of the square of this amplitude:

$$I(\mathbf{k}) = \langle |F(\mathbf{k})|^2 \rangle = \sum_{i=1}^{N} \sum_{j=1}^{N} \langle B_i(\mathbf{k}) B_j(\mathbf{k}) \exp\left(i\mathbf{k} \cdot (\mathbf{r}_i - \mathbf{r}_j)\right) \rangle \qquad (3.104)$$

In the monodisperse case, where all N particles are identical, and in the absence of any orientational order, linked to the particle shape (which would introduce correlations between particle positions $\mathbf{r}_i$ and amplitudes $B_i(\mathbf{k})$), the intensity simplifies to:

$$I(\mathbf{k}) = N B(0)^2 P(\mathbf{k}) S(\mathbf{k}) \tag{3.105}$$

where $S(\mathbf{k})$ is the centre-of-mass structure factor, defined in equation (3.46), while $P(\mathbf{k})$ is the form factor:

$$P(\mathbf{k}) = \left[\frac{B(\mathbf{k})}{B(0)} \right]^2 \tag{3.106}$$

normalized such that $P(k = 0) = 1$.

Spherical colloidal particles

Consider first the case of a suspension of homogeneous, spherical colloidal particles of radius R, and refractive index n_c. $B(r)$ reduces then to $n_c - n_0$ ($|\mathbf{r}| < R$) and, according to equation (3.100):

$$B(k) = 4\pi(n_c - n_0) \int_0^R \frac{\sin(kr)}{kr} r^2 \, dr = v_c(n_c - n_0)\frac{3 j_1(kR)}{kR} \tag{3.107}$$

where $j_1(x) = (\sin x - x \cos x)/x^2$ is the first-order spherical Bessel function, and $v_c = 4\pi R^3/3$. $B(0) = v_c(n_c - n_0)$, and the form factor is:

$$P(k) = \left[\frac{3 j_1(kR)}{kR} \right]^2 \tag{3.108}$$

For very dilute suspensions $S(k) = 1$ for all k, so that the form factor may be directly extracted from a measurement of the intensity of scattered light as a function of scattering angle. If the particles are known to be spherical and monodisperse, such a measurement allows a direct determination of the particle radius R. The form factor decreases rapidly with k and has a first zero for $kR \simeq 4.5$; it is negligibly small at larger k. This means that the structure factor $S(k)$ is practically inaccessible for $k > 4/R$ by light-scattering experiments.

The form factor is the internal structure factor of a single composite particle characterized by a discrete or a continuous distribution of scatterers. As long as this distribution is spherically symmetric, the form factor will only depend on the modulus k of the wavenumber, as in equation (3.108). As another example, the form factor for a spherical shell of radius R and thickness $\Delta R \ll R$ (which may be a crude model for a vesicle) is easily calculated to be

$$P(k) = \left[\frac{\sin(kR)}{kR} \right]^2 \tag{3.109}$$

which decays much more slowly with k than the form factor (3.108) for a homogeneous sphere.

Rods and platelets

In the case of non-spherical particles of well defined orientation, the form factor will depend on the relative orientation of the scattering vector and of the symmetry axes of the particle. As an example consider the case of a homogeneous cylinder of radius R, length L and unit vector $\mathbf{u}$ along its axis. The corresponding form factor is:

$$P(\mathbf{k}) = \left[\frac{\sin(\mathbf{k} \cdot \mathbf{u} L/2)}{\mathbf{k} \cdot \mathbf{u} L/2} \right]^2 \left[\frac{J_1(k_\perp R)}{k_\perp R} \right]^2 \cdot \tag{3.110}$$

which, as expected, depends on the angle between $\mathbf{u}$ and $\mathbf{k}$; $k_\perp = |\mathbf{k} \wedge \mathbf{u}|$. Two limits are particularly interesting: that of an infinitely thin rod ($R \to 0$), as a model for a very stiff polymer, for which $P(\mathbf{k})$ reduces to the first factor on the right-hand side of equation (3.110); and that of an infinitely thin disc ($L \to 0$), which may serve as a model for clay platelets, where $P(\mathbf{k})$ reduces to the second factor.

An isolated rod or disc suspended in a solvent will rotate unhindered in the course of its Brownian motion, so that the intensity of the light scattered by a very dilute suspension will be proportional to the angular average of the form factor over all orientations:

$$\langle P(k) \rangle_\mathbf{u} = \frac{1}{4\pi} \int P(\mathbf{k}) \, d\Omega_\mathbf{u} \tag{3.111}$$

In the cases of thin rods and discs, this leads to the following orientation-averaged form factors:

$$\langle P(k) \rangle_\mathbf{u} = \frac{\pi}{kL} \qquad kL \gg 2 \qquad \text{rods} \tag{3.112}$$

$$\langle P(k) \rangle_\mathbf{u} = \frac{2}{k^2 R^2} \left[1 - \frac{J_1(2kR)}{kR} \right] \qquad \text{discs} \tag{3.113}$$

Gaussian polymer

Another important form factor, or internal structure factor, is that of a single polymer coil. Consider a linear polymer comprising M identical monomers positioned at $\mathbf{r}_\nu$ ($1 \le \nu \le M$). The single polymer structure factor (corresponding to an infinite dilution limit) is defined as in equation (3.46):

$$S_0(\mathbf{k}) = \frac{1}{M} \langle \rho_\mathbf{k} \rho_{-\mathbf{k}} \rangle \tag{3.114}$$

where the collective coordinate is once more given by equation (3.44), the sum being over the monomers. $\rho_\mathbf{k}$ is proportional to the amplitude $B(\mathbf{k})$ introduced in equation (3.100) with $b_\nu(\mathbf{r} - \mathbf{r}_\nu) = \delta(\mathbf{r} - \mathbf{r}_\nu)$. The statistical average in equation (3.114) is over all polymer conformations compatible with the internal constraints of the macromolecule. We consider here the case of the Gaussian chain introduced in section 1.5, in which case the statistical weight of a configuration is given by the Gaussian weight (1.32). Using this Gaussian weight, the average of each of the terms contributing to the structure factor is easily computed as

$$\langle \exp(i\mathbf{k} \cdot (\mathbf{r}_\nu - \mathbf{r}_\mu)) \rangle = \exp(-\mathbf{k}^2 |\mu - \nu| b^2/6) \tag{3.115}$$

The final expression for $S_0(k)$ then follows by summing over all $1 \le \mu, \nu \le M$; for large M, the discrete summations can be replaced by continuous integrations:

$$S_0(k) = \frac{1}{M} \int_0^M d\mu \int_0^M d\nu \, \exp(-\mathbf{k}^2 |\mu - \nu| b^2/6) = M f(k R_G) \tag{3.116}$$

where $f(x) = 2(\exp(-x^2) - 1 + x^2)/x^4$ is called the Debye function, while $R_G^2 = Mb^2/6$ is the radius of gyration (equation (1.24)). For large arguments, $f(x) \sim 2/x^2$ so that

$$S_0(k) = \frac{2M}{k^2 R_G^2} \quad \text{for} \quad k R_G \gg 1 \tag{3.117}$$

Finally, a useful approximation to the structure factor, valid within 15% over the whole k range, and which incorporates the correct asymptotic behaviour, is given by

$$S_0(k) \simeq \frac{M}{1 + k^2 R_G^2/2} \tag{3.118}$$

From the inverse Fourier transform of this expression one obtains the approximate monomer pair distribution function inside the coil,

$$g_0(r) \simeq \frac{R_G}{r} \exp(-\sqrt{2} r/R_G) \tag{3.119}$$

Expanding $S_0(k)$ (equation (3.116)) for small wavevectors, it is easily seen that

$$S_0(k) = S_0(0)(1 - k^2 R_G^2/3) \quad \text{for} \quad k R_G \ll 1 \tag{3.120}$$

This is an illustration of the general Guinier theorem, valid for any scattering object, which allows extraction of the radius of gyration (defined by equation (1.24)) from the small k behaviour of the structure factor. For any system made of M point scatterers, one finds, upon expanding the general expression for small k, and using the fact that $\mathbf{r}_\mu - \mathbf{r}_\nu$ has an isotropic distribution

$$S_0(k) = \frac{1}{M} \sum_{\mu=1}^M \sum_{\nu=1}^M \langle \exp(i\mathbf{k} \cdot (\mathbf{r}_\mu - \mathbf{r}_\nu)) \rangle = \frac{1}{M} \sum_{\mu=1}^M \sum_{\nu=1}^M \left[1 - k^2 \langle (\mathbf{r}_\mu - \mathbf{r}_\nu)^2 \rangle + \mathcal{O}(k^4) \right] \tag{3.121}$$

which, when using the definition (1.25) of the radius of gyration, reduces to (3.120).

Finally, we note that the expression (3.116) of the structure factor is, as should be the case for a measurable property of the polymer chain, expressible in terms of measurable quantities only (in this case the radius of gyration and the molecular weight M), without explicit reference to the underlying microscopic model.

Although the structure factor calculated here describes correctly ideal chains only, this is a general feature that will remain true when we consider the effect of excluded volume interactions.

Fractal objects

Scatterers in a composite object are not always distributed according to standard geometric patterns, such as surfaces, lines or dense objects. The ideal polymer coil, for example, has its beads distributed along a random walk, which is a *fractal object* of fractal (or Hausdorff) dimension $d_f = 2$. Generally speaking, the fractal dimension can be defined as follows. For an object (ensemble of points) of typical size R, one counts the number of d-dimensional cubes of lateral size ℓ that are needed to cover the object completely. For $\ell \ll R$, this number scales as

$$N_c(\ell) \sim (R/\ell)^{d_f} \tag{3.122}$$

which defines the fractal dimension d_f. Obviously, for standard objects such as lines or surfaces d_f coincides with the usual definition of the dimension ($d_f = 1$ for a line and $d_f = 2$ for a surface). For the ideal polymer chain, the fractal dimension $d_f = 2$ can be determined from the scaling relation $R \sim M^{1/2} b$. If p is the number of monomers that leads to a coil of size ℓ, $\ell = p^{1/2} b$. Now, the number of cubes needed to cover the whole chain is $M/p = (R/\ell)^2$. Therefore the fractal dimension of the chain is 2, independently of the dimension of space. In three dimensions, the chain is a tenuous object, with an average mass density that decreases with size like $1/M^{1/2}$. In two dimensions, however, the chain is essentially a compact object.

Exercise: Show that the pair correlation function inside a fractal object behaves as $g(r) \sim 1/r^{d-d_f}$ for $r \ll R$.

Quite generally, the form factor of a fractal object behaves as

$$P(k) \sim k^{-d_f} \tag{3.123}$$

for $1/b \gg k \gg 1/R$ where b is some molecular size. That this is the case can be checked explicitly for the various objects (rod, plate or polymer) that were considered above. To establish (3.123) in the general case, we use a simple argument due to T. Witten. The scattering at wavevector k is considered as arising from cubes of volume k^{-d}. Inside a cube, scattering is fully coherent, so that the intensity scattered by one cube is proportional to the *square* of the number $N_s(k)$ of scatterers in the cube. Different cubes, in contrast, are separated by more than one wavelength and scatter independently. The total scattering intensity is therefore

$$P(k) \sim N_s(k)^2 N_c(1/k) \tag{3.124}$$

with N_c given by (3.122). N_c and N_s are related by $M = N_s(k)N_c(1/k)$, where M is the number of scatterers, so that one finally obtains (leaving out k independent factors)

$$P(k) \sim 1/N_c(1/k) \sim (kL)^{-d_f} \tag{3.125}$$

.

The reader may have noted that the result for the spherical colloid, equation (3.108), behaves as $1/k^4$ at large k, which seems to contradict the general statement (3.123). There is a subtle difference, however, between scattering by a fractal object and scattering by a d-dimensional object bounded by a (possibly fractal) boundary. In the latter case, only the cubes that cover the boundary of the object actually contribute to the scattering, since the other cubes are completely homogeneous. If d_B is the fractal dimension of the boundary, the number of boundary cubes that contribute to the scattering at wavevector k is therefore $N_c(1/k) \sim (kL)^{d_B}$. The number of scatterers in a cube, however, is now independent of d_B, and of order ρk^{-d}, where ρ is the density of the bulk object[5]. As a result, the scattering intensity is now proportional to

$$P(k) \sim \left[\rho k^{-d}\right]^2 (kL)^{d_B} \sim k^{2d-d_B}. \tag{3.126}$$

For the sphere, $d = 3$ and $d_B = 2$ so that the k^{-4} behaviour is obtained. This k^{-4} scaling is more generally characteristic of scattering by a three-dimensional object bounded by a smooth surface, and is known as Porod's law.

3.10   Application 4: random phase approximation

The so-called 'random phase approximation' (or RPA)[6] provides a systematic procedure to calculate the local structure and thermodynamic properties of systems of particles interacting via weak or long-range forces, based on a combination of a mean field approximation and of linear response theory.

The starting point is the separation of the total potential energy of N interacting particles into a reference part, involving only short-range interactions, and a perturbation, accounting for weak and/or long-range forces, as in equation (3.59).

[5] Typically a boundary cube will be half filled by scatterers, independently of the shape of the boundary.

[6] Historically, the 'random phase' term was introduced to describe a closely related approximation used to calculate the response function of quantum many-body systems. In the context of fluids, the RPA implies, as mentioned at the end of this section, a Gaussian distribution for the Fourier components of the density; such a distribution implies obviously that the *phases* of these complex numbers are randomly distributed and independent.

The simpler case of point particles is considered first. Putting aside the frequent collisions experienced by any one particle, due to the short-range (excluded volume) forces, the local potential $\phi(\mathbf{r})$ felt by a particle at $\mathbf{r}$ is the sum of the external potential $\phi_{\text{ext}}(\mathbf{r})$ and of the 'molecular field', or mean potential energy due to slowly varying, long-range interactions $w(\mathbf{r} - \mathbf{r}')$ with all other particles in the fluid. If $\delta\rho(\mathbf{r})$ denotes the deviation of the local density from its mean, the potential may hence be expressed as:

$$\phi(\mathbf{r}) = \phi_{\text{ext}}(\mathbf{r}) + \int w(\mathbf{r} - \mathbf{r}')\delta\rho(\mathbf{r}') \, d\mathbf{r}' \qquad (3.127)$$

Note that the second term incorporates a mean field assumption, since it neglects any correlations between particles at $\mathbf{r}$ and $\mathbf{r}'$. Taking Fourier transforms of both sides of equation (3.127), and using the notation of section 3.6, one arrives at:

$$\phi_{\mathbf{k}} = \delta\phi_{\mathbf{k}}^{\text{ext}} + w(\mathbf{k})\delta\langle\rho_{\mathbf{k}}\rangle \qquad (3.128)$$

According to the linear response result (3.6)

$$\delta\langle\rho_{\mathbf{k}}\rangle = \chi_0(\mathbf{k})\left[\delta\phi_{\mathbf{k}}^{\text{ext}} + w(\mathbf{k})\delta\langle\rho_{\mathbf{k}}\rangle\right] \qquad (3.129)$$

where $\chi_0(\mathbf{k})$ is related to the static structure factor $S_0(\mathbf{k})$ of the reference system (i.e. an auxiliary system of particles interacting only via the short-range potential energy) by equation (3.66). Solving equation (3.129) for $\delta\langle\rho_{\mathbf{k}}\rangle$ we obtain:

$$\delta\langle\rho_{\mathbf{k}}\rangle = \frac{\chi_0(\mathbf{k})}{1 - \chi_0(\mathbf{k})w(\mathbf{k})}\delta\phi_{\mathbf{k}}^{\text{ext}} \qquad (3.130)$$

Now, according to (3.66), $\delta\langle\rho_{\mathbf{k}}\rangle = \chi(\mathbf{k})\delta\phi_{\mathbf{k}}^{\text{ext}}$, where $\chi(\mathbf{k})$ is the static response function (or susceptibility) of the actual system, i.e. a system of particles interacting through the full potential energy (3.59). Identification leads to the required RPA result:

$$\chi(\mathbf{k}) = \frac{\chi_0(\mathbf{k})}{1 - \chi_0(\mathbf{k})w(\mathbf{k})} \qquad (3.131)$$

or in terms of the structure factors $S(\mathbf{k})$ and $S_0(\mathbf{k})$ of the actual and reference systems:

$$\frac{1}{S(\mathbf{k})} = \frac{1}{S_0(\mathbf{k})} + \frac{\rho w(\mathbf{k})}{k_B T} \qquad (3.132)$$

Obviously, a necessary requirement for this kind of approximation is that the Fourier transform $w(\mathbf{k})$ of the 'long range' potential $w(r)$ exists.

Using the OZ relation (3.49), an alternative expression for the RPA relates the direct correlation functions of the actual and reference systems:

$$c(r) = c_0(r) - \frac{w(r)}{k_B T} \qquad (3.133)$$

Note that the RPA result (3.133) is compatible with the general assumption (3.41).

In the absence of strong excluded volume interactions, or for low density fluids or dilute solutions, the reference system may be taken to be an assembly of non-interacting particles, i.e. an ideal gas, for which $c_0(r) = 0$ and $S_0(k) = 1$. This leads to the crudest RPA expression for $S(k)$,

$$S^{RPA}(\mathbf{k}) = \frac{1}{1 + \beta \rho w(\mathbf{k})} \tag{3.134}$$

Two classic applications of the RPA are to ionic solutions and to polymer solutions and melts.

RPA for electrolytes and Debye–Hückel limiting law

For the description of ionic solutions, we adopt the so-called 'primitive model' whereby the aqueous solvent is treated as a dielectric continuum of permittivity ϵ ($\epsilon = 78$ for water at room temperature), while cations and anions are treated as hard spheres of diameters $\sigma_{\pm}$ carrying charges $q_{\pm} = z_{\pm}e$, where $z_{\pm}$ are the valences. If ρ_+ and ρ_- denote the number densities of cations and anions, overall charge neutrality requires that:

$$\rho_+ z_+ + \rho_- z_- = 0 \tag{3.135}$$

For the sake of simplicity, we consider explicitly the 'restricted' version of the primitive model, for which $z_+ = -z_- = z$ (and hence $\rho_+ = \rho_- = \rho$) and $\sigma_+ = \sigma_- = \sigma$. Moreover, restriction will be made to very dilute solutions (say a solution of NaCl of concentration less than 0.1 mol/l), such that $\sigma \ll d = \rho^{-1/3}$. Under these conditions the ionic cores are not expected to contribute significantly to the long-range pair correlations between ions, and to the resulting thermodynamic properties, so that ions may be considered to be point particles in first approximation[7]. Within the RPA description, the reference system is hence an ideal gas, while the interaction potential between ions is simply given by Coulomb's law (inside a uniform dielectric):

$$w_{\alpha\beta}(r) = \frac{z_\alpha z_\beta e^2}{4\pi\epsilon_0\epsilon r} \qquad \alpha, \beta = +, - \tag{3.136}$$

The RPA with an ideal gas reference system follows from an immediate generalization of equation (3.133) (with $c_0(r) = 0$) to a two-component system, i.e.

$$c_{\alpha\beta}(r) = -\frac{w_{\alpha\beta}(r)}{k_B T} = -z_\alpha z_\beta \frac{\ell_B}{r} \tag{3.137}$$

where $\ell_B = e^2/(4\pi\epsilon_0\epsilon k_B T)$ is the Bjerrum length, approximately 0.72 nm in water at room temperature. The correlation functions $h_{\alpha\beta}(k)$, and hence the partial

---

[7] Strictly speaking, a non-zero core diameter σ is required between oppositely charged ions, to prevent the 'Coulomb collapse' between anion–cation pairs, which would lead to an infinite attractive energy as $r \to 0$.

structure factors $S_{\alpha\beta}(k)$, (as defined in equation (3.53)) are then easily calculated from the multi-component generalization of the OZ relation (3.39), namely, for ν components:

$$h_{\alpha\beta}(r) = c_{\alpha\beta}(r) + \sum_{\gamma=1}^{\nu} \rho_\gamma \int c_{\alpha\gamma}(\mathbf{r}')h_{\gamma\beta}(\mathbf{r}-\mathbf{r}')\,d\mathbf{r}' \qquad (3.138)$$

In the present symmetric two-component case, there are only two independent correlation functions, $h_{++}(r) = h_{--}(r)$ and $h_{+-}(r)$, and the OZ relations (3.138) reduce to a set of two equations between h_{++}, h_{+-} and c_{++}, c_{+-}. Moreover these equations decouple if one considers the linear combinations:

$$h_N = h_{++} + h_{+-} \qquad h_C = h_{++} - h_{+-} \qquad (3.139)$$

with similar definitions for c_N and c_C. $h_N(r)$ is the number density correlation function, which characterizes pair correlations between ions independently of their charge, while $h_C(r)$ is the charge density correlation function. The decoupled OZ relations read in Fourier space:

$$h_N(k) = c_N(k)\,[1 + \rho h_N(k)] \qquad h_C(k) = c_C(k)\,[1 + \rho h_C(k)] \qquad (3.140)$$

$c_N(k)$ vanishes identically within the RPA (equation (3.137)), so that there are no number density correlations in the restricted primitive model at the RPA level. However (3.137) also implies that $c_C(r) = -2z^2 \ell_B/r$, and hence $c_C(k) = -8\pi z^2 \ell_B/k^2$. Substitution into equation (3.140) then yields the following charge structure factor:

$$S_c(k) = 1 + \rho h_c(k) = \frac{k^2}{k^2 + \kappa_D^2} \qquad (3.141)$$

where $\kappa_D = 1/\lambda_D = \sqrt{8\pi\rho z^2 \ell_B}$ is the inverse of the Debye screening length, the physical significance of which becomes clear when one considers the charge correlation function $h_c(r)$ obtained by Fourier transformation of $h_c(k)$:

$$h_c(r) = 2h_{++}(r) = -2h_{+-}(r) = -\frac{2z^2 \ell_B}{r}\exp(-r/\lambda_D) \qquad (3.142)$$

This result clearly demonstrates the physical significance of the Debye screening length: correlations between ions decrease exponentially, with a decay length equal to λ_D. Charges of equal sign repel each other ($h_{++}(r) < 0$), while opposite charges attract ($h_{+-}(r) > 0$), in such a way that a 'screening cloud' of opposite charge forms around each ion; at distances $r > \lambda_D$, the total charge inside the screening cloud (including that of the central ion) vanishes exponentially. This 'perfect screening' is embodied in the small k behaviour of $S_c(k)$ deduced from

(3.142), namely [8]:

$$S_c(k) = \frac{k^2}{\kappa_D^2} + \mathcal{O}(k^4) \tag{3.143}$$

Physically the vanishing of $S_c(k)$ at small wavevectors is a consequence of elec-troneutrality. Large-scale charge fluctuations have an extensive energy cost, and are therefore ruled out in the thermodynamic limit. However, the RPA result (3.142) is unphysical at short range, where it predicts divergent, negative val-ues of $g_{++} = h_{++} - 1$ as $r \to 0$; this is a direct consequence of the neglect of correlations within the mean field picture.

The RPA results obtained for the restricted primitive model are easily general-ized to primitive models of electrolytes with arbitrary numbers v of ionic species of arbitrary valences z_α. The general definition of the Debye screening length is:

$$\kappa_D = \lambda_D^{-1} = \sqrt{4\pi\ell_B \sum_{\alpha=1}^{v} \rho_\alpha z_\alpha^2} \tag{3.144}$$

while the partial pair correlation functions $h_{\alpha\beta}(r)$ are given by an obvious gener-alization of (3.142), namely:

$$h_{\alpha\beta}(r) = -\frac{z_\alpha z_\beta \ell_B}{r} \exp(-r/\lambda_D) \tag{3.145}$$

These pair correlation functions may then be used to calculate the osmotic prop-erties of ionic solutions, with the help of the standard relations between thermo-dynamics and pair structure established in section 3.5. In particular, the osmotic pressure Π may be calculated from the multicomponent generalization of the virial relation (2.44).

$$\Pi = k_B T \sum_\alpha \rho_\alpha - \frac{2\pi}{3} \sum_\alpha \sum_\beta \rho_\alpha \rho_\beta \int_0^\infty g_{\alpha\beta}(r) \frac{dw_{\alpha\beta}(r)}{dr} r^2 dr = k_B T \rho \left[1 - \frac{\kappa_D^3}{24\pi\rho} \right] \tag{3.146}$$

Note that in going from the first to the second equality in (3.146), the $g_{\alpha\beta}$ may be replaced by the $h_{\alpha\beta}$, due to the overall charge neutrality condition $\sum_\alpha z_\alpha \rho_\alpha = 0$. The first term is the usual van't Hoff term for ideal solutions (non-interacting ions), while the second term leads to a lowering of the osmotic pressure, which shows that the attraction between oppositely charged ions more than compen-sates for the repulsion between ions of equal sign. Equation (3.146) is called the 'limiting law', valid for dilute electrolyte solutions, which was established by Debye and Hückel in 1923, using a different but equivalent approach, based upon a linearized version of Poisson–Boltzmann theory, to which we shall return in section 7.6 in the study of electric double-layers near surfaces. Equation (3.146) allows one to assess the range of validity of the RPA approach: for $\rho > 4/\pi\ell_B^3$

---

[8] The validity of equation (3.143) goes beyond the RPA, and was shown by Stillinger and Lovett to be an exact condition for ionic fluids.

it would predict a negative osmotic compressibility, and the 'correction' term would become larger than the term that corresponds to the reference system.

It is important to notice that the Debye–Hückel correction to the osmotic pressure of an ideal solution does not correspond to second order in the virial expansion (2.79). Indeed, the correction to van't Hoff's law is proportional to $\rho^{3/2}$ instead of ρ. This is a consequence of the infinite range of the Coulomb interaction potential (3.136), which leads to a divergent second virial coefficient (2.80), and in fact a divergence of all higher order virial coefficients as well.

Experimental measurements of the osmotic pressure are in agreement with the Debye–Hückel limiting law at low ionic strength. As the ionic concentration increases, finite core ($\sigma \neq 0$) effects become important, and the osmotic pressure starts to increase. Finite core effects may be approximately accounted for within a generalization of the RPA, called the 'mean spherical approximation' (MSA), whereby the RPA ansatz is kept for $r > \sigma_{\alpha\beta}$, while the $h_{\alpha\beta}$ are required to satisfy the exact core condition $h_{\alpha\beta}(r) = -1$ for $r < \sigma_{\alpha\beta}$. Together with the OZ relations (3.138), these closure relations form a closed set, which was solved analytically [9]. Their osmotic equation of state reduces to the Debye–Hückel law in the limit $\sigma_{\alpha\beta} \to 0$.

Structure factor of dense polymer solutions

As a second application of the RPA, we consider the structure factor of a solution or melt of non-ideal polymers. The reference system will now be an assembly of non-interacting, or ideal, polymer chains. As the different chains are statistically independent, the structure factor $S_0(k)$ of this assembly is simply the form factor of the Gaussian chain calculated in section 3.9 (equation (3.116)).

This reference system ignores all interactions between monomers on the same or different chains, except for the connectivity embodied in the harmonic spring force between successive monomers on the same polymer. In reality, any two monomers that are not directly connected will interact if they come sufficiently close in the course of the conformational fluctuations of the polymer chains. The interaction is generally short ranged (except in the case of polyelectrolytes) compared to the characteristic dimension of a chain (e.g. its radius of gyration R_G), and is identical for monomers (assumed to be all of the same chemical species) on the same or on different chains. The interaction is repulsive for polymers in a 'good' solvent, and attractive in a 'poor' solvent. In view of the short-range nature of the (effective) interaction one may adopt a simple 'contact' potential [10]

[9] E. Waisman and J-L. Lebowitz, *J. Chem. Phys.* **56**, 3086, 3093 (1972).
[10] Although the δ function appearing in equation (3.147) may seem a very singular interaction, it is actually very well behaved, in the sense that it admits a Fourier transform. This feature allows a direct application of the RPA method.

between monomers, of the form:

$$w(\mathbf{r} - \mathbf{r}') = v k_{\mathrm{B}} T \delta(\mathbf{r} - \mathbf{r}') \tag{3.147}$$

such that the total potential energy of interaction between all monomers (N chains of M monomers each) is:

$$W_N = \frac{v k_{\mathrm{B}} T}{2} \sum_{i=1}^{N} \sum_{v=1}^{M} \sum_{j=1}^{N} \sum_{\mu=1}^{M} \delta(\mathbf{r}_{i,v} - \mathbf{r}_{j,\mu}) \tag{3.148}$$

v, called the 'excluded volume parameter' has the dimension of a volume, and plays, in fact, the role of a second virial coefficient. The above model implicitly assumes that the monomer concentration is sufficiently low for triplet and higher order interactions to be negligible.

With the non-interacting gaussian chains as reference system, the general RPA result (3.132) leads to the following simple expression for the structure factor of N interacting polymers:

$$\frac{1}{S(k)} = \frac{1}{S_0(k)} + \rho v \tag{3.149}$$

where $S_0(k)$ is given by (3.116) and ρ is the monomer concentration. Substituting the Lorentzian approximation (3.118) for $S_0(k)$ we find:

$$S(k) = \frac{12}{b^2}(k^2 + 1/\xi^2) \tag{3.150}$$

where ξ is given by

$$\xi^2 = \frac{M b^2}{12(1 + M \rho v)} \simeq \frac{b^2}{12 \rho v} \tag{3.151}$$

and the last equality follows from taking the large M limit. The monomer–monomer correlation function finally follows from inverse Fourier transformation of $S(k)$, with the result:

$$h(r) = \frac{3}{\pi \rho b^2 r} \exp(-r/\xi) \tag{3.152}$$

ξ therefore defines a *correlation length* for the monomer density fluctuations. This correlation length decreases with increasing monomer concentration.

The correlation length ξ is also a *screening length* for the excluded volume interaction (3.147). This can be seen as follows. Suppose we introduce a monomer at the origin ($\mathbf{r} = 0$) and compute the effective potential felt by another monomer at distance r from the origin. This effective potential is the sum of the potential created by the monomer at the origin, and of the contribution from the response of the solution to this external perturbation. Using the usual linear response argument, the density in the perturbed solution has Fourier components

$$\delta \rho_{\mathbf{k}} = -S(\mathbf{k}) \rho v \tag{3.153}$$

or, in real space

$$\rho(r) = \rho - \frac{3\rho v}{\pi b^2 r} \exp(-r/\xi) \tag{3.154}$$

and one obtains an effective potential [11] from equation (3.127)

$$v_{\text{eff}}(r) = k_B T v \left[\delta(\mathbf{r}) - \frac{1}{4\pi\xi^2 r} \exp(-r/\xi) \right] \tag{3.155}$$

The remarkable property of this effective potential is that $\int v_{\text{eff}}(r) \, d\mathbf{r} = 0$. The attractive contribution exactly compensates the repulsion. An important consequence of this exact screening is that a chain in a dense solution or melt behaves, on large scales (i.e. on scales larger than ξ), as an ideal chain, with Gaussian statistics, $R_G \sim \sqrt{M} b$. This is the case because, when seen on scales larger than ξ, the effective interaction between two monomers on the same chain is neither repulsive nor attractive [12].

Finally, it is important to understand the limit of validity of the RPA approach for polymers. To this end, we will consider the osmotic equation of state for the polymer solution. As the polymers are composite molecules, it is not possible to use directly the virial formula as in the electrolyte case, and we will admit without proof that the equation of state at the RPA level is given by

$$\Pi = k_B T \left[\frac{\rho}{M} + \frac{1}{2}\rho v^2 - \frac{1}{24\pi\xi^3} \right] \tag{3.156}$$

The origin of the three terms in (3.156) is easily understood. The first term is just the ideal gas term for a gas of chains with density ρ/M. The second term accounts for the repulsive interactions between chains, computed at the mean field level, i.e. taking $g(r) = 1$. The last term, of the form $k_B T/\xi^3$, is typical of a 'fluctuation' correction, and accounts for the fact that the correlations between monomers act to make the repulsion less effective. Note the similarity of this term with the corresponding correction computed for electrolytes, the Debye limiting law (3.146).

From equation (3.156) it is easily seen that as $\rho \to 0$ the osmotic compressibility becomes negative, which indicates a failure of the approach. The RPA theory is valid only if the third term in (3.156) is a correction compared to the mean field contribution, which is true if $\rho > v/b^6$. The RPA for polymer solutions is valid for high densities only, when the solution is sufficiently uniform for the neglect of correlations embodied in equation (3.127) to be valid. When the solution becomes more dilute, large correlated fluctuations take place on the scale ξ, which can no longer be treated using this simple assumption (see section 5.5).

---

[11] In the definition of v_{eff} the mean contribution $\rho v k_B T$ coming from the uniform density is subtracted.

[12] The argument can of course be made rigorous by computing, at the level of perturbation theory, the swelling of a chain induced by the effective potential (3.155).

Finally, let us mention, without describing the technical justification of this statement, that the RPA approximation can be shown to imply a Gaussian distribution of the density fluctuations. The probability function for the collective coordinates ρ_k which is compatible with the RPA form of the structure factor is of the form

$$\mathcal{P}(\{\rho_k\}) = \prod_k \frac{b^2}{12\pi}(k^2 + \xi^{-2}) \exp\left(-\frac{b^2}{12}(k^2 + \xi^{-2})\rho_k \rho_{-k}\right) \tag{3.157}$$

which of course leads back to the expression (3.150) for the structure factor upon integration over ρ_k. Note that this distribution implies that the phases of the complex numbers ρ_k are uncorrelated, uniformly distributed random numbers, consistent with the 'RPA' denomination.

Further reading

Fluctuations are discussed in many textbooks on statistical physics. L. Landau and E. Lifshitz's *Statistical Physics*, Butterworth-Heinemann, London, 1980, is probably one of the oldest references on the subject, and in spite of a rather outdated presentation remains worth reading. More modern presentations, with special emphasis on liquid state structure, can be found in D. Chandler's textbook *Introduction to Modern Statistical Physics*, Oxford University Press, Oxford, 1987, and, at a more advanced level, in J.-P. Hansen and I.R. McDonald, *Theory of Simple Liquids*, Academic Press, New York, 1986.

A (difficult) reference text on the physics of dielectric media, including both macroscopic and microscopic aspects, is Fröhlich's *Theory of Dielectrics*, Oxford University Press, Oxford, 1989.

B. Berne and R. Pecora, *Dynamic Light Scattering*, Wiley, New York, 1976, reprinted by Dover, is an excellent textbook on scattering techniques, with discussions of structure and form factors that are of course not limited to light scattering experiments.

The RPA approximation is presented in several books on polymer physics, in particular P.G. de Gennes, *Scaling Concepts in Polymer Physics*, Cornell University Press, Ithaca, NY, 1979, and M. Doi and S.F. Edwards, *Theory of Polymer Dynamics*, Clarendon Press, Oxford, 1986. Both books also discuss form factors of single polymer chains. Application of the RPA to charged systems, in the form of the Debye–Hückel approximation, is presented in one way or another in essentially all textbooks on statistical physics (e.g. Landau and Lifshitz).

II Phase transitions

4 Mean field approaches

4.1 Lattice models and mean field treatment

Changes in temperature, pressure or chemical composition can lead to continuous or discontinuous phase changes, as illustrated schematically in figure 1.1. Phase transitions are ubiquitous in simple and complex fluids, and the thermodynamic conditions for phase equilibria have been spelled out in section 2.3. The most common phase transitions, like vapour condensation, freezing or the isotropic to nematic transition in liquid crystals, are first-order transitions, characterized by discontinuities of first derivatives of the free energy, like the entropy or the molar volume. Two (or more phases) can coexist over a range of temperatures or pressures and *metastable* thermodynamic states of one phase can exist for appreciable times under conditions where another phase has a lower free energy. Most first-order transitions can be reasonably well described within mean field theory, which is the leitmotiv of the present chapter. A key quantity in the description of first-order transitions is the *order parameter*, which generally characterizes some broken symmetry, taking a finite value in the phase of lower symmetry, and vanishing discontinuously at the transition towards the phase of higher symmetry. In many cases the order parameter can be associated with some microscopic variable, like the orientation vector of mesogenic molecules, but in the case of a transition between two isotropic fluids, with full rotational and translational invariance, like a liquid and its coexisting vapour, or two liquid mixtures of different compositions, the order parameter is a macroscopic characteristic, conveniently chosen to be the difference in density or concentration between the two phases.

Second-order phase transitions, however, correspond to a continuous vanishing of the order parameter at a well defined temperature, called the critical temperature, below which the transition is first order. They are characterized by strong fluctuations and consequently cannot be correctly described within mean field theory, which neglects fluctuations entirely. Critical fluctuations will be the object of the following chapter.

If molecular details of a fluid are not essential for the understanding of large-scale behaviour, as is the case near second-order phase transitions (where the characteristic correlation length of fluctuations will be shown to diverge), or when considering global characteristics of macromolecules in solution, a greatly simplified, coarse-grained description of the system, as embodied in the lattice gas model introduced in section 1.4 (cf. figure 1.11), is generally sufficient. Molecular positions are restricted to a discrete set on a regular space-filling array of cells of volume v_0, and a single occupancy constraint is imposed, whereby each cell can contain at most one molecule.

We first consider the lattice model of an incompressible binary mixture of two molecular species, A and B. Each cell is occupied by either an A or a B molecule, so that the total number of molecules, N, equals the total number of sites, M (this can be easily generalized to a compressible mixture, by allowing some cells to be empty, so that $N < M$). A single, discrete variable n_i is associated with each cell $1 \leq i \leq N$, which can take on two possible values only, $n_i = 0$ (when the cell contains an A molecule) and $n_i = 1$ (when the cell contains a B molecule).

Only molecules in nearest-neighbour cells interact (although this may be easily generalized), with pair energies $-\epsilon_{AA}$, $-\epsilon_{BB}$ and $-\epsilon_{AB}$ corresponding to each of the three possible neighbour pairs.

A configuration is completely specified by the set of N occupation numbers $\{n_i\}$. The total number of possible configurations is 2^N, and for any configuration, the total interaction energy is:

$$E(\{n_i\}) = -\sum_{(i,j)} \epsilon_{AA}(1 - n_i)(1 - n_j) + \epsilon_{AB}(1 - n_i)n_j + \epsilon_{AB}n_i(1 - n_j) + \epsilon_{BB}n_i n_j$$

(4.1)

where the sum (i, j) is over all $N\nu/2$ nearest-neighbour pairs of cells (ν being the lattice coordination number). Equation (4.1) is easily recast in the form:

$$E(\{n_i\}) = \epsilon \sum_{(i,j)} n_i(1 - n_j) - 2B \sum_i n_i + C \qquad (4.2)$$

where $\epsilon = \epsilon_{AA} + \epsilon_{BB} - 2\epsilon_{AB}$, $B = -\nu(\epsilon_{AA} - \epsilon_{BB})/4$ and $C = -N\nu\epsilon_{AA}/2$.

The above form clearly shows that occupation of neighbouring sites by molecules of different species (AB pairs) will lower the energy, provided $\epsilon < 0$. It may hence be anticipated that $\epsilon_{AA} + \epsilon_{BB} < 2\epsilon_{AB}$ ($\epsilon < 0$) favours mixing, while $\epsilon_{AA} + \epsilon_{BB} > 2\epsilon_{AB}$ favours segregation. In practice, the latter is the most common situation, as unlike atoms or molecules tend to have a weaker attraction than similar species.

A convenient order parameter to characterize a possible phase separation is the concentration ϕ of B molecules:

$$\frac{1}{N}\left\langle \sum_{i=1}^{N} n_i \right\rangle = \frac{N_B}{N} = \phi \qquad (4.3)$$

Since $N = N_A + N_B$ is fixed, the appropriate statistical ensemble is a semi-grand canonical ensemble characterized by a single (relative) chemical potential

μ, conjugate to the total number N_B of B molecules, which can vary between 0 and N. The corresponding grand partition function of the binary mixture is:

$$\Xi(\mu, N, T) = \prod_{i=1}^{N} \sum_{n_i=0,1} \exp\left(-\beta \left[E(\{n_i\}) - \mu \sum_j n_j\right]\right) \qquad (4.4)$$

The resulting grand potential is

$$\Omega(\mu, N, T) = -k_B T \ln \Xi \qquad (4.5)$$

and the Helmholtz free energy per site follows from the standard thermodynamic relation (2.19):

$$f(\phi, T) = \frac{1}{N} F(\phi, N, T) = \frac{1}{N} \Omega(\mu, N, T) + \mu \phi \qquad (4.6)$$

f is the sum of energy and entropy contributions:

$$f = \frac{1}{N}(U - TS) = u - Ts \qquad (4.7)$$

Both terms are easily evaluated within the mean field approximation which neglects correlations between occupation numbers n_i of neighbouring cells. A quick derivation is based on the Bragg–Williams argument. Since sites are assumed to be uncorrelated, S reduces to the entropy of mixing of two ideal gases, which is easily calculated from the total number of configurations, given by equation (1.22), using Stirling's formula:

$$S = k_B \ln \Omega = -Nk_B [\phi \ln \phi + (1 - \phi) \ln(1 - \phi)] \qquad (4.8)$$

The internal energy U is the statistical average of the energy (4.2), evaluated by assuming that the n_i are independent variables. In the symmetric case, where $\epsilon_{AA} = \epsilon_{BB}$ (and hence $B = 0$ in equation (4.2)), one finds, within a state-independent constant:

$$U \simeq \epsilon \sum_{(i,j)} \langle n_i \rangle \langle (1 - n_j) \rangle = \frac{N\nu\epsilon}{2} \phi(1 - \phi) \qquad (4.9)$$

Gathering results:

$$f = \epsilon \frac{\nu}{2} \phi(1 - \phi) + k_B T [\phi \ln \phi + (1 - \phi) \ln(1 - \phi)] \qquad (4.10)$$

If $\epsilon < 0$, the two contributions to this free energy of mixing are negative, so that for any T, f will be a convex function of ϕ, with a single minimum for $\phi = 0.5$ (the symmetric model must be invariant under the $\phi \leftrightarrow 1 - \phi$ transformation), corresponding to complete miscibility.

If, however, $\epsilon < 0$, the energy and entropy terms compete, and at sufficiently low temperature, f will exhibit two minima at $\phi = \phi_1$ and $\phi = \phi_2 = 1 - \phi_1$, separated by a concave region, with a maximum at $\phi = 1/2$. The concavity signals phase separation (cf. figure 4.1): for any concentration $\phi_1 < \phi < \phi_2$, the system can lower its free energy by decomposing into an 'A-rich' phase with

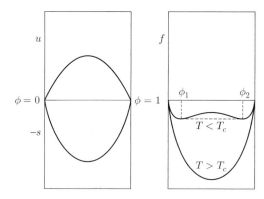

Figure 4.1. Free energy of a symmetric mixture. The left panel shows the entropic (convex) and energetic (concave) contributions, $-s$ and u, to the free energy. The resulting free energy per particle, $f = u - Ts$ is shown on the right for $T > T_c$ and $T < T_c$. For $T < T_c$, systems with $\phi_1 < \phi < \phi_2$ are not thermodynamically stable.

composition ϕ_1 and a 'B-rich' phase with composition ϕ_2. The composition ϕ of the mixture is fully determined by specifying the temperature T and relative chemical potential μ. For any given T, ϕ and μ are linked by the thermodynamic relation:

$$\mu = \left(\frac{\partial f}{\partial \phi} \right)_T \tag{4.11}$$

Substituting equation (4.10) one arrives at the following implicit relation for ϕ:

$$\phi = \frac{1}{2} \left[1 + \tanh\left(\frac{\beta}{2} \left(\mu + \epsilon \nu \left(\phi - \frac{1}{2} \right) \right) \right) \right] \tag{4.12}$$

which may be solved numerically or graphically. Equation (4.12) admits a solution for all temperatures, provided $\mu < \epsilon \nu / 2$, which is the energetic cost of interchanging an A and a B molecule.

Due to the A $\leftrightarrow$ B interchange symmetry of the model, $\mu = \mu_A - \mu_B$ must vanish at coexistence when phase separation occurs, i.e. the compositions of the coexisting phases are determined by a standard double-tangent construction, which is horizontal in the symmetric model, as illustrated in figure 4.1. Setting $\mu = 0$ in equation (4.12) the composition variable $x = \phi - 1/2$ satisfies the following condition at coexistence:

$$x = \frac{1}{2} \tanh\left(\frac{\beta \epsilon \nu}{2} x \right) \tag{4.13}$$

which admits a non-zero solution only provided the initial slope of the right-hand side exceeds 1, i.e.:

$$T < T_c = \epsilon \nu / 4 k_B \tag{4.14}$$

where T_c is the critical temperature, above which the free energy is a convex function for all values of ϕ and hence no demixing occurs. The result (4.14) may also be recovered by noting that for the symmetric mixture, the critical composition is necessarily $\phi_c = 1/2$, and that at the critical temperature, the minimum of $f(\phi)$ for $T > T_c$ turns into a maximum for $T < T_c$, so that:

$$\left.\frac{\partial^2 f}{\partial \phi^2}\right|_{T=T_c,\phi=1/2} = 0 \tag{4.15}$$

From the definition (4.3) of ϕ, and from equations (4.4) and (4.5), it is immediately clear that:

$$\phi = -\frac{1}{N}\left(\frac{\partial \Omega}{\partial \mu}\right)_{N,T} \tag{4.16}$$

Taking the second derivative, one arrives at the following fluctuation relation, which is a special case of the general result (2.51):

$$k_B T \left(\frac{\partial \phi}{\partial \mu}\right)_{N,T} = -\frac{k_B T}{N}\left(\frac{\partial^2 \Omega}{\partial \mu^2}\right)_{N,T} = \frac{1}{N}\left(\langle N_B^2 \rangle - \langle N_B \rangle^2\right) \tag{4.17}$$

which immediately implies that $(\frac{\partial \phi}{\partial \mu})_{N,T} > 0$, and, in view of relation (4.11), the thermodynamic stability condition of the mixture:

$$\left(\frac{\partial^2 f}{\partial \phi^2}\right)_{N,T} > 0 \tag{4.18}$$

The locus of points in the (T, ϕ) plane where $\frac{\partial^2 f}{\partial \phi^2}$, called the spinodal line, terminates at the critical point (T_c, ϕ_c). From figure 4.1 it is clear that the spinodal lies inside the coexistence line, or binodal, and that the two curves only meet at their maximum, which is the critical point, as illustrated in figure 4.2. Thermodynamic states inside the spinodal line are unstable against low amplitude fluctuations, leading to rapid spinodal decomposition into stable coexisting (demixed) phases, a kinetic process described in section 9.3. Thermodynamic states between the binodal and the spinodal are metastable, and eventually phase separate via rare nucleation events, to be described in section 10.6.

The scenario just described for a symmetric mixture carries over to the physically more relevant asymmetric case, where $\epsilon_{AA} \neq \epsilon_{BB}$, with only quantitative changes, as illustrated in the following exercise.

Exercise: Calculate within mean field theory (Bragg–Williams approximation) the free energy per site $f(\phi, T)$ in the general, asymmetric case where $\epsilon_{AA} \neq \epsilon_{BB}$. Derive from it the free energy of mixing, defined by $\Delta f(\phi) = f(\phi) - f(\phi = 0) - f(\phi = 1)$. Show explicitly that this expression is no longer invariant under the substitution $\phi \leftrightarrow 1 - \phi$. What immediate consequence for the critical point follows from this lack of invariance? Defining $\Delta\epsilon = \epsilon_{AA} - \epsilon_{BB}$, write down the two equations that determine the critical concentration ϕ_c and temperature T_c. Solve these equations numerically for the case $\Delta\epsilon/\epsilon = 0.1$.

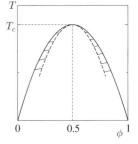

Figure 4.2. Phase diagram of a symmetric binary mixture. The full line is the binodal line, the dashed line is the mean field spinodal line.

The 'lattice gas', illustrated in figure 1.11, is a special case of the above model for binary mixtures, where only one species of molecules is present, say B, occupying a fraction ϕ of the available cells, whereas the remaining fraction $1 - \phi$ of cells are empty; in other words the molecules A are replaced by vacancies, such that $\epsilon_{AA} = \epsilon_{AB} = 0$. Due to the vacancies, the lattice gas is compressible, and is the simplest possible model to study condensation, i.e. the gas–liquid transition, which will be investigated within a continuous model in section 4.3.

As noted earlier, the binary mixture may also be made compressible by allowing for three types of sites, namely sites occupied by either A or B molecules, and vacancies. The composition and overall density (or fraction of occupied cells) are now independent thermodynamic variables controlled by two freely varying chemical potentials μ_A and μ_B (rather than merely by their difference as in the incompressible case). The single critical point is now replaced by a critical line as the overall density (or equivalently the pressure) is varied. Note that due to the competition between demixing and condensation, binary mixtures can exhibit a rich variety of topologically different phase diagrams [1].

Finally it is easily shown that the lattice model for incompressible binary mixtures is isomorphous to the familiar Ising spin model of magnetism. Indeed, the occupation variables n_i can be mapped into spin variables S_i, taking values ± 1 (up or down), by the transformation:

$$S_i = 2n_i - 1 \tag{4.19}$$

The energy function (4.2) may then be cast, within a constant, into the standard Ising form:

$$H(\{S_i\}) = -B \sum_i S_i - J \sum_{(i,j)} S_i S_j \tag{4.20}$$

where $J = \epsilon/4$, and the coefficient B is seen to play the role of the external magnetic field (multiplied by the magnetic dipole moment of the spin-carrying particles). $J < 0$ favours antiparallel neighbouring spins (antiferromagnetic order). The natural order parameter for the Ising model is obviously the mean magnetization per spin:

$$\zeta = \frac{1}{N} \left\langle \sum_i S_i \right\rangle \tag{4.21}$$

which plays the role of the composition variable ϕ in the binary mixture; ζ takes non-zero values in the ferromagnetic phase, below the critical (Curie) temperature. The appropriate ensemble to describe a system with a fixed number of

---

[1] R.L. Scott and P.H. Van Konynenburg, *Discuss. Faraday Soc.* **49**, 87 (1970). G. Pittion-Rossillon, *J. Chem. Phys.* **73**, 3398 (1980).

two-state spins is the canonical ensemble. The partition function reads:

$$Q = \prod_{i=1}^{N} \sum_{S_i = \pm 1} \exp(-\beta H_0 - \beta H_1) \tag{4.22}$$

where H_0 is the first term on the right-hand side of the Hamiltonian (4.20) (coupling to the external field), while H_1 is the term describing the coupling between nearest-neighbour spins. The Helmholtz free energy per spin is:

$$f = -\frac{k_B T}{N} \ln Q_N \tag{4.23}$$

while the order parameter is given by a relation similar to equation (4.16) for the binary mixture (with the applied field playing the role of the chemical potential) namely:

$$\zeta = -\left(\frac{\partial f(T, B)}{\partial B}\right)_T \tag{4.24}$$

If spin–spin interactions can be neglected, i.e. if $H_1 = 0$ as is physically reasonable in the high temperature paramagnetic phase, the partition function trivially factorizes into N identical one-spin partition functions q:

$$Q_N^{(0)} = q^N = 2\left[\cosh(\beta B)\right]^N \tag{4.25}$$

Substitution into equations (4.23) and (4.24) then leads to

$$\zeta = \tanh(\beta B) \tag{4.26}$$

i.e. the order parameter is non-zero only in a finite applied field (paramagnetic phase).

At lower temperatures H_1 can no longer be neglected. Mean field theory will be formulated here in a more systematic way than the Bragg–Williams approximation (while being physically equivalent), by invoking the Gibbs–Bogoliubov inequality (3.62). To that purpose, the Hamiltonian (4.20) is split into two terms, $H = H_0' + H_1'$, where:

$$H_0' = -(B + B_1) \sum_i S_i \tag{4.27}$$

$$H_1' = -J \sum_{(i,j)} S_i S_j + B_1 \sum_i S_i \tag{4.28}$$

Where B_1 is an auxiliary 'molecular' field, as first introduced by Pierre Weiss in 1907. H_0' characterizes the reference system of non-interacting spins, while the perturbation H_1' (equivalent to W_N in section 3.5) accounts for the interactions between spins. According to the inequality (3.62), the free energy of the system of fully interacting spins is bounded above by:

$$f \leq f^{(0)}(B_1) + \frac{1}{N} \langle H_1 \rangle^{(0)}(B_1) \tag{4.29}$$

where both the reference free energy $f^{(0)}$ and the first-order perturbation correction, calculated within the independent spin ensemble of the reference system

$$\frac{1}{N}\langle H_1\rangle^{(0)}(B_1) = -J\frac{\nu}{2}\langle S\rangle^2 + B_1\langle S\rangle = -J\frac{\nu}{2}\zeta^2 + B_1\zeta \tag{4.30}$$

depend on the auxiliary field B_1. The optimum value of the latter is determined by minimizing the right-hand side of (4.29) with respect to B_1. Remembering equations (4.25) and (4.26) (with B replaced by B_1), this immediately leads to the following implicit equation for B_1:

$$B_1 = \nu J \tanh[\beta(B + B_1)] \tag{4.31}$$

The corresponding order parameter follows from (4.24), with f given by (4.29); taking due account of the extremum condition, one easily verifies that:

$$\zeta = \tanh[\beta(B + B_1)] = \frac{B_1}{J\nu} \tag{4.32}$$

which reflects the intuitively satisfying result that the 'molecular field' felt by any one spin is in fact the field of its nearest neighbours. In the absence of an applied field B the order parameter is hence a solution of the implicit equation:

$$\zeta = \tanh(\beta\nu J\zeta) = \tanh(\theta\zeta/T) \tag{4.33}$$

where $\theta = \nu J/k_B$ is a characteristic temperature related to the coupling between spins. Equation (4.33) is strongly reminiscent of the corresponding self-consistency equation (4.13) for a binary mixture. Using a graphical solution of equation (4.33), there will be an intersection of the diagonal $y = \zeta$ and of the curve $y = \tanh(\theta\zeta/T)$ at a non-zero value of ζ, provided the initial slope of the latter is larger than one, i.e. when $T < \theta$. A spontaneous magnetization ($\zeta \neq 0$) thus appears in zero applied field, at all temperatures below the critical (or Curie) temperature:

$$T_c = \theta = \frac{\nu J}{k_B} \tag{4.34}$$

This corresponds to the ferromagnetic phase, while above T_c no spontaneous magnetization occurs (paramagnetic phase). Just below T_c, ζ is small, so that $\tanh(T_c\zeta/T)$ may be expanded in a Taylor series of its argument. Retaining the first two terms, and substituting in equation (4.33), one arrives at a quadratic equation for the order parameter, which is easily solved to yield:

$$\zeta = \pm\sqrt{3}\left(\frac{T_c - T}{T_c}\right)^{1/2} \tag{4.35}$$

This simple power law is a first example of a scaling relation which characterizes the behaviour of the system in the vicinity of the critical point. The exponent $1/2$ is a general mean field result for the order parameter, which is modified when critical fluctuations are properly taken into account (cf. section 5.2).

The isomorphism of the Ising model and of the lattice gas has far-reaching consequences. It implies that near the critical point, where molecular details are unimportant, magnetic systems and fluids have similar critical behaviour. In particular, the scaling laws which characterize this behaviour admit the same exponents for these widely different physical systems, which are said to belong to the same *universality class*.

4.2 Landau theory of phase transitions

The previous section provides a classic example of a mean field theory of a first-order phase transition terminating in a critical point. A unified framework for the study of first- and second-order phase transitions within the mean field spirit is provided by Landau's phenomenological theory which, by ignoring microscopic details of the system under consideration, applies to any phase transition characterized by a macroscopic order parameter ζ (which would be the composition variable ϕ in the example of section 4.1).

An order parameter is a tensorial quantity of rank 0 (scalar), 1 (vector) or higher which provides an average measure of the global symmetry of a thermo-dynamic phase of a substance. In most cases high temperature phases are fully disordered on the molecular scale, both translationally and orientationally, and this disorder reflects itself in a high degree of symmetry of macroscopic properties which are fully invariant under arbitrary translations (homogeneity) and rotations (isotropy). Low temperature phases (like crystals or liquid crystal mesophases) are characterized by the spontaneous appearance of long-range translational or orientational order, which corresponds to a reduction in symmetry; the symmetry of the high temperature phase is said to be broken, and a suitably defined order parameter will take on non-zero values, which depend on the thermodynamic state variables. In the simple magnetic case examined in the previous section, magnets will spontaneously align along a preferred direction even in zero field to provide a finite magnetization in the low temperature ferromagnetic phase, below a critical (Curie) temperature T_c. Clearly the full rotational invariance of the high temperature paramagnetic phase is broken below T_c; the ferromagnetic phase is invariant only under rotations around the direction of the spontaneous magnetization (cylindrical symmetry). Similar considerations apply to the isotropic to nematic transition of liquid crystals.

In a number of cases, particularly relevant for liquid matter, the two coexisting phases have the same symmetry. For example an isotropic liquid and its coexisting vapour are fully homogeneous and isotropic; the same is true of coexisting fluid mixtures of different chemical compositions, separated by a miscibility gap. Since there is no order parameter associated with a broken symmetry, it is convenient to characterize such transitions by a thermodynamic order parameter, which is the difference in density $\Delta\rho = \rho_l - \rho_g$ between the coexisting liquid and gas phases, or the difference in concentration of one of the species in a binary mixture.

At a phase transition, the order parameter may vanish either discontinuously (i.e. jump from a finite value to zero) or continuously, as the temperature is increased. The first situation corresponds to a first-order phase transition, which is also characterized by discontinuities of first derivatives of the appropriate thermodynamic potential (cf. section 2.3), while a continuous vanishing of the order parameter signals a second-order transition, where second derivatives of the thermodynamic potential exhibit singularities. First-order transitions between phases of the same symmetry generally terminate at a critical point, or critical line, where the transition becomes second order; the most common example is illustrated in figure 1.1, where the gas–liquid transition is first order when the temperature is lowered, except along the critical isochore $\rho = \rho_c$ where the transition is second order at the critical temperature, which coincides with the top of the coexistence curve.

For the sake of simplicity, only scalar order parameters will be considered in the subsequent discussion. The order parameter ζ is assumed to couple linearly to a conjugate 'ordering' field, which may be an external force field (like a magnetic field) or a thermodynamic control variable (like the chemical potential). Landau theory casts all phase transition scenarios (some of which have been alluded to above) in a unified mathematical framework. To capture the link between Landau's phenomenological theory, and a statistical mechanics formulation, it proves convenient to partition the microstates of the system according to the corresponding value of ζ. Let $E(\zeta)$ be the mean energy of the system associated with a prescribed value of ζ, in the absence of the ordering field (which will henceforth be denoted by B), and let $\omega(\zeta)$ denote the corresponding density of states, i.e. the number of microstates (or the volume in phase space) compatible with a value of the order parameter between ζ and $\zeta + d\zeta$.

In the presence of the ordering field, the partition function may then be cast in the form (remembering that ζ is intensive)

$$
\begin{aligned}
Q(T, B) &= \int_{-\infty}^{+\infty} \omega(\zeta) \exp\left[-\beta\left(E(\zeta) - BV\zeta\right)\right] d\zeta \\
&= \int_{-\infty}^{+\infty} \exp\left[-\beta\left(A(T, \zeta) - BV\zeta\right)\right] d\zeta
\end{aligned}
\tag{4.36}
$$

where V is the volume, $S(\zeta) = k_B \ln \omega(\zeta)$ is the entropy of the system with the order parameter in the range, $\zeta, \zeta + d\zeta$, and $A(T, \zeta) = E(\zeta) - TS(\zeta)$ denotes the *Landau free energy*. The resulting thermodynamic potential (Helmholtz or Gibbs free energy, or grand potential, depending on the choice of independent thermodynamic variables) is finally given by the standard definition:

$$
F(T, B) = -k_B T \ln Q(T, B)
\tag{4.37}
$$

The probability of observing a value ζ of the order parameter in a given thermo-dynamic state is:

$$P(\zeta) = \frac{1}{Q} \exp\left[-\beta\left(A(T,\zeta) - BV\zeta\right)\right] \tag{4.38}$$

while the mean (or equilibrium) value of ζ follows from:

$$\bar{\zeta} = \int_{-\infty}^{+\infty} P(\zeta)\zeta \, d\zeta = \left(\frac{-\partial f}{\partial B}\right)_T \tag{4.39}$$

where $f = F/V$ is now a free energy per unit volume, or free energy density.

The approximation inherent in Landau's theory is to replace the integral in equation (4.36) by the maximum value of the integrand, which corresponds to the most probable, or equilibrium, value of ζ. The maximum is reached when the argument of the Boltzmann factor, $A - BV\zeta$, is at its minimum. If $a = A/V$ denotes the Landau free energy density, the extremum conditions are:

$$\frac{\partial}{\partial\zeta}\left(a(T,\zeta) - B\zeta\right)_{\zeta=\bar{\zeta}} = 0 \tag{4.40}$$

$$\frac{\partial^2}{\partial\zeta^2}\left(a(T,\zeta) - B\zeta\right)_{\zeta=\bar{\zeta}} > 0 \tag{4.41}$$

and the resulting estimate of the free energy is:

$$f(T, B) = a(T, \bar{\zeta}) - B\bar{\zeta} \tag{4.42}$$

The two free energies coincide in zero field, but are functions of different variables related by a Legendre transformation. Landau's approximation, which restricts ζ to the contribution from the most probable value, $\bar{\zeta}$, neglects any fluctuations around this value, which is the essence of mean field theory.

Since the most probable value of the order parameter vanishes in the high temperature, high symmetry phase, and remains small in the lower symmetry phase in the vicinity of a second-order phase transition, or near a weakly first-order transition (characterized by a small discontinuity of the order parameter), it seems natural to write down a Taylor expansion of the Landau free energy density in powers of ζ:

$$a(T, \zeta) = a_0(T) + \sum_{\ell=2}^{n} \frac{a_\ell(T)}{\ell}\zeta^\ell + \mathcal{O}(\zeta^{\ell+1}) \tag{4.43}$$

The coefficients $a_\ell(T)$ are functions of the temperature, and of any other in-dependent state variables, like pressure or composition, not explicitly shown in equation (4.43). Note that a does not contain a term linear in ζ, as is clear from the extremum condition (4.40) in zero field, when $\bar{\zeta} = 0$ in the high symmetry phase. Underlying the Landau expansion (4.43) is the fundamental assumption of analyticity of the free energy function, which in fact turns out not to be true near second-order phase transitions, as will be discussed in chapter 5.

Symmetry considerations may lead to a reduction of the number of non-vanishing coefficients a_ℓ. In particular, in the case of a scalar order parameter, if the system properties are invariant under the inversion $\zeta \leftrightarrow -\zeta$, only even powers of ζ appear in equation (4.43). This is true for the Ising model, and the closely related incompressible symmetric binary mixture model (where $\zeta = \phi - 1/2$), considered in section 4.1. The predictive power of Landau's theory is illustrated next by investigating the phase behaviour resulting from three choices for low order Taylor series of the form (4.43).

(a) $\ell = 2$ and 4

This series in even powers, truncated at $\ell = 4$, is appropriate for the description of the aforementioned Ising and symmetric binary mixture models, and more generally for systems invariant under the $\zeta \leftrightarrow -\zeta$ inversion:

$$a(T, \zeta) = a_0(T) + \frac{a_2(T)}{2}\zeta^2 + \frac{a_4(T)}{4}\zeta^4 \tag{4.44}$$

In order to ensure convergence of the integral (4.36) (stability), $a_4(T)$ must be positive. The extremum condition (4.40) in zero ordering field leads to $a_2(T)\bar{\zeta} + a_4(T)\bar{\zeta}^3 = 0$, which admits the following roots:

$$\bar{\zeta} = 0 \qquad \bar{\zeta} = \sqrt{-a_2(T)/a_4(T)} \tag{4.45}$$

For $\bar{\zeta} = 0$ to be the only real and stable root above some critical temperature T_c, $a_2(T)$ must be positive for $T > T_c$. There will be a non-zero, stable root below T_c, provided $a_2(T) < 0$ for $T < T_c$. Hence $a_2(T)$ must change sign at $T = T_c$. Since the theory is restricted to the vicinity of the phase transition, the simplest assumption compatible with these requirements is:

$$a_2(T) = \alpha_2(T - T_c) \tag{4.46}$$

$$a_4(T) = a_4(T_c) = \alpha_4 \tag{4.47}$$

where both α_2 and α_4 are positive constants. The resulting free energy curves are plotted in figure 4.3 Notice the characteristic flattening of the free energy curve at the critical temperature, where the single minimum at $\zeta = 0$ ($T > T_c$) breaks up into two symmetric minima at $\pm\bar{\zeta} \neq 0$ ($T < T_c$); below T_c the appearance of a non-vanishing order parameter signals spontaneous symmetry breaking[2]. Substitution of the expressions (4.46) and (4.47) into (4.45) shows that the order parameter vanishes continuously at T_c, with a critical exponent $\beta = 1/2$:

$$\bar{\zeta} = \left(\frac{\alpha_2 T_c}{\alpha_4}\right)^{1/2} \left(\frac{T_c - T}{T_c}\right)^{1/2} \tag{4.48}$$

---

[2] That is, the minimum free energy state of the system does not have the full symmetry of the original Hamiltonian or free energy.

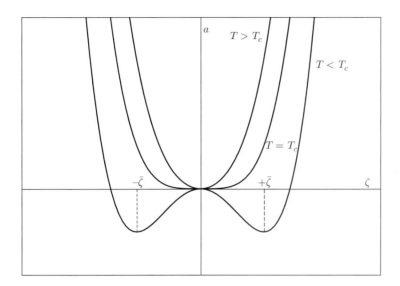

Figure 4.3. Landau free energy curves, $a(T, \zeta)$ versus ζ, for the a_2, a_4 model. For $T < T_c$, two symmetric minima at $\zeta = \pm\bar{\zeta} \neq 0$ develop.

Exercise: Show that a Taylor expansion of the Bragg–Williams approximation (4.10) for the free energy of the symmetric binary mixture, in powers of $\phi - \phi_c$ leads back precisely to the Landau form (4.43); work out the expressions for the coefficients a_0, $a_2(T)$ and $a_4(T)$ in terms of microscopic parameters of the model.

The susceptibility χ in zero field is defined as the response of the order parameter to a vanishingly small external field B:

$$\chi = \left(\frac{\partial \bar{\zeta}}{\partial B}\right)_T \Bigg|_{B=0} \tag{4.49}$$

where $\bar{\zeta}$ satisfies the extremum condition (4.40) when $B \neq 0$, namely:

$$a_2(T)\bar{\zeta} + a_4(T)\bar{\zeta}^3 = B \tag{4.50}$$

Differentiating both sides of (4.50) with respect to B, and substituting the equilibrium values of $\bar{\zeta}$ above and below T_c, one arrives at:

$$\chi = \frac{1}{\alpha_2(T - T_c)} \quad \text{for } T > T_c; \qquad \chi = \frac{1}{2\alpha_2(T_c - T)} \quad \text{for } T < T_c \tag{4.51}$$

The susceptibility, a second derivative of the free energy, is seen to diverge at $T = T_c$, and the divergence is characterized by the critical exponent $\gamma = 1$. Note that the prefactors (or 'critical amplitudes') differ by a factor of 2 above and below T_c. In the case of a binary mixture, the ordering field is the difference in chemical potentials, and the susceptibility is the inverse of the second derivative of the free energy with respect to the composition variable. Along the critical isotherm, $a_2(T = T_c) = 0$, and $B \sim \overline{\zeta}^{\delta}$, with the critical exponent $\delta = 3$ within Landau theory.

Exercise: Using the Landau free energy (4.44), show that in zero field, the entropy per unit volume s is continuous at T_c, while the specific heat per unit volume, c, has a finite discontinuity.

In summary, the form (4.44) of the Landau free energy provides an adequate phenomenological framework for the study of second-order phase transitions of highly symmetric systems, which turn into a first-order transition below T_c, where two phases with opposite, finite values of the order parameter can coexist. The form of the Landau free energy considered next extends these considerations to non-symmetric systems, for which a cubic term must be included in the Taylor expansion.

(b) $\ell = 2, 3$ and 4

Consider a Landau free energy of the form:

$$a(T, \zeta) = a_0(T) + \frac{a_2(T)}{2}\zeta^2 + \frac{a_3(T)}{3}\zeta^3 + \frac{a_4(T)}{4}\zeta^4 \qquad (4.52)$$

For the system to be stable against large fluctuations of the order parameter, $a_4(T)$ must be positive at all temperatures, thus ensuring convergence of the integral (4.36). The Landau free energy (4.52) leads to two very different scenarios according to the dependence of a_2 and a_3 on temperature.

(b) (i) Gas–liquid transition. The phase diagram of this familiar transition is qualitatively similar to that of the Ising model or the equivalent symmetric binary mixture examined under case (a), but there is a marked asymmetry between the gas and liquid portions of the coexistence curve below the critical point, as illustrated in figure 1.1; this asymmetry requires the presence of the cubic term in equation (4.52).

The natural independent variables are the temperature T and the chemical potential of the single molecular species; at coexistence the chemical potential must be the same in both phases at any given temperature. The appropriate thermodynamic potential is the grand potential per unit volume (cf. Equation (2.19)).

$$\omega(\mu, T) = \frac{\Omega(\mu, T)}{V} = f(\rho, T) - \mu\rho \qquad (4.53)$$

where f is the Helmholtz free energy density. In the vicinity of the critical point, the relevant order parameter is $\zeta = \rho - \rho_c$, where ρ_c is the critical density. Expanding $f(\rho, T)$ in a Taylor series in powers of ζ, it is immediately clear that the grand potential can be written, for arbitrary values of ζ, as:

$$\omega(T, \mu; \zeta) = a(T, \zeta) - B\zeta \qquad (4.54)$$

with a of the form (4.52), and the ordering field defined by:

$$B = \mu - \left.\frac{\partial f(\rho, T)}{\partial \rho}\right|_{\rho=\rho_c} = \mu - \mu_c(T) \qquad (4.55)$$

while the coefficients $a_\ell(T)$ $(\ell \geq 2)$ are $\frac{1}{(\ell-1)!} \frac{\partial^\ell f(\rho,T)}{\partial \rho^\ell}$.

The ordering field, equal to the chemical potential minus its value $\mu_c(T)$ on the critical isochore, vanishes on the latter, and plays a role very similar to the applied magnetic field for the Ising model. The critical point is reached by lowering the temperature along the critical isochore, just as the Curie point of a ferromagnet is reached by lowering the temperature in zero magnetic field.

For any given value of μ (or B) and T, the equilibrium value(s) $\bar{\zeta}$ of ζ must once more satisfy the extremum conditions (4.40) and (4.41). In zero field (i.e. along the critical isochore), these lead to:

$$\zeta(a_2 + a_3\zeta + a_4\zeta^2) = 0 \qquad (4.56)$$

$$a_2 + 2a_3\zeta + 3a_4\zeta^2 > 0 \qquad (4.57)$$

The root $\bar{\zeta} = 0$ is stable in the range of temperatures where $a_2(T) > 0$. For equations (4.56) and (4.57) to describe a continuous phase transition at some critical temperature T_c, two non-zero roots $\bar{\zeta}$ below T_c must continuously go to zero, such that at $T = T_c$, the root is triply degenerate. According to (4.56) this requires that $a_2(T_c) = a_3(T_c) = 0$, i.e. both coefficients, which are assumed to be continuous functions of T, must change sign at T_c. The immediate vicinity of the critical point may hence be studied by setting $a_2(T) = a_2(T - T_c) = \alpha_2\Delta T$, $a_3(T) = \alpha_3\Delta T$ and neglecting the temperature dependence of the positive coefficient $a_4(T)$ which is replaced by its value $\alpha_4 = a_4(T_c)$. α_2 and α_4 are positive constants, while α_3 may, a priori, be of either sign. Similarly the ordering field B, which depends on both μ and T, may be expanded to lowest order as $B = \Delta\mu - \alpha_1\Delta T$, with $\Delta\mu = \mu - \mu_c(T_c)$[3] and $\alpha_1 = \frac{\partial^2 f}{\partial \rho \partial T}|_{T=T_c, \rho=\rho_c}$. The pressure along the $\zeta = 0$ isochore follows directly from $P = -\omega(\mu_c(T), T)$ which may be expanded to lowest order in $\Delta\mu$ and ΔT, around $P_c = -\omega(\mu_c(T_c), T_c)$, resulting in:

$$P - P_c = D(T - T_c) + \mathcal{O}(\Delta T^2) \qquad (4.58)$$

where $D = \alpha_1\rho_c - \alpha_0$, with $\alpha_0 = \frac{\partial f(\rho_c, T)}{\partial T}|_{T=T_c}$; the pressure varies linearly along the critical isochore. The inverse isothermal compressibility $\chi_T = \rho^2(\partial^2 f/\partial\rho^2)_T$

---

[3] $\mu_c(T_c)$ is the value of the chemical potential *at the critical point*. Along the critical isochore, $\mu_c(T) = \mu_c(T_c) + \alpha_1\Delta T$.

reduces to $\rho_c^2 a_2(T) = \rho_c^2 \alpha_2 \Delta T$ along the critical isochore, so that the compressibility diverges at the critical point:

$$\chi_T = \frac{C}{T - T_c} \qquad T > T_c \tag{4.59}$$

i.e. the critical exponent $\gamma = 1$ is identical to that characterizing the divergence of the magnetic susceptibility in equation (4.51), as expected.

For $T < T_c$, coexistence of two phases (liquid and gas) with respective densities $\rho_\ell(T)$ and $\rho_g(T)$ is possible for a single value of the chemical potential, $\mu_{coex}(T)$. The densities of the coexisting phases correspond to the minima $\zeta_\ell(T) = \rho_\ell(T) - \rho_c$ and $\zeta_g(T) = \rho_g(T) - \rho_c$ of the potential

$$\omega(\mu, T; \zeta) = \frac{a_2(T)}{2}\zeta^2 + \frac{a_3(T)}{3}\zeta^3 + \frac{a_4(T)}{4}\zeta^4 - B(\mu, T)\zeta \tag{4.60}$$

At coexistence, the two minima have equal depth. The potential ω for $\mu = \mu_{coex}(T)$ can therefore be written in the form

$$\omega(\mu, T; \zeta) = \omega_0(T) + \frac{A(T)}{4}(\zeta - \zeta_\ell(T))^2(\zeta - \zeta_g(T))^2 \tag{4.61}$$

which is the generic form for a fourth-order polynomial with equal minima at ζ_ℓ and ζ_g. $\omega_0(T)$ is, by construction, the opposite of the coexistence pressure. Identifying term by term the polynomials (4.60) and (4.61), one has $A(T) = a_4(T)$, $-A(T)(\zeta_\ell(T) + \zeta_g(T))/2 = a_3(T)/3$, and $A(T)(\zeta_\ell(T)^2 + \zeta_g(T)^2 + 4\zeta_\ell(T)\zeta_g(T))/4 = a_2(T)/2$. Hence

$$\frac{1}{2}(\zeta_\ell(T) + \zeta_g(T)) = -\frac{a_3(T)}{3a_4(T)} \tag{4.62}$$

and

$$(\zeta_\ell(T) - \zeta_g(T))^2 = \frac{4a_2(T)}{a_4(T)} + \frac{2a_3(T)^2}{3a_4(T)^2} \tag{4.63}$$

Close to the critical point the expansion of the coefficients to first order in ΔT may be used, which results in

$$\frac{1}{2}(\zeta_\ell(T) + \zeta_g(T)) = -\frac{\alpha_3}{3\alpha_4}\Delta T \tag{4.64}$$

which expresses the law of rectilinear diameters, i.e. the observation that the average density of the two coexisting phases varies linearly with ΔT. Since experimental data show that this average density increases with ΔT, the coefficient α_3 is necessarily positive for the gas–liquid transition. At the same order in ΔT, the density difference follows a scaling relation with an exponent $\beta = 1/2$, identical to the result (4.51) for the Ising model:

$$\zeta_\ell(T) - \zeta_g(T) = -\left(\frac{4\alpha_2 T_c}{\alpha_4}\right)^{1/2}\left(\frac{\Delta T}{T_c}\right)^{1/2} \tag{4.65}$$

Finally, the coefficient of the linear term in (4.61), $-A(\zeta_\ell + \zeta_g)\zeta_\ell\zeta_g$, is easily seen to be of order $(\Delta T)^2$. This implies that the coefficient $B(\mu, T)$ in equation (4.60) is of order $(\Delta T)^2$ on the coexistence line near the critical point. In other words, the coexistence line $(\mu = \mu_{coex}(T))$, in the (μ, T) plane, has the same slope as the critical isochore $(\mu = \mu_c(T))$ at the critical point.

(b) (ii) First-order transition without a critical point.

The first-order liquid–gas transition is between two phases of identical (fully isotropic) symmetry, and the two phases continuously merge into a single fluid phase at the critical point. This scenario is not generally observed when the two coexisting phases are of different symmetry, as is the case for the transition from the isotropic to nematic phase in liquid crystals, or in the freezing transition from an isotropic or nematic liquid to a crystalline solid. A natural order parameter is one which characterizes the broken symmetry of the low temperature phase. The order parameter for the nematic phase will be defined in section 4.5, but meanwhile the possibility of describing a discontinuous transition is considered within Landau's framework, assuming that the order parameter is a scalar, and that the Landau free energy is of the form (4.52). As in the previous case, the coefficient $a_2(T)$ of ζ^2 is assumed to change sign at some temperature, say T_0 (which will turn out *not to be* the transition temperature), but contrarily to the previous case, $a_3(T)$ is now assumed to remain of the same sign throughout the relevant temperature range, and will be chosen to be negative; $a_4(T)$ must of course remain positive throughout. The chosen thermodynamic potential will be the free energy, represented by the expansion (4.52) in powers of ζ, and no ordering field B is required, as for the Curie transition of a ferromagnet in zero field. The extremum conditions (4.56) and (4.57) show that $\zeta = 0$ is always a solution, which becomes unstable for $a_2(T) < 0$, i.e. $T < T_0$, when the extremum is no longer a minimum, but rather a maximum in the free energy curves. The non-zero roots of the quadratic factor become real when $a_3 - 4a_2a_4 \geq 0$; with $a_2(T) = \alpha_2(T - T_0)$, $a_3(T) = \alpha_3$ and $a_4(T) = \alpha_4$ (where the α_i are constants), an inflection point in the free energy curve occurs at $T = T_0' = T_0 + \alpha_3^2/4\alpha_2\alpha_4$, and $\zeta = |\alpha_3|/2\alpha_4$. For $T < T_0'$, a minimum at $\bar{\zeta} = |\alpha_3|/2\alpha_4 + \sqrt{\alpha_2(T_0 - T)/\alpha_4}$ develops in the free energy curves, as illustrated in figure 4.4 At temperatures close to T_0', the minimum in the free energy lies above that at $\zeta = 0$, and hence corresponds to a *metastable* phase with broken symmetry. As the temperature decreases, the value of $a(T, \bar{\zeta}) - a(T, \zeta = 0)$ drops and goes through zero at a critical temperature T_c and order parameter ζ_c determined by the conditions $a(T_c, \zeta_c) = a(T_c, \zeta = 0)$ and $(\partial a/\partial \zeta)|_{\zeta=\zeta_c} = 0$ which yield:

$$\zeta_c = \frac{2|a_3|}{3a_4} \qquad T_c = T_0 + \frac{2\alpha_1^2}{9\alpha_3\alpha_4} \qquad (4.66)$$

Figure 4.4. Same as figure 4.3 for the a_2, a_3, a_4 model with $a_3 \neq 0$ (nematic type). At $T = T_c$, a minimum for $\zeta \neq 0$ becomes more stable than the $\zeta = 0$ minimum. The minimum at $\zeta = 0$ becomes unstable at $T = T_0$.

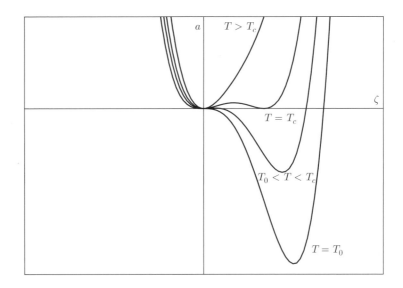

For $T < T_c$, the free energy minimum at finite ζ is lower than that at $\zeta = 0$. In other words, at the critical temperature T_c, the order parameter jumps discontinuously from $\zeta = 0$ to $\zeta = \zeta_c$, and below that temperature, the high symmetry phase is metastable, and eventually becomes unstable for $T < T_0$. At $T = T_c$ the two phases coexist, but do not merge continuously, as in the liquid–gas transition. The discontinuous transition is characterized by a jump in entropy density equal to:

$$\Delta s = -\left.\frac{\partial a(T, \zeta_c)}{\partial T}\right|_{T=T_c} + \left.\frac{\partial a(T, 0)}{\partial T}\right|_{T=T_c} = \frac{-2\alpha_2\alpha_3^2}{9\alpha_4^2} \tag{4.67}$$

which is negative, as one may expect in going from a disordered (symmetric) to an ordered (broken symmetry) phase. The corresponding latent heat per unit volume is $l = T_c \Delta s$. In general the coefficients in the Landau free energy (4.52) will not only depend on temperature, but also on an additional intensive thermodynamic state variable, e.g. pressure or density (in the simplest case of a one-component system). The critical temperature T_c will hence be a function of this additional variable, and upon varying the latter, a phase coexistence line will be mapped out, as shown in figure 1.1. The presence of metastable states in the range $T_0' > T > T_0$ means that the ordered, low temperature phase may be overheated, while the disordered, high temperature phase may be supercooled, as routinely observed, e.g. along the melting and freezing lines. Finally, a limitation of the Landau theory of first-order phase transitions must be borne in mind: since the Landau free energy is a truncated Taylor expansion, and since the order parameter undergoes a finite jump, the theory will only provide meaningful results if the transition is weakly first order, i.e. if the jump ζ_c is sufficiently small so that the terms beyond ζ^4 in the Landau expansion are negligible.

(c) $\ell = 2, 4$ and 6

An interesting combination of the second-order scenarios of cases (a) and (b) (i) and of the first-order scenario (b) (ii) is achieved by considering the following Landau free energy in even powers of the order parameter, valid for systems invariant under the inversion $\zeta \leftrightarrow -\zeta$:

$$a(T, \zeta) = a_0(T) + \frac{a_2(T)}{2}\zeta^2 + \frac{a_4(T)}{2}\zeta^4 + \frac{a_6(T)}{2}\zeta^6 \qquad (4.68)$$

The coefficient of the largest power, a_6, must be positive for stability, and will be assumed to be independent of T. $a_2(T)$ is once more assumed to change sign at some temperature T_0, i.e. $a_2(T) = \alpha_2(T - T_0)$. As long as $a_4(T)$ is positive, the presence of the term makes no qualitative change to the second-order scenario described in case (a) (Ising or symmetric mixture models); the continuous phase transition takes place at a temperature $T_c = T_0$. In fact, since T_0 generally depends on a second intensive state variable (e.g. the pressure), a line of second-order phase transitions is mapped out as long as $a_4 > 0$.

If $a_4 < 0$, however, the scenario is that of a first-order phase transition, similar to that described under case (b) (ii). In the absence of an ordering field, the extremum condition now leads to:

$$\alpha_2(T - T_0)\zeta + a_4\zeta^3 + a_6\zeta^5 = 0 \qquad (4.69)$$

As long as $T > T_c$, where

$$T_c = T_0 + \frac{3a_4^2}{16\alpha_2 a_6} \qquad (4.70)$$

the solution of (4.69) with the lowest free energy is $\overline{\zeta} = 0$. At $T = T_c$, the order parameter jumps discontinuously from zero to a finite value:

$$\overline{\zeta} = \pm\left(\frac{3|a_4|}{4a_6}\right)^{1/2} \qquad (4.71)$$

For $T < T_c$ the ordered phase, characterized by a finite value of $\overline{\zeta}$, has the lowest free energy, and is hence the stable phase, but as long as $T > T_0$, the disordered phase $\overline{\zeta} = 0$ remains metastable, since it belongs to a local minimum of $a(T, \zeta)$.

In summary, the critical temperature at which the phase transition occurs is given by:

$$\begin{aligned} T_c - T_0 &= 0 & a_4 > 0 \text{ (second order)} \\ T_c - T_0 &= \frac{3a_4^2}{16\alpha_2 a_6} & a_4 < 0 \text{ (first order)} \end{aligned} \qquad (4.72)$$

When $a_4 = 0$, the line of second-order phase transitions merges into a line of first-order transitions, at a tricritical point, which is defined by the two requirements $a_2 = a_4 = 0$. Since a_2 and a_4 depend on temperature and on the additional state variable, say pressure, the two equations $a_2(T, P) = 0$ and $a_4(T, P) = 0$

determine the coordinates of the tricritical point. Note that at the tricritical point, the extremum condition, with $a_4 = 0$, leads to the following scaling relation for the vanishing of the order parameter:

$$\bar{\zeta} = \left(\frac{\alpha_2}{a_6}\right)^{1/4} (T_c - T)^{1/4} \tag{4.73}$$

i.e. the tricritical exponent $\beta = 1/4$, rather than $\beta = 1/2$ for a normal critical point.

4.3   Application 1: van der Waals theory of condensation

The Landau theory of phase transitions is phenomenological in the sense that it provides no a priori knowledge of the coefficients a_ℓ in the expansion (4.43) in powers of the order parameter. These must be determined by independent experimental measurements, or from microscopic models for the free energy, like the Bragg–Williams approximation (4.10). This section presents a related microscopic theory of the gas–liquid transition (or condensation), which goes back to van der Waals (1873), and is the earliest example of a mean field theory.

Van der Waals' key observation is that typical pair potentials between quasi-spherical molecules, like the Lennard-Jones potential (1.9), naturally split into a short-range part $v_0(r)$, giving rise to a steeply repulsive force (over the range $0 < r < 2^{1/6}\sigma$ in the case of the Lennard-Jones potential), and a long-range part $w(r)$ corresponding to the attractive dispersion interaction, as illustrated in figure 1.4. The total potential energy (1.47) of the fluid accordingly splits into two terms, as in equation (3.59). The repulsive component $V_N^{(0)}$ determines the short-range order of the fluid, which is dominated by excluded volume effects, while the more smoothly varying W_N, which controls the cohesion of the fluid, may be regarded as a perturbation, which hardly affects the local structure, but contributes significantly to the thermodynamic properties. According to the Gibbs–Bogoliubov inequality (3.62), the free energy F of the fully interacting system is bounded above by the sum of the free energy of the 'reference system' of particles interacting only via $v_0(r)$, and of the first-order correction due to the 'perturbation'. The mean field approximation amounts to replacing the inequality by an equality, thus neglecting all higher order terms in a systematic expansion in powers of the perturbation; in terms of the free energy density:

$$f(\rho, T) = f^{(0)}(\rho, T) + \frac{\rho^2}{2} \int g^{(0)}(r)w(r)\,d\mathbf{r} \tag{4.74}$$

where $g^{(0)}(r)$ denotes the pair distribution function of the reference system. Starting from the result (4.74), further steps are required to arrive at a tractable expression for the free energy. Since the reference potential $v_0(r)$ is steeply repulsive, it may be reasonably approximated by the infinitely repulsive hard sphere potential, with an appropriate diameter σ. The latter may roughly be chosen such that $v_0(r = \sigma) = k_B T$; a more satisfactory procedure would be to consider σ as a variational parameter in a Gibbs–Bogoliubov minimization of the free energy of the actual reference system in terms of an underlying hard sphere system. The free energy of the hard sphere fluid is easily derived from the accurate Carnahan–Starling equation of state (2.92) by integrating the thermodynamic relation (2.12). The perturbation term in (4.74) is approximated by replacing $g^{(0)}(r)$ by its low density limit (1.2) which, for hard spheres, reduces to the step function $\theta(r - \sigma)$. The mean field expression for the free energy then reads:

$$f(\rho, T) = f^{(0)}(\rho, T) - a\rho^2 \tag{4.75}$$

where $a = -2\pi \int_\sigma^\infty w(r) r^2 \, dr$ is a positive constant (since $w(r)$ is negative), independent of density and temperature. This expression for the free energy can be used to determine the liquid–vapour coexistence curve, by enforcing the equality of pressures and chemical potentials of the low and high density phases. Following van der Waals' original work, a further simplification amounts to evaluating $f^{(0)}(\rho, T)$ within the simple free volume approximation (1.17), which immediately leads to:

$$f^{(0)}(\rho, T) = \rho k_B T \left[\ln(\rho \Lambda^3) - 1 - \ln(1 - \rho b) \right] \tag{4.76}$$

where $b = v_{ex}/2 = 2\pi\sigma^3/3$ is one half the excluded volume. The resulting equation of state is that originally proposed by van der Waals, namely:

$$\frac{P}{\rho k_B T} = \frac{1}{1 - \rho b} - \frac{a\rho}{k_B T} \tag{4.77}$$

or equivalently, in terms of the volume per molecule:

$$\left(P + \frac{a}{v^2} \right) (v - b) = k_B T \tag{4.78}$$

The resulting P–v isotherms increase monotonically upon compression at high temperatures but exhibit characteristic van der Waals 'loops' below a critical temperature T_c, determined by the conditions:

$$\left(\frac{\partial P}{\partial v} \right)_{T=T_c, v=v_c} = \left(\frac{\partial^2 P}{\partial v^2} \right)_{T=T_c, v=v_c} = 0 \tag{4.79}$$

which, together with the equation of state (4.77), yield the following critical parameters: $v_c = 3b$, $P_c = a/27b^2$ and $k_B T_c = 8a/27b$. The critical ratio $P_c v_c / k_B T_c$ is universal, i.e. independent of the molecular parameters a and b, equal to $3/8$, which is reasonably close to the experimental value $\simeq 0.3$ observed for most simple fluids.

For $T < T_c$, the P–v isotherms exhibit, within the van der Waals loops, regions where P decreases with v; this unphysical region, which violates the stability condition of a non-negative compressibility χ_T, is eliminated by the condition of equal chemical potentials in the coexisting gas and liquid phases, which translates into the familiar Maxwell equal area construction whereby, for each $T < T_c$, the unphysical loop, a direct consequence of the mean field approximation, is replaced by a horizontal tie-line, reflecting the condition of equal pressures. The locus of points in the P–v plane where $(\partial P/\partial v)_T$ vanishes, i.e. where the compressibility diverges, is the spinodal line, which lies inside the binodal; the two curves meet only at the critical point. Since the van der Waals free energy (4.75) is fully analytic, it can be Taylor expanded in powers of $\zeta = \rho - \rho_c$ in the vicinity of the critical point, and one may easily check that the corresponding Landau free energy is precisely of the form (4.52), so that the whole scenario (b) (i) discussed in section 4.2 applies directly to van der Waals' microscopic theory; in particular, the critical exponents take their 'classical' mean field values ($\beta = 1/2$, $\gamma = 1$, $\delta = 3$) as expected.

A final interesting feature of the van der Waals equation of state is that it obeys a 'law of corresponding states'. Introducing the reduced variables $t = T/T_c$, $p = P/P_c$, $\phi = v/v_c$, (4.78) may be cast in the universal form:

$$(p + 3/\phi^2)(3\phi - 1) = 8t \tag{4.80}$$

which contains no reference to any specific substance; chemical specificity is entirely contained in the values of T_c and v_c. Such a universality is reasonably well obeyed by experimental data on many fluids, although the functional form of the equation of state is not as simple as (4.80).

4.4 Application 2: Flory–Huggins theory of polymer blends

As shown in figure 1.11, the lattice model for a binary mixture, discussed in section 4.1, can be generalized to the case where one or both of the species A and B are monomers belonging to polymer chains. Consider first the case of a polymer solution. Of the N lattice sites, N_A are occupied by solvent molecules, and N_B by monomers. M monomers are connected into a polymer chain, so that the total number of chains is $N_p = N_B/M$. The solvent and monomer concentrations are $(1 - \phi)$ and ϕ, while the polymer concentration is ϕ/M. In the Bragg–Williams approximation, and within a constant, the internal energy per site $u = U/N$ follows from equation (4.2):

$$u = \frac{\nu\epsilon}{2}\phi(1 - \phi) + \frac{\nu}{2}\phi\Delta\epsilon \tag{4.81}$$

where $\Delta\epsilon = \epsilon_{AA} - \epsilon_{BB}$. The entropy of mixing of free (unconnected) monomers and solvent molecules would still be given by equation (4.8), but the connectivity constraint, whereby each polymer sequentially links M monomers, strongly reduces the contribution of the polymer chains to the entropy. Intuitively, the reduction due to the connectivity is expected to be a factor $1/M$ compared to independent monomers. Assuming statistical independence of the intramolecular bonding constraints, Flory and Huggins [4] used a simple combinatorial analysis to estimate the number of ways of placing N_p polymer chains, of M connected monomers each, on a lattice of N sites, to confirm this intuitive guess, i.e.:

$$s = -k_B \left[\frac{1}{M} \phi \ln \phi + (1 - \phi) \ln(1 - \phi) \right] \tag{4.82}$$

The resulting reduced free energy of mixing is then given by the celebrated Flory–Huggins form:

$$f_m^*(\phi) = [f(\phi) - \phi f(1) - (1 - \phi) f(0)]/k_B T$$
$$= \frac{\phi}{M} \ln \phi + (1 - \phi) \ln(1 - \phi) + \chi(T)\phi(1 - \phi) \tag{4.83}$$

where $\chi(T) = v\epsilon/2k_B T$ is the dimensionless Flory–Huggins parameter.

The osmotic pressure of the polymer may be calculated from the general expression (2.59) (valid for an incompressible solvent), which may be re-expressed in terms of $f_m^*(\phi)$ to yield:

$$\beta \Pi v = \frac{\partial f_m^*}{\partial \phi} - f_m^*(\phi) = \frac{\phi}{M} - \ln(1 - \phi) - \phi - \chi \phi^2 \tag{4.84}$$

where $v = V/N$ is the volume per lattice site. In the low monomer concentration limit ($\phi \ll 1$), this may be expanded to yield

$$\beta \Pi v = \frac{\phi}{M} + \frac{1}{2}(1 - 2\chi(T))\phi^2 + \mathcal{O}(\phi^3) \tag{4.85}$$

or, in terms of the polymer density $\rho_p = N_p/V$:

$$\beta \Pi = \rho_p + B_2(T)\rho_p^2 + \mathcal{O}(\rho_p^3) \tag{4.86}$$

where the second virial coefficient is:

$$B_2(T) = \frac{M^2 v}{2}(1 - 2\chi(T)) \tag{4.87}$$

In the very dilute limit, equation (4.86) leads back to van't Hoff's result for an ideal solution. However the range of validity of van't Hoff's law shrinks with increasing molecular weight (or degree of polymerization), since the second virial correction is proportional to M^2. Note that if the overall monomer concentration is known, the molecular weight of the polymers may be directly determined by a measurement of the osmotic pressure in the dilute regime (osmometry): the higher the degree of polymerization, the lower the osmotic pressure. With increasing polymer concentration, the second term in equation (4.86) rapidly takes over, and the osmotic pressure scales essentially as ρ_p^2. It should be noted that if the virial expansion is written in powers of the mole fraction of monomers ϕ, rather than ρ_p, the second virial correction is independent of molecular weight

[4] P.J. Flory, *Principles of Polymer Chemistry*, Cornell University Press, New York, 1953.

(cf. equation (4.85)). The second virial coefficient is positive at high temperatures, corresponding to the regime of *good solvent* (i.e. the monomers repel each
other and prefer to be surrounded by solvent molecules), but changes sign at the
θ-temperature $\theta = \epsilon v/k_B$, where $\chi = 1/2$; at that temperature the excluded volume repulsion (embodied in the single occupancy constraint), is exactly cancelled
by the effective attraction between monomers; at $T = \theta$, the polymers behave like
an ideal solution, apart from higher order virial corrections.

As for the symmetric binary mixture studied in section 4.1, the Flory–
Huggins free energy (4.83) will drive a phase separation into dilute and concentrated polymer solutions below a critical temperature. The critical coordi-
nates T_c and ϕ_c are easily calculated from the conditions $(\partial^2 f_m(\phi)/\partial\phi^2)_{\phi=\phi_c} = (\partial^3 f_m(\phi)/\partial\phi^3)_{\phi=\phi_c} = 0$, with the result:

$$\phi_c = \frac{1}{1 + \sqrt{M}} \simeq \frac{1}{\sqrt{M}} \qquad T_c = \frac{\theta}{(1 + 1/\sqrt{M})^2} \simeq \theta \qquad (4.88)$$

The critical concentration decreases, while the critical temperature increases with
molecular mass, in agreement with experimental observation.

Although the Flory–Huggins approach is originally based on a simple lattice
model, it should be noted that the expression for the mixing free energy (4.83) is
often used more generally to reproduce experimental results, by allowing $\chi(T)$ to
be an adjustable parameter, often represented in the form $\chi(T) = A + B/T$. The
constant term in $\chi(T)$ accounts then for an entropic contribution to the complex,
effective interaction between the monomers. By re-expressing the theory in terms
of the θ-temperature, such that $\chi(\theta) = 1/2$, the previous results stay unchanged.

The Flory–Huggins analysis may be extended to polymer blends, i.e. melts
containing two polymer species of molecular weights M_A and M_B. If the fractions
of lattice sites occupied by A and B monomers are ϕ_A and $\phi_B = 1 - \phi_A$, the
generalization of the free energy of mixing (4.83) reads:

$$f_m^*(\phi_A, \phi_B) = \frac{1}{M_A}\phi_A \ln \phi_A + \frac{1}{M_B}\phi_B \ln \phi_B + \chi(T)\phi_A\phi_B \qquad (4.89)$$

and the resulting critical coordinates are now:

$$\phi_c = \frac{1}{1 + \sqrt{M_A/M_B}} \qquad T_c = \frac{\theta}{(1/\sqrt{M_A} + 1/\sqrt{M_B})^2} \qquad (4.90)$$

The critical temperature clearly becomes very large when both species have large
molecular weights, so that polymer blends will demix at almost any temperature,
except when $\chi \sim \epsilon \simeq 0$ (compatible polymers). In view of this strong tendency
to demixing, it is of interest to consider the case of AB block copolymers obtained by chemically joining incompatible A and B polymers into a single chain.
Macroscopic phase separation is now prevented by the chemical bond, but the A
ends of different chains will tend to cluster locally, while the B ends will have the
same tendency, resulting in mesoscopic domains rich in either A or B polymers.

Such local phase separation, described in section 7.5, will drive self-assembly of copolymers into spherical, lamellar, cylindrical or more complicated structures, reminiscent of surfactant self-assembly.

4.5 Application 3: isotropic–nematic transition

As already briefly mentioned in section 1.2, certain organic compounds do not melt directly from a crystalline solid into an isotropic liquid phase upon heating, but exhibit, over a relatively broad range of temperatures (typically tens of degrees kelvin), intermediate phases (or mesophases) with features characteristic both of a liquid (e.g. fluidity) and of a crystal (e.g. optical birefringence). Mesophases are only observed with organic compounds (called liquid crystals) involving molecules with highly anisotropic shapes; these may be rod-like molecules, with a length to width aspect ratio of typically 4 or more, or discoid (flat, plate-like) molecules. A widely studied example of a rod-like mesogenic molecule is p-azoxyanisole (or PAA), $H_3C-O-C_6H_6-N=N-C_6H_6-O-CH_3$, which is approximately 2 nm long and 0.5 nm wide. The benzene rings ensure the rigidity of the molecule. PAA melts at $T = 118.2\,^{\circ}C$ into a nematic mesophase, which goes over into an isotropic liquid phase at $T = 135.3\,^{\circ}C$. The melting transition is strongly first order, with a latent heat of 29.57 kJ/mol, while the second transition is only weakly first order (latent heat of 0.57 kJ/mol).

The phase behaviour of materials of mesogenic molecules is essentially controlled by temperature, and the corresponding liquid crystals are called thermotropic. Liquid crystal behaviour is also observed for suspensions of elongated colloidal particles. A widely studied example is the tobacco mosaic virus (TMV) suspended in water, which is a 300 nm long and 20 nm wide rigid rod. For such suspensions, phase behaviour is controlled by particle concentration rather than by temperature (lyotropic liquid crystals).

All mesophases (a partial list of which is given in section 1.2) are anisotropic, i.e. their optical, dielectric, magnetic and elastic properties are orientation dependent, and must hence be characterized by tensors rather than by mere scalars. This anisotropy may be traced back to the partial alignment of the molecular axis of the mesogenic particles along a direction materialized by a unit vector $\mathbf{n}$, called the director. This alignment gives rise to long-range orientational order, extending over macroscopic distances under favourable conditions (e.g. by the application of a weak magnetic field).

Restricting the discussion to uniaxial nematics, an order parameter must be defined to distinguish the isotropic and nematic phases. Let $\mathbf{u}_i$ denote the unit vector specifying the orientation of a rod-like or disc-like molecule, which generally has inversion ($\mathbf{u}_i \leftrightarrow -\mathbf{u}_i$) symmetry. The order parameter must hence be quadrupolar (rather than dipolar) in nature, i.e. it must be a second-rank tensor, quadratic in the $\mathbf{u}_i$. It is convenient to associate the following symmetric traceless tensor with any orientational configuration of the N molecules:

$$\underline{Q} = \frac{1}{N} \sum_{i=1}^{N} \mathbf{u}_i \otimes \mathbf{u}_i - \frac{1}{3}\underline{I} \tag{4.91}$$

Keeping in mind that the N molecules are identical, the tensorial order parameter may then be defined as:

$$\underline{\zeta} = \langle \underline{Q} \rangle = \langle \mathbf{u}_i \otimes \mathbf{u}_j \rangle - \frac{1}{3}\underline{I} \tag{4.92}$$

The statistical average may be expressed in terms of the director $\mathbf{n}$ according to:

$$\underline{\zeta} = \zeta(\mathbf{n} \otimes \mathbf{n} - \frac{1}{3}\underline{I}) = \begin{pmatrix} -\zeta/3 & 0 & 0 \\ 0 & -\zeta/3 & 0 \\ 0 & 0 & 2\zeta/3 \end{pmatrix} \tag{4.93}$$

where the z-axis is chosen along $\mathbf{n}$, and the scalar quantity ζ is:

$$\zeta = \frac{1}{2}\left[3\langle(\mathbf{u}_i \cdot \mathbf{n})^2\rangle - 1\right] \tag{4.94}$$

ζ is conveniently expressed in terms of the orientational distribution function $p(\Omega)$, where $p(\Omega)\,d\Omega$ is the probability of finding a molecule with polar angles $\Omega = (\theta, \phi)$ in a coordinate system with the polar axis along $\mathbf{n}$. Since a uniaxial nematic has full cylindrical symmetry around the director, $p(\Omega)$ is independent of the azimuthal angle ϕ. ζ is hence given by:

$$\zeta = \frac{1}{2}\langle 3\cos^2\theta - 1\rangle \equiv \langle P_2(\cos\theta)\rangle = 2\pi \int_0^\pi P_2(\cos\theta)p(\theta)\sin\theta\,d\theta \tag{4.95}$$

where $P_2(x)$ is the second-order Legendre polynomial; $p(\Omega)$ is normalized, i.e. $2\pi \int_0^\pi p(\theta)\sin\theta\,d\theta = 1$. In the isotropic phase, all orientations are equally probable, so that $p(\Omega) = 1/4\pi$, and $\zeta = 0$. In the uniaxial nematic phase, the molecules align preferentially along $\mathbf{n}$ or $-\mathbf{n}$, and $p(\theta)$ is peaked around $\theta = 0$ and $\theta = \pi$; ζ is then positive. The Landau free energy appropriate for the description of the isotropic–nematic transition may be written in terms of scalar combinations of the tensor $\underline{\zeta}$. Since the free energy must be invariant under all rotations of the director, it may be shown that the free energy reduces to a series in powers of the scalar (4.94), of the form (4.43) (scenario (b) (ii) of section 4.2).

We now turn our attention to a microscopic theory of the isotropic–nematic transition in lyotropic liquid crystals, put forward by L. Onsager in 1942 [5]. The theory accounts only for excluded volume effects, by considering long cylindrical rods of length L and diameter D, in the limit where the aspect ratio $x = L/D \gg 1$. The packing fraction of such rods is $\phi = \rho\pi L(D/2)^2$. Clearly, the excluded volume of two rods, i.e. the volume around one rod, from which the centre of another rod is excluded, depends on their relative orientation. When the two rods

---

[5] A detailed account is given by L. Onsager, *Ann. NY Acad. Sci.* **51**, 627 (1949).

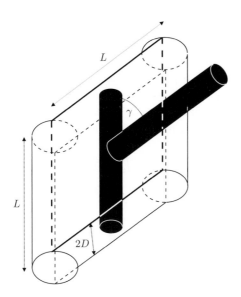

Figure 4.5. Schematic representation of the excluded volume between two rod-like molecules with axis forming an angle γ.

are parallel, this volume is of order LD^2, while for perpendicular rods it is much larger, of order $L^2D = LD^2x$. The packing of rods at high values of ϕ is thus much more efficient when they align along one direction, like matches in a box. More generally, as may be checked from figure 4.5, the excluded volume of two rods depends on the angle between their axes. For sufficiently large aspect ratios, when end effects may be neglected, the excluded volume is:

$$v_{\mathrm{ex}} = 2L^2D|\sin\gamma| \tag{4.96}$$

Onsager considers the dilute limit of suspensions of very long rods. At low concentration, the free energy may be approximated by the sum of the ideal contribution and the second virial coefficient approximation of the excess part, derived from (2.79), i.e.:

$$f^* = \frac{\beta F}{N} = \left[\ln(\rho\Lambda^3) - 1 + B_2(T)\rho + \mathcal{O}(\rho^2)\right] \tag{4.97}$$

where the second virial coefficient (2.80) must be generalized to the case of non-spherical particles, according to:

$$B_2(T) = -\frac{1}{2V}\int d\mathbf{r}\int d\mathbf{r}'\int d\Omega\int d\Omega'\,\Phi(\mathbf{r} - \mathbf{r}', \Omega, \Omega') \tag{4.98}$$

For hard core particles (like the hard rods under consideration here), the Mayer function $\Phi = -1$ when two particles overlap, and $\Phi = 0$ otherwise, so that $B_2(T)$ is one half the excluded volume, averaged over all possible orientations Ω and Ω' of the two particles.

Expression (4.97) of the free energy applies to the isotropic case. In the ordered nematic phase, orientations with respect to the director are not equally probable, but are distributed according to $p(\Omega)$. Onsager formally considers the nematic suspension to be a multicomponent mixture, where each species is associated with a given orientation Ω; the mean number of rods per unit volume with orientations in the range $d\Omega$ about Ω is $p(\Omega)\,d\Omega$. Generalizing the free energy (2.40) of a mixture to such a continuous distribution, and remembering that the excluded volume for two rods of orientations Ω and Ω' making an angle γ [6] is given by (4.96), the free energy of the nematic phase may be written as:

$$f^*[p(\Omega)] = \ln(\rho\Lambda^3) - 1 + \int_{4\pi} p(\Omega)\ln(4\pi p(\Omega))\,d\Omega$$

$$+ \rho L^2 D \int_{4\pi} d\Omega \int_{4\pi} d\Omega'\, p(\Omega)p(\Omega')\sin\gamma + \mathcal{O}(\rho^2) \qquad (4.99)$$

The reduced free energy f^* is a functional of the (yet unknown) distribution function $p(\Omega)$. In the isotropic phase, where $p(\Omega) = 1/4\pi$, (4.99) reduces to (4.97). The ideal and excess contributions to the free energy (4.99) are of entropic nature (apart from $k_B T$, there is no energy scale in the problem). The ideal term is lowest for an isotropic distribution, and hence favours the orientationally disordered isotropic phase. The excess term in (4.99) obviously decreases as the orientational order increases, i.e. when the rods are increasingly aligned, so that $|\sin\gamma|$ is small; it hence favours the nematic phase. The weight of the excess contribution increases with increasing density ρ, relative to the ideal term. Physically, the loss of orientational entropy will be more than compensated by the enhanced translational freedom which is favoured when the rods have a common orientation, thus minimizing their mutual hindrance. The competition between the two contributions is settled by the general variational principle, which requires the free energy to be at its overall minimum with respect to variations of the orientational distribution function $p(\Omega)$, subject to the normalization constraint $\int_{4\pi} p(\Omega)\,d\Omega = 1$. If $p'(\Omega)$ denotes a trial distribution function, the extremum condition reads:

$$\left.\frac{\delta f^*[p'(\Omega)]}{\delta p'(\Omega)}\right|_{p'=p} = \mu^* \qquad (4.100)$$

where the left-hand side denotes functional differentiation with respect to $p'(\Omega)$, $p(\Omega)$ is the equilibrium distribution function, while μ^* is the Lagrange multiplier associated with the normalization constraint (in fact the chemical potential of the rods). Substituting (4.99) into (4.100), and using the rules of functional differentiation (to which we shall briefly return in section 7.1), one arrives at the following Euler–Lagrange equation for $p(\Omega)$:

$$\ln(4\pi p(\Omega)) = \mu^* - 2\rho L^2 D \int p(\Omega')\sin\gamma\,d\Omega' = \mu^* - \frac{8}{\pi}\phi x \int p(\Omega')\sin\gamma\,d\Omega' \qquad (4.101)$$

[6] If $\Omega = (\theta, \phi)$ and $\Omega' = (\theta', \phi')$, then $\cos\gamma = \cos\theta\cos\theta' + \sin\theta\sin\theta'\cos(\phi - \phi')$.

This integral equation always admits the isotropic solution $p(\Omega) = 1/4\pi$, with $\mu^* = \beta\mu_{ex} = 2B_2\rho$, B_2 being the second virial coefficient (4.98) averaged over all orientations of two rods:

$$B_2 = \frac{1}{2(4\pi)^2} \int_{4\pi} d\Omega \int_{4\pi} d\Omega' v_{ex} = \frac{1}{2} \frac{2L^2 D}{4\pi} \int_{4\pi} \sin\gamma \, d\Omega' = \frac{\pi L^2 D}{4} \qquad (4.102)$$

However, for a sufficiently high packing fraction ϕ, or large aspect ratio x, (4.101) also admits an anisotropic solution which, when substituted back into (4.99), eventually leads to a lower free energy. The search for the ansiotropic solution is most conveniently carried out by the use of a simple trial function. Following Onsager, we choose the properly normalized trial function:

$$p'(\Omega) \equiv p'(\theta) = \frac{\alpha \cosh(\alpha \cos\theta)}{4\pi \sinh(\alpha)} \qquad (4.103)$$

where α is a variational parameter which controls the width of the distribution function. The corresponding order parameter is easily calculated from equation (4.94):

$$\zeta(\alpha) = 1 + \frac{3}{\alpha^2} - 3\frac{\coth\alpha}{\alpha} \qquad (4.104)$$

When $\alpha \to 0$, $p(\theta) \to 1/4\pi$ and $\zeta \to 0$ (isotropic phase), while when $\alpha \to \infty$, $\zeta \to 1$ (perfect orientational order).

When (4.103) is substituted into (4.99), the integrals may be calculated explicitly as functions of the width parameter α (details can be found in Onsager's original paper); the reduced free energy (4.99) is now of the form:

$$f^*(\alpha) = \left[\ln(\rho\Lambda^3) - 1\right] + f_1(\alpha) + \phi x f_2(\alpha) \qquad (4.105)$$

which is to be minimized with respect to α for any given value of the control parameter ϕx. In the isotropic phase ($\alpha = 0$), $f_1(\alpha) = 0$ and $f_2(\alpha) = 1$, leading back to the result (4.97) (with B_2 given by (4.102)). An explicit calculation shows that a minimum occurs at a non-zero value of α, i.e. for a finite value of the order parameter, when $\phi x > 4$. The bifurcation is discontinuous, as in the Landau scenario of a first-order transition. Phase coexistence between the isotropic and nematic phases is determined by the usual conditions of equality of chemical potentials and osmotic pressures:

$$\beta\mu = f^* + \rho\left(\frac{\partial f^*}{\partial\rho}\right)_\alpha = \ln(\rho\Lambda^3) + f_1(\alpha) + 2\phi x f_2(\alpha) \qquad (4.106)$$

$$\beta\Pi = \rho^2\left(\frac{\partial f^*}{\partial\rho}\right)_\alpha = \rho(1 + \phi x f_2(\alpha)) \qquad (4.107)$$

The values of α to be inserted into these relations are those which satisfy $(\partial f^*/\partial\alpha)_\rho = 0$, i.e. $\alpha = 0$ (isotropic phase), and the non-zero value of α which satisfies the extremum condition for $\phi x > 4$; this latter value of α depends on density, but this dependence does not contribute to the density derivatives in the relations (4.106) and (4.107), because of the extremum condition.

The resulting values of the control parameter ϕx for the coexisting isotropic and nematic phases are $(\phi x)_i = 3.34$, while $(\phi x)_n = 4.49$, which translates, according to (4.104), into a value $\zeta_n = 0.848$ of the order parameter in the nematic phase at coexistence. Thus Onsager theory predicts a strongly first-order phase transition $((\Delta\phi x)/(\phi x) = 34\%$, in contrast to experimental observation of a discontinuity of the order of 1%). The nematic phase is also much more strongly ordered than suggested by experiment ($\zeta \simeq 0.5$). These shortcomings of the theory can be traced back to the use of a low density approximation for the excess part of the free energy. This approximation is only justified in the limit of very long rods ($x \gg 1$) since the packing fractions of the two coexisting phases decrease like $1/x$; in that limit the contributions of higher order terms in the virial expansion may be expected to become negligible. In real systems the aspect ratio is never large enough to justify truncation of the virial series after second order.

Despite its quantitative shortcomings, Onsager theory remains a milestone in modern understanding of first-order phase transitions, since it is the first example of a purely entropic phase transition which is not driven by a competition between energy and entropy contributions to the free energy (as is the case for liquid–vapour coexistence), but rather by the competition between orientational and correlational contributions to the entropy (there is no interaction energy involved in purely hard core or athermal systems). Onsager theory is also one of the first examples of density functional theory (DFT) of inhomogeneous fluids, the inhomogeneity pertaining here to molecular orientation. DFT will be examined in more detail in chapter 7. Modern versions of DFT, which are better adapted to dense fluids than Onsager's theory, have been applied to the isotropic–nematic transition, and have shown that the predictions of Onsager theory can only be expected to be quantitavely reliable for $x > 100$[7].

For physically relevant values of the aspect ratio x, accurate information on the phase diagram of model liquid crystals, like spherocylinders, is provided by computer simulations[8]. An example of a phase diagram in the density–aspect ratio (ρ, x) plane is shown in figure 4.6.

4.6 Application 4: freezing

The isotropic to nematic transition examined in the preceding section is a good example of spontaneously broken symmetry: the full rotational invariance of the

---

[7] For a review of improvements to Onsager's theory, see G.J. Vroege and H.N.W. Lekkerkerker, *Rep. Progr. Phys.* **55**, 1241 (1992).

[8] For a review, see M.P. Allen, G.T. Evans, D. Frenkel and B.M. Mulder, *Adv. Chem. Phys.* **86**, 1 (1993).

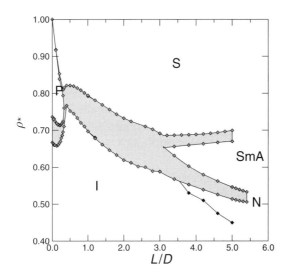

Figure 4.6. Phase diagram of a liquid crystal model (hard spherocylinders) in the density/aspect ratio plane. I, isotropic phase; N, nematic phase; P, plastic phase (FCC crystal, with free rotation of the molecules); SmA, smectic A phase; S, orientationally ordered crystal. The shaded areas correspond to two-phase coexistence. (Courtesy P. Bolhuis, University of Amsterdam.)

isotropic phase gives way to a nematic state with long-range orientational order, which has only cylindrical symmetry around the director. In the smectic phase, translational invariance is moreover broken in the direction normal to the layers, while in a columnar liquid crystal phase, translational invariance is broken in the two directions inside the plane orthogonal to the columnar stacks. In these phases long-range translational order thus appears spontaneously in one or two directions, in addition to the orientational order of the molecular axes. These partially ordered phases are, however, only observed for mesogenic (e.g. rod-like or disc-like) molecules. On the contrary, at sufficiently low temperatures, nearly all substances will undergo a much more common symmetry-breaking transition, namely they will freeze into a three-dimensional crystal. In other words almost any fluid will, under adequate experimental conditions (e.g. under a sufficiently slow cooling rate), crystallize into a periodic, three-dimensional macroscopic structure, characterized by a Bravais lattice. The spontaneous self-assembly of molecules of any shape, and even of mixtures, into macroscopic periodic crystal structures is one of the most striking and universal phenomena in condensed matter, and it is fair to say that it is far from being fully understood.

But crystallization is not restricted to molecular systems; it is also observed in colloidal assemblies of mesoscopic particles suspended in some molecular solvent. Colloidal crystals are not mere curiosities, but they have been shedding new light on the freezing transition, because they are easily visualized by optical or electron microscopy, and their kinetics (e.g. of nucleation) are more readily resolved, because of the much longer time scales associated with mesoscopic particles. From a practical standpoint, they are also a promising route towards the elaboration of three-dimensional 'photonic crystals'[9]. Moreover, colloidal crystals have very low rigidity; a typical elastic modulus, an energy per unit volume, may be roughly estimated from the characteristic energy scale, $k_B T$ (the same as for molecular crystals), and from the colloidal size σ, to be $k_B T/\sigma^3$, which yields values many orders of magnitude smaller than for the molecular counterparts, since colloidal diameters are typically 10^2–10^3 times larger than molecular size. Colloidal crystals, which may be melted by the application of a very modest shear force ('shear melting'), are thus a good example of 'soft matter'.

A convenient order parameter to characterize the broken translational symmetry is the statistical average $\rho_G = \langle \tilde{\rho}_G \rangle$ of a collective coordinate

$$\tilde{\rho}_G = \sum_{i=1}^{N} \exp(i\mathbf{G} \cdot \mathbf{r}_i) \qquad (4.108)$$

where the $\mathbf{r}_i$ are the instantaneous positions of the centres of mass of the N molecules (or colloidal particles) in the crystal, and $\mathbf{G}$ is a reciprocal lattice vector of the periodic Bravais lattice. In a crystal, when molecular positions deviate little from lattice sites, $\mathbf{G} \cdot \mathbf{r}_i$ is always close to a multiple of 2π, so that the real part of ρ_G will be of order N, whereas in a liquid or in a glass, long-range positional disorder leads to 'random' phases $\mathbf{G} \cdot \mathbf{r}_i$, so that $\rho_G = 0$. In fact $\tilde{\rho}_G$ and ρ_G are just the Fourier transforms of the microscopic density (3.30), and the local density (3.31), and the latter may be expanded as:

$$\rho(\mathbf{r}) = \frac{1}{V} \sum_{\mathbf{G}} \rho_G \exp(-i\mathbf{G} \cdot \mathbf{r}) \qquad (4.109)$$

This yields a spatially modulated local density with the periodicity of the underlying Bravais lattice. In the liquid, only the $\mathbf{G} = 0$ component contributes, and $\rho(\mathbf{r})$ reduces to N/V as expected.

A theory of freezing, reminiscent of Onsager's theory of the previous section, may be formulated, whereby the free energy of the crystal phase is related to that of the fluid by a generalized Landau expansion in powers of the order parameters ρ_G[10]. Using the density functional formalism of chapter 7, the coefficients of the expansion may be expressed in terms of the direct correlation functions of the fluid phase. A subsequent minimization of the free energy leads to a bifurcation scenario similar to that of Onsager's description of the isotropic–nematic transition in liquid crystals.

The periodicity of the crystal gives rise to Bragg peaks in the diffraction pattern of X-rays (in the case of molecular crystals) or of visible light (in the case

[9] See e.g. J. Wijnhoven and W.L. Vos, *Science* **281**, 802 (1998).
[10] T.V. Ramakrishnan and M. Youssouff, *Phys. Rev. B* **19**, 2275 (1979). For a recent review, see P.A. Monson and D.A. Kofke, *Adv. Chem. Phys.* **115**, 113 (2000).

of colloidal crystals, where the lattice spacing is comparable to the wavelength of light). The Bragg peaks are broadened by the thermal vibrations of the molecules, and their width provides a measure of the mean square deviation of molecules from their equilibrium (lattice) positions, $\mathbf{R}_i$, $\langle \mathbf{u}^2 \rangle = \langle (\mathbf{r} - \mathbf{R}_i)^2 \rangle$. The peaks in measured fluid structure factors $S(k)$ (see figure 3.1), whose amplitude decreases rapidly with wavenumber k, are all that remains of the Bragg peaks once the crystal has melted.

The ratio of the root mean square displacement of the molecules over the nearest-neighbour distance d in the crystal lattice, $\delta = \sqrt{\langle \mathbf{u}^2 \rangle}/d$ provides a convenient melting criterion due to Lindemann (1910), which states that upon heating, a crystal will melt when the ratio δ reaches a critical value, independent of the material. In other words melting occurs when the amplitude of the thermal vibrations of the molecules reaches a well defined fraction of the lattice spacing. A compilation of experimental data on a variety of substances, and of computer simulations of various models, confirms the semi-quantitative validity of Lindemann's criterion, with a 'universal' critical value $\delta_L \simeq 0.15$. The usefulness of the criterion lies in the fact that δ is a rapidly varying function of temperature or density. Overheated (metastable) states of the crystal are routinely observed and characterized by values of $\delta > \delta_L$; however, for values of δ significantly larger than δ_L, the crystal becomes mechanically unstable [11].

One of the early achievements of computer simulations in the 1950s was to show that a fluid of hard spheres crystallizes into an FCC solid when compressed beyond a packing fraction $\phi \simeq 0.5$ [12]. Since hard sphere systems are athermal (the excess free energy is purely entropic), this remarkable discovery led to the realization that, at least for simple substances, the freezing transition is essentially entropy driven. This is reminiscent of Onsager's theory of the isotropic–nematic transition of hard rods. The somewhat counter-intuitive conclusion to be drawn from the existence of a freezing transition for hard spheres is that at sufficiently high packing fractions, the entropy of the ordered solid is larger than the entropy of the disordered fluid. Contrary to the case of hard rods, hard spheres have only translational degrees of freedom, and the entropy balance is controlled by a subtle competition between configurational and correlational contributions to the entropy. The former is strongly reduced in crystalline solids, because of long-range order, and hence always favours the disordered fluid. However, the very disorder of fluid-like configurations means that the latter are easily 'jammed', as shown in figure 4.7, i.e. the available (free) volume within which the centre of one

---

[11] The thermodynamic melting transition is not, in general, associated with a mechanical instability of the crystal, with vanishing elastic constant (Born instability) which takes place only at much higher temperatures.

[12] B. J. Alder and T.E. Wainwright, *J. Chem. Phys.* **27**, 1208 (1957). W.W. Wood and J.D. Jacobson, *J. Chem. Phys.* **27**, 1207 (1957).

Figure 4.7. Comparison of free volume in a dense fluid and in a crystalline solid, at the same density. The crystalline configuration is clearly less jammed than the disordered one.

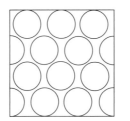 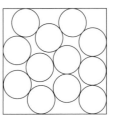

sphere may move when the centres of the surrounding spheres are fixed is strongly reduced compared to the case of the ordered crystal phase, where each sphere still has substantial freedom to move within the regular cavity formed by its shell of nearest neighbours; this enhanced free volume increases the entropy of the crystal and eventually favours the latter over the disordered fluid. The fluid–solid phase diagram of a hard sphere fluid reduces to two vertical coexistence lines in the density–temperature plane, with the packing fractions of the coexisting fluid and solid phases $\phi_f = 0.49$ and $\phi_s = 0.54$; the corresponding relative volume change on freezing, $\Delta V / V = 10\%$, is typical of many simple substances (e.g. rare gases). So is the change in entropy per particle, $\Delta S / N k_B \simeq 1$.

The influence of weak long-range attractions between molecules on phase coexistence can be estimated with the help of thermodynamic perturbation theory (cf. section 3.5) and shown to lead to a modest broadening of the coexistence region at low temperatures. In contrast, the softness of the intermolecular repulsion leads to coexistence curves with a finite (rather than infinite) slope, as shown schematically in the phase diagram in figure 1.1.

The predominance of hard core packing effects at the freezing transition is illustrated by a simple structural criterion [13] which states that a liquid will freeze when the amplitude of the main peak in the fluid structure factor $S(k)$ reaches a universal value equal to that observed for the hard sphere system ($S(k)_{max} \simeq 2.85$); this criterion is indeed well verified by many simple molecular and colloidal systems and, in conjunction with its melting counterpart (the Lindemann criterion), may be used to obtain quick estimates of the fluid–solid coexistence curves.

However, the hard sphere scenario of freezing does not apply to all fluids, and there can be a number of complications.

The structure of ionic crystals, like NaCl and $CaCl_2$, is dominated by charge ordering, a regular alternation of anions and cations, imposed by the strong constraints of local charge neutrality. In this case enthalpic effects, associated with the long-range Coulombic interactions, compete with packing requirements. The relative volume change on melting is generally significantly larger than for non-ionic systems.

---

[13] J.P. Hansen and L. Verlet, *Phys. Rev.* **184**, 151 (1969).

In binary mixtures or alloys, the efficient packing of molecules or colloidal particles may be frustrated by size mismatch, except for restricted ranges of size ratios and concentrations. According to the classic Hume–Rothery rule, metals cannot form substitutional binary alloys for size ratios $\sigma_2/\sigma_1 \leq 0.85$. In fact the topology of the phase diagram of binary hard sphere mixtures varies dramatically with size ratio; eutectic behaviour, with phase separation in the solid, but not in the fluid, occurs at about the Hume–Rothery ratio [14]. At smaller size ratios, say $\sigma_2/\sigma_1 \simeq 0.5$, stoichiometrically ordered binary alloys, like AB_2 and AB_{13} where B denotes the smaller species, are predicted by the binary hard sphere model, and have indeed been observed in binary colloidal mixtures [15]. The macroscopic self-assembly of binary systems into complex superlattices (the unit cell of the AB_{13} structure contains no less than 112 particles!) on the basis of entropic considerations alone, is a truly remarkable finding.

An additional twist arises in highly asymmetric colloidal systems involving large and small colloidal particles, or colloidal particles and non-adsorbing polymer coils. The range of the effective depletion attraction between the large particles induced by the smaller particles (as discussed in section 2.7), is of the order of the size of the latter, which may only be a small fraction of the hard core diameter of the former. Such 'extreme' potentials, involving a very narrow attractive well, can lead to phase diagrams which differ qualitatively from those of simple molecular substances. In particular, for sufficiently narrow wells, the 'liquid–gas' transition (which mirrors a fluid–fluid transition of the underlying binary system) can be 'pre-empted' by the freezing transition, so that both the critical and the triple points vanish and the freezing transition broadens dramatically at large concentrations of small particles, which lead to a deeper potential well, and hence to a lower effective temperature [16].

Finally network-forming systems, like water or silicate, have peculiar freezing behaviour, because the spatial networks due to strong intermolecular bonds (e.g. the hydrogen-bond network of water) exist in both liquid and solid phases, and are only partially broken upon melting. The strong network leads to rather open (as opposed to closely packed) structures, like the four-fold coordinated diamond structure of ice. The hard sphere paradigm clearly does not apply to such systems, where highly directional interactions replace purely excluded volume considerations, which cannot explain, among other effects, the well known contraction of water upon melting (i.e. $\Delta V < 0$).

[14] J-L. Barrat, J-P. Hansen and M. Baus, *J. Phys. C* **20**, 1413 (1987).

[15] J.V. Sanders, *Philos. Mag. A* **42**, 705 (1980). P. Bartlett, R.H. Ottewill and P.N. Pusey, *Phys. Rev. Lett.* **68**, 3801 (1992). M.D. Eldridge, P.A. Madden, P.N. Pusey and P. Bartlett, *Mol. Phys.* **84**, 395 (1995).

[16] M. Dijkstra, R. van Roij and R. Evans, *Phys. Rev. E* **59**, 5744 (1999); this paper contains an excellent bibliography of earlier work.

Exercise: Show that for systems of molecules interacting via the inverse-power pair potential:

$$v(r) = \epsilon \left(\frac{\sigma}{r}\right)^n \qquad n > 3 \tag{4.110}$$

The freezing and melting lines in the density–temperature plane are necessarily of the form:

$$T_f = C_f \rho_f^{n/3} \qquad T_m = C_m \rho_m^{n/3} \tag{4.111}$$

where C_f and C_m are constants.

Hint: Show first, using an appropriate change of variables in the configuration integral, that the reduced excess free energy of the system $f^* = F^{ex}/Nk_BT$, is a function of $\rho^{*n/3}/T^*$ only (where $\rho^* = \rho\sigma^3$ and $T^* = k_BT/\epsilon$), rather than of ρ^* and T^* separately.

5 Critical fluctuations and scaling

The mean field theories of phase transitions considered in the previous chapter explicitly or implicitly neglect fluctuations, i.e. local deviations of the order parameter from its most probable, macroscopic value. This neglect is epitomized in the basic assumption underpinning Landau's theory, i.e. that the order parameter is uniformly replaced by the value which minimizes the phenomenological Landau free energy; fluctuations around this most probable value do not contribute to the partition function (4.36). The fact that local deviations of the order parameter from its average are not built into mean field considerations means, a fortiori, that correlations between fluctuations occurring in different subsystems are neglected. Correlations on the molecular scale, as quantified by the density correlation functions or static structure factors introduced in section 3.4, dominate first-order phase transitions like freezing, and can be accounted for quantitatively within standard structural theories of dense fluids, like the mean spherical approximation (MSA) and Percus–Yevick (PY) approximation mentioned in chapter 3. The range of these spatial correlations is typically a few molecular diameters. However, upon approaching the critical point of a second-order phase transition, e.g. the liquid–vapour critical point, density fluctuations become correlated over mesoscopic scales, while the macroscopic response of the system (e.g. the compressibility) diverges. A quantitative description of large-scale fluctuations is a highly challenging problem which has only been understood since the 1970s in terms of the idea of scale invariance, one of the key concepts of this chapter, which may also be applied to other, apparently unrelated, 'critical' objects like polymers.

The present chapter introduces the concept of correlation length, shows how Landau theory can be extended to include fluctuations and how the postulate of scale invariance can be exploited to generate scaling laws describing critical behaviour. Similar ideas are put to good use to describe polymer solutions and blends in the applications sections.

5.1 The correlation length

Consider a one-component fluid undergoing a vapour–liquid transition. The microscopic structure is characterized by the static structure factor $S(k)$. According to (3.52), the long wavelength limit of $S(k)$ diverges at the critical point, since the isothermal compressibility is singular there (cf. the result (4.51), valid within the mean field approximation). On the other hand $S(k)$ is related to the direct correlation function via the Ornstein–Zernike (OZ) relation (3.49). Now, while the $k \rightarrow 0$ singularity of $S(k)$ reflects the divergent range of density fluctuations at the critical point, there is no a priori reason why $c(k)$ should exhibit a small k singularity, since the range of $c(r)$ is expected to remain of the order of the range of the intermolecular potential. Hence, following Ornstein and Zernike (1914), we make the simplest possible assumption that $c(k)$ is a regular function of k as $k \rightarrow 0$, and may hence be expanded in powers of k; due to isotropy, the lowest order terms in the Taylor expansion are of the form:

$$\rho c(k) = c_0(\rho, T) + c_2(\rho, T)k^2 + \mathcal{O}(k^4) \tag{5.1}$$

According to (3.52) and (3.49):

$$\lim_{k \rightarrow 0} \rho c(k) = c_0(\rho, T) = 1 - \chi_T^{\mathrm{id}}/\chi_T \tag{5.2}$$

where χ_T^{id} is the compressibility of an ideal gas; c_0 clearly remains finite at the critical point. Expanding the Fourier transform of $c(r)$ in powers of k, one easily verifies that:

$$c_2(\rho, T) = -2\pi\rho \int_0^\infty c(r)r^4 \, dr = -R^2 \tag{5.3}$$

where R is a characteristic length, sometimes referred to as the Debye persistence length; note that the integral is expected to be positive, if the asymptotic form (3.41) is assumed, since $v(r)$ must be attractive at sufficiently long range to induce condensation. Substitution of (5.1) and (5.2) into (3.52) leads to the following small k behaviour of $S(k)$:

$$S(k) = \frac{R^{-2}}{K^2 + k^2} \tag{5.4}$$

where $K^2 = R^{-2}\chi_T^{\mathrm{id}}/\chi_T$, which vanishes at the critical point. Fourier transformation of (5.4) back to r space shows that the asymptotic (large r) behaviour of the correlation function $h(r)$ is:

$$h(r) = \frac{1}{4\pi\rho R^2} \frac{\exp(-r/\xi)}{r} \tag{5.5}$$

which introduces the correlation length:

$$\xi = K^{-1} = R\sqrt{(\chi_T/\chi_T^{\mathrm{id}})} \tag{5.6}$$

Away from the critical point spatial correlations decay exponentially; as the critical point is approached, ξ increases with a critical exponent $\nu = \gamma/2$, one half
the exponent which characterizes the divergence of χ_T. Mean field theory predicts $\gamma = 1$ (cf. (4.51)) and hence $\nu = 1/2$. The universal experimental values
are closer to 1.22 and 0.61. At the critical point correlations become extremely
long ranged, since they then decay only like $1/r$. Note that a power-law decay is
characteristic of self-similar behaviour, since it involves no intrinsic length scale,
similar to ξ in the exponential decay (5.5).

The result (5.6) for the decay of spatial correlations may be derived from a
different viewpoint, based on the so-called Landau–Ginzburg generalization of
the Landau free energy introduced in section 4.2. The Landau–Ginzburg free
energy is conveniently derived by taking the continuum limit of a microscopic
lattice model, like the binary mixture model examined in section 4.1. The mean
occupation number (i.e. the order parameter) $\langle n_i \rangle = \phi$ is now allowed to vary
smoothly between lattice sites, so that the mean field approximation to the internal
energy, equation (4.9), generalizes to:

$$U = \frac{1}{2} \sum_i \sum_{j \neq i} \epsilon_{ij} \phi_i (1 - \phi_j) \tag{5.7}$$

where $\epsilon_{ij} = \epsilon(\mathbf{r}_i - \mathbf{r}_j)$ depends on the relative positions $\mathbf{r}_i$ and $\mathbf{r}_j$ of lattice sites.
Similarly the mean field entropy (4.8) generalizes to:

$$S = -k_B \sum_i [\phi_i \ln \phi_i + (1 - \phi_i) \ln(1 - \phi_i)] \tag{5.8}$$

By writing $\phi_i(1 - \phi_j) = [(\phi_i - \phi_j)^2 - (\phi_i^2 + \phi_j^2) + 2\phi_i]/2$, the internal energy
(5.7) may be cast into the sum of non-local and local contributions:

$$U = \frac{1}{4} \sum_i \sum_{j \neq i} \epsilon_{ij} (\phi_i - \phi_j)^2 - \frac{\epsilon}{2} \sum_i \left(\phi_i^2 - \phi_i \right) \tag{5.9}$$

where $\epsilon = \sum_{j \neq i} \epsilon_{ij}$ and the coupling between sites is assumed to be translationally invariant.

In the continuum limit, the discrete variables ϕ_i turn into a continuously vary-
ing order parameter field $\phi(\mathbf{r})$, and $(\phi_i - \phi_j)^2 \rightarrow d^2 (\nabla \phi)^2$, where d is the lattice
parameter (the distance between nearest-neighbour lattice sites); for simplicity it
is assumed that in the initial lattice model, $\epsilon_{ij} \neq 0$ only between nearest neighbours. Denoting $\epsilon d^2/2$ by bd^3, the continuum limit of the energy (5.9) may be
cast in the form:

$$U = \frac{1}{2} \int_V \left[bd^3 (\nabla \phi)^2 + \epsilon \phi(\mathbf{r})(1 - \phi(\mathbf{r})) \right] \frac{d\mathbf{r}}{d^3} \tag{5.10}$$

Similarly, the continuum expression of the entropy (5.8) reads:

$$S = -k_B \int_V [\phi(\mathbf{r}) \ln \phi(\mathbf{r}) + (1 - \phi(\mathbf{r})) \ln(1 - \phi(\mathbf{r}))] \frac{d\mathbf{r}}{d^3} \tag{5.11}$$

The free energy $F = U - TS$ may finally be cast in the form:

$$F = \int_V \left[\frac{1}{2} b(\nabla \phi)^2 + f(\phi(\mathbf{r})) \right] d\mathbf{r} \qquad (5.12)$$

where $f(\phi(\mathbf{r}))$ is the local free energy density:

$$f(\phi(\mathbf{r})) = \frac{\epsilon}{d^3} \phi(\mathbf{r})(1 - \phi(\mathbf{r})) + \frac{k_B T}{d^3} [\phi(\mathbf{r}) \ln \phi(\mathbf{r}) + (1 - \phi(\mathbf{r})) \ln(1 - \phi(\mathbf{r}))] \qquad (5.13)$$

In the vicinity of the critical point, the next step is to expand the local free energy density in a Taylor series in $\zeta(\mathbf{r}) = \phi(\mathbf{r}) - \phi_c$, along the lines of the Landau expansion (4.43); only even terms contribute in the symmetric case (cf. (4.44)), so that the Landau–Ginzburg free energy finally reads:

$$F[\zeta(\mathbf{r})] = \int_V \left[\frac{1}{2} b(\nabla \zeta)^2 + \frac{a_2}{2} \zeta^2(\mathbf{r}) + \frac{a_4}{4} \zeta^4(\mathbf{r}) \right] d\mathbf{r} \qquad (5.14)$$

The temperature-dependent coefficients $a_2(T)$ and $a_4(T)$ may be related to the microscopic parameters ϵ and d appearing in (5.13) by carrying out the Taylor expansion on that expression. In Landau theory, the free energy F is a function of the mean order parameter ζ, while the Landau–Ginzburg free energy allows for local inhomogeneities of ζ, which become important near the critical point. The expression (5.14) is intimately related to the 'square gradient' approximation which will be encountered in the density functional theory of inhomogeneous fluids (section 7.2). The Landau–Ginzburg free energy (5.14) may be used to explore density fluctuations near the critical point of the gas–liquid transition. If the latter is approached from above ($T > T_c$) along the critical isochore ($\rho = \rho_c$), the change in free energy due to a fluctuation in local density, $\Delta \rho(\mathbf{r}) = \rho(\mathbf{r}) - \rho_c$ is:

$$\Delta F = F[\rho(\mathbf{r})] - F(\rho_c) = \int_V \left(\frac{a}{2} (\delta \rho(\mathbf{r}))^2 + \frac{b}{2} (\nabla \rho(\mathbf{r}))^2 \right) d\mathbf{r} \qquad (5.15)$$

where $a = a_2(T) > 0$ (since $T > T_c$), and the quartic term in (5.14) is neglected, as long as the density fluctuation is of sufficiently small amplitude. The coefficient a is the second derivative of the thermodynamic free energy density

$$a = \left(\frac{\partial^2 f(\rho, T)}{\partial \rho^2} \right)_{T, \rho=\rho_c} = \left(\frac{\partial \mu}{\partial \rho} \right)_{T, \rho=\rho_c} = \frac{1}{\rho_c^2 \chi_T} \qquad (5.16)$$

Note that the coefficient b in (5.15) must be positive for F to be at its minimum for a uniform density. Expanding $\Delta \rho(\mathbf{r})$ in a Fourier series:

$$\Delta \rho(\mathbf{r}) = \frac{1}{V} \sum_{\mathbf{k}} \rho_{\mathbf{k}} \exp(-i\mathbf{k} \cdot \mathbf{r}) \qquad (5.17)$$

ΔF may be cast in the form:

$$\Delta F = \frac{1}{V^2} \sum_{\mathbf{k}} \sum_{\mathbf{k'}} \left(\frac{a}{2} - \frac{b}{2} \mathbf{k} \cdot \mathbf{k'} \right) \rho_{\mathbf{k}} \rho_{\mathbf{k'}} \int_V d^3\mathbf{r} \exp\left(i(\mathbf{k} + \mathbf{k'}) \cdot \mathbf{r} \right)$$

$$= \frac{1}{2V} \sum_{\mathbf{k}} (a + b\mathbf{k}^2) |\rho_{\mathbf{k}}|^2$$

(5.18)

The probability of a density fluctuation characterized by the set of Fourier components $\{\rho_{\mathbf{k}}\}$ factorizes into a product of Gaussians, in agreement with the general result (3.7):

$$w(\{\rho_{\mathbf{k}}\}) \sim \exp\left(-\frac{\Delta F}{k_B T} \right) = \prod_{\mathbf{k}} w(|\rho_{\mathbf{k}}|)$$

(5.19)

where:

$$w(\{\rho_{\mathbf{k}}\}) \sim \exp\left(-\frac{1}{2V k_B T} (a + b\mathbf{k}^2) |\rho_{\mathbf{k}}|^2 \right)$$

(5.20)

This Gaussian distribution leads directly to the following form of the structure factor:

$$S(\mathbf{k}) = \frac{1}{N} \langle |\rho_{\mathbf{k}}|^2 \rangle = \frac{k_B T}{a + b\mathbf{k}^2}$$

(5.21)

which has the same Lorentzian form as the OZ result (5.4). Like the latter, the expression (5.21) is expected to be valid for $k \ll 1/d$, where $d = \rho^{-1/3}$ is the mean spacing between molecules, since only fluctuations on a mesoscopic scale have been considered. Note that the correlation length associated with (5.21) is $\xi = \sqrt{b/a} = \rho_c \sqrt{b \chi_T}$ which diverges like the square root of the compressibility at the critical point, in agreement with the result (5.6) of the OZ analysis.

The above Landau–Ginzburg analysis of density fluctuations in simple fluids may be extended to more complex fluids like oil–water microemulsions in the presence of surfactants [1] (cf. section 2.6). In that case the order parameter $\zeta(\mathbf{r})$ may be associated with the local molar water-to-oil ratio.

Due to the formation of local structures (oil in water or water in oil droplets, lamellar or bicontinuous phases) driven by the surface-active amphiphilic molecules, the order parameter will have rapid spatial variations, so that higher powers in the gradient of $\zeta(\mathbf{r})$ become important. If the microemulsion is isotropic, the term beyond the square gradient term in (5.15) may be cast in the form of the square of the Laplacian of $\zeta(\mathbf{r})$, so that the free energy functional reads:

$$F[\zeta(\mathbf{r})] = \frac{1}{2} \int \left[a\zeta(\mathbf{r})^2 + b|\nabla\zeta(\mathbf{r})|^2 + c(\nabla^2\zeta(\mathbf{r}))^2 \right] d\mathbf{r}$$

(5.22)

[1] M. Teubner and R. Strey, *J. Chem. Phys.* **87**, 3195 (1987).

the Fourier space representation of which is:

$$\Delta F = \frac{1}{2V} \sum_{\mathbf{k}} (a + bk^2 + ck^4)|\zeta_{\mathbf{k}}|^2 \tag{5.23}$$

which is still a quadratic function in the fluctuating variable $\zeta_{\mathbf{k}}$. While in simple liquids b must be positive to ensure stability, it can be driven to negative values by the surface tension reducing action of the surfactant; in that case $c > 0$, and stability against large amplitude fluctuations is guaranteed provided the coefficient of $|\zeta_{\mathbf{k}}|^2$ is positive for all wavenumbers, which requires $4ac - b^2 > 0$.

Proceeding as for the derivation of (5.21), one finds that the structure factor associated with the fluctuations of ζ is:

$$S(k) \sim \langle |\zeta_{\mathbf{k}}|^2 \rangle \sim \frac{1}{a + bk^2 + ck^4} \tag{5.24}$$

which has a peak at $k = \sqrt{-b/2c}$ before decaying like $1/k^4$ at large k, in agreement with small-angle neutron scattering data from microemulsions. The asymptotic behaviour of the corresponding space correlation function is:

$$h_{\zeta\zeta}(r) \sim \frac{\exp(-r/\xi_1)}{r} \sin\left(\frac{2\pi r}{\xi_2}\right) \tag{5.25}$$

where the two length scales are:

$$\xi_1 = \sqrt{\sqrt{\frac{a}{4c}} + \frac{b}{4c}} \qquad \xi_2 = 2\pi \sqrt{\sqrt{\frac{a}{4c}} - \frac{b}{4c}} \tag{5.26}$$

ξ_1 characterizes the decay of the correlations, and is analogous to the OZ correlation length, while ξ_2 is a measure of the 'periodicity' of the oil-rich and water-rich domains in the microemulsion.

5.2 Fluctuations and dimensionality

The growing importance of fluctuations in the local order parameter as the critical point is approached has been stressed repeatedly. It is important to obtain a quantitative estimate of the amplitude of the fluctuations of the order parameter, compared to its mean, in order to assess the internal consistency of mean field theory. It will become clear that this consistency test is intimately related to the dimensionality d of the system, and the subsequent discussion will be formulated for arbitrary d. Moreover, since the correlation length ξ becomes much larger than molecular size, or the lattice constant a, a coarse-grained description, along the lines of the continuum limit leading to the Landau–Ginzburg free energy (5.12) will be adopted. Since molecular details are irrelevant near T_c, one might as well forget about the exact physical significance of the local order parameter $\phi(\mathbf{r})$, and consider it to be a field variable in some effective Hamiltonian H, similar to (5.12)

$$H[\phi] = \int_V \left[\frac{1}{2} r\phi^2 + u\phi^4 + \frac{1}{2} c(\nabla \phi)^2 \right] d\mathbf{r} \qquad (5.27)$$

where the integration is over a d-dimensional volume, and rather standard field-theoretical notation has been adopted. The field is moreover coupled to some spatially variable external field $B(\mathbf{r})$, reminiscent of the magnetic field in the Ising spin Hamiltonian (4.20). The partition function is given by a spatially varying generalization of (4.36):

$$Q[B] = \int \mathcal{D}\phi \exp\left\{-\beta H[\phi] + \beta \int d^d\mathbf{r} B(\mathbf{r})\phi(\mathbf{r})\right\} = \exp\left\{-\beta \Omega[B]\right\} \qquad (5.28)$$

Q and the corresponding thermodynamic potential Ω are functionals of the external field B. Q is obtained by functional integration over all possible realizations of the field $\phi(\mathbf{r})$, denoted by $\mathcal{D}\phi$; this is in fact an infinite-dimensional integral which can be understood as the limit of a multidimensional integral over all possible values of the variables attached to discrete lattice sites, when the continuum limit ($a \to 0$, $N \to \infty$, $Na^d = $ constant) is taken.

$$\mathcal{D}\phi = \lim_{N \to \infty} \prod_{i=1}^{N} d\phi_i \qquad (5.29)$$

The obvious generalization of standard statistical mechanics relations like (2.50) and (2.51) (where the conjugate variables μ and N play roles similar to the conjugate fields $B(\mathbf{r})$ and $\phi(\mathbf{r})$) allows one to express the mean value of $\phi(\mathbf{r})$, and the correlation of its fluctuations at two different points of the volume V, as:

$$\langle \phi(\mathbf{r}) \rangle = -\frac{\delta \Omega}{\delta B(\mathbf{r})} \qquad (5.30)$$

$$G(\mathbf{r}, \mathbf{r}') = \langle (\phi(\mathbf{r}) - \langle \phi \rangle)(\phi(\mathbf{r}') - \langle \phi \rangle) \rangle = k_B T \frac{\delta^2 \Omega}{\delta B(\mathbf{r})\delta B(\mathbf{r}')} \qquad (5.31)$$

$$= -k_B T \frac{\delta \langle \phi(\mathbf{r}) \rangle}{\delta B(\mathbf{r}')} = -k_B T \chi(\mathbf{r}, \mathbf{r}') \qquad (5.32)$$

The derivatives appearing in these equations are functional derivatives which generalize the standard partial derivative of functions of several variables to the case of functionals, which involve an infinite number of variables, namely the values of a function at every point in space (see also section 7.1). $G(\mathbf{r}, \mathbf{r}')$ is of course analogous to the two-point correlation function introduced in section 3.4 (equation (3.32)), while $\chi(\mathbf{r}, \mathbf{r}')$ is the susceptibility (or linear response function) which is the $\mathbf{r}$-space equivalent of the susceptibility introduced in (3.66) (the latter relation pertains to a homogeneous fluid). Relations very similar to (5.30) and (5.31) will be re-derived within the framework of the density functional theory of inhomogeneous fluids (chapter 7).

It proves convenient (as it will in chapter 7) to switch to a thermodynamic potential (in fact a Helmholtz free energy) which is a functional of $\langle\phi(\mathbf{r})\rangle$, rather than of the conjugate field $B(\mathbf{r})$, by a standard Legendre transformation (see e.g. (2.19)):

$$F[\langle\phi(\mathbf{r})\rangle] = \Omega[B] + \int_V B(\mathbf{r})\langle\phi(\mathbf{r})\rangle d^d\mathbf{r} \tag{5.33}$$

The relation conjugate to (5.30) is:

$$\frac{\delta F}{\delta\langle\phi(\mathbf{r})\rangle} = B(\mathbf{r}) \tag{5.34}$$

Starting from the fundamental identity for functional differentiation:

$$\frac{\delta\langle\phi(\mathbf{r})\rangle}{\delta\langle\phi(\mathbf{r}')\rangle} = \delta^d(\mathbf{r} - \mathbf{r}') \tag{5.35}$$

where δ^d denotes the d-dimensional Dirac δ distribution, and using the chain rule:

$$\frac{\delta\langle\phi(\mathbf{r})\rangle}{\delta\langle\phi(\mathbf{r}')\rangle} = \int_V \frac{\delta\langle\phi(\mathbf{r})\rangle}{\delta B(\mathbf{r}'')} \frac{\delta B(\mathbf{r}'')}{\delta\langle\phi(\mathbf{r}')\rangle} d^d\mathbf{r}'' \tag{5.36}$$

one arrives, upon substituting the last equality of (5.31) and (5.34) into the right-hand side of (5.36)

$$\int \chi(\mathbf{r}', \mathbf{r}'') \frac{\delta^2 F}{\delta\langle\phi(\mathbf{r}'')\rangle\delta\langle\phi(\mathbf{r}')\rangle} d^d\mathbf{r}'' = \delta^d(\mathbf{r} - \mathbf{r}') \tag{5.37}$$

which shows that:

$$\frac{\delta^2 F}{\delta\langle\phi(\mathbf{r}'')\rangle\delta\langle\phi(\mathbf{r}')\rangle} = \chi^{-1}(\mathbf{r}, \mathbf{r}') \tag{5.38}$$

where $\chi^{-1}(\mathbf{r}, \mathbf{r}')$ is the inverse, in a matrix sense (with continuous indices), of the susceptibility $\chi(\mathbf{r}, \mathbf{r}')$. In an obvious generalization of Landau theory (where the order parameter is homogeneous), the mean field approximation now amounts to replacing the functional integral in (5.28) by the value of the integrand corresponding to the field $\phi_0(\mathbf{r})$ which maximizes this integrand, i.e. satisfies the extremum condition:

$$\left.\frac{\delta H[\phi]}{\delta\phi(\mathbf{r})}\right|_{\phi(\mathbf{r})=\phi_0(\mathbf{r})} = B(\mathbf{r}) \tag{5.39}$$

The partition function and the thermodynamic potential now reduce to:

$$Q[B] = \exp\left\{-\beta H[\phi_0] + \beta \int B(\mathbf{r})\phi_0(\mathbf{r}) d^d\mathbf{r}\right\} \tag{5.40}$$

$$\Omega[B] = H[\phi_0] - \int B(\mathbf{r})\phi_0(\mathbf{r}) d^d\mathbf{r} \tag{5.41}$$

Combination of (5.30), (5.39) and (5.41) shows that, as expected, $\langle \phi(\mathbf{r}) \rangle = \phi_0(\mathbf{r})$, while substitution of (5.41) into (5.33) yields:

$$F[\langle \phi(\mathbf{r}) \rangle] = H[\langle \phi(\mathbf{r}) \rangle] \tag{5.42}$$

In the absence of external field $B(\mathbf{r})$, the extremum condition (5.39) leads to a homogeneous field $\langle \phi \rangle = \phi_0$, which may be identified with the homogeneous order parameter of Landau theory. The results from Landau theory, with the free energy density (4.44), may then be directly carried over, with the correspondence $r(T) = a_2(T) = \alpha_2(T - T_c)$ and $u = a_4/4$. In particular, $\langle \phi \rangle = 0, T > T_c$; $\langle \phi \rangle = \pm\sqrt{-r/4u}, T < T_c$, and

$$\chi = \frac{\partial \langle \phi \rangle}{\partial B} = \frac{1}{r + 12u \langle \phi \rangle^2} \tag{5.43}$$

while the specific heat per unit volume is

$$c_v = -T \frac{\partial^2 (F/V)}{\partial T^2} = 0 \qquad T > T_c \tag{5.44}$$

$$c_v = T\alpha_2^2/8u \qquad T < T_c \tag{5.45}$$

The two-point correlation function, or the susceptibility (5.30) may also be calculated within the mean field approximation, along lines similar to the calculation in section 5.1. Substitution of (5.27) into (5.39) yields:

$$r\phi_0(\mathbf{r}) + 4u\phi_0^3(\mathbf{r}) - c\nabla^2 \phi_0(\mathbf{r}) = B(\mathbf{r}) \tag{5.46}$$

which is then functionally differentiated with respect to $B(\mathbf{r}')$. Remembering (5.31) and (5.35):

$$r\chi(\mathbf{r}, \mathbf{r}') + 12u\phi_0^2 \chi(\mathbf{r}, \mathbf{r}') - c\nabla^2 \chi(\mathbf{r}, \mathbf{r}') = \delta^d(\mathbf{r} - \mathbf{r}') \tag{5.47}$$

Going to the homogeneous (zero field) limit, where $\phi_0 = \langle \phi \rangle$ is constant, $\chi(\mathbf{r}, \mathbf{r}') = \chi(\mathbf{r} - \mathbf{r}')$, and taking the d-dimensional Fourier transform, one finds, as expected (cf. (5.21)) a Lorentzian form:

$$\chi(\mathbf{k}) = \frac{1}{r + 12u \langle \phi \rangle^2 + ck^2} = \frac{\chi_0}{1 + k^2 \xi^2} \tag{5.48}$$

where the correlation length ξ is given by:

$$\xi^2 = \frac{c}{r + 12u \langle \phi \rangle^2} \sim \xi_0^2 \frac{T_c}{|T - T_c|} \tag{5.49}$$

with $\xi_0 = \sqrt{c/(\alpha_2 T_c)}$ for $T > T_c$ and $\xi_0 = \sqrt{c/(2\alpha_2 T_c)}$ for $T < T_c$. Note that all mean field critical exponents are independent of dimensionality d. Taking the inverse Fourier transform of (5.48), one arrives at the d-dimensional generalization of (5.5):

$$\chi(\mathbf{r}, \mathbf{r}') = \frac{1}{4\pi c|\mathbf{r} - \mathbf{r}'|^{d-2}} \exp\left(-\frac{|\mathbf{r} - \mathbf{r}'|}{\xi}\right) \tag{5.50}$$

In order to examine the internal consistency of the mean field approximation, consider the Ginzburg ratio R_G, which is the ratio of the mean square fluctuation of the order parameter, coarse-grained over a volume V_ξ of dimension ξ, over the square of the mean value of the same coarse-grained order parameter in the ordered phase ($T < T_c$):

$$\overline{\phi} = \frac{1}{V_\xi} \int_{V_\xi} \phi(\mathbf{r}) \, d^d\mathbf{r} \tag{5.51}$$

$$R_G = \frac{\langle \overline{\phi}^2 \rangle - \langle \overline{\phi} \rangle^2}{\langle \overline{\phi} \rangle} \tag{5.52}$$

Using (5.31) for a homogeneous system, R_G may be cast in the form:

$$R_G = \xi^{-d} \langle \phi \rangle^{-2} \int_{V_\xi} G(\mathbf{r}) \, d^d\mathbf{r} \tag{5.53}$$

valid in the homogeneous limit. Substituting the mean field results for $\langle \phi \rangle = \sqrt{-r/4u}$ and $\int_{V_\xi} G(\mathbf{r}) \, d^d\mathbf{r} \simeq k_B T r^{-1}$, one finds that:

$$R_G \sim uc^{-d/2} r^{d/2-2} \sim uc^{-d/2} \alpha_2^{d/2-2}(T - T_c)^{d/2-2} \tag{5.54}$$

Since the coefficients are all finite near T_c, it is clear from (5.54) that R_G is well behaved near the critical point for $d \geq 4$. $d = 4$ is the *upper critical dimensionality*. For $d < 4$, R_G diverges at T_c, so that the mean field approximation is inconsistent, since the fluctuations of the order parameter become much larger than its mean value. For $d < 4$, there is a range of temperatures near T_c, called the fluctuation range, where $R_G \gg 1$, signalling the breakdown of mean field theory. For temperatures well outside that range, however, R_G becomes sufficiently small for mean field theory to be applicable, in agreement with experimental observation. Note that in view of the scaling with $T - T_c$ in (5.54), the fluctuation range is expected to increase as the dimensionality decreases. In particular, second-order phase transitions are totally suppressed by large fluctuations for $d < 2$, contrary to the mean field scenario whose predictions are independent of d; $d = 2$ is referred to as the *lower critical dimensionality*.

The cross-over from the mean field to the fluctuation regime may be expressed in more quantitative terms by considering the discontinuity in specific heat per unit volume, at the critical point predicted by mean field theory. Combination of (5.44) and (5.54) yields:

$$R_G = \frac{k_B}{\xi_0^d \Delta c_v} \left(\frac{|T - T_c|}{T_c} \right)^{(d-4)/2} \tag{5.55}$$

where $\xi_0 = \sqrt{c/\alpha_2 T_c}$ is a characteristic length, equal to the prefactor of the mean field correlation length in (5.49). If a is a molecular length scale, the jump in specific heat may be estimated to be $\Delta c_v = k_B/a^d$ and (5.55) then shows that

Table 5.1. *Definition of the main critical exponents*

Specific heat	$C \sim \|T - T_c\|^{-\alpha}$		
Order parameter	$\zeta \sim (T_c - T)^{\beta}$	for	$T < T_c$
Susceptibility	$\chi \sim \|T - T_c\|^{-\gamma}$		
Critical isotherm	$B \sim \zeta^{\delta}$	at	$T = T_c$
Correlation function	$G(k) \sim k^{-2+\eta}$	at	$T = T_c$
Correlation length	$\xi \sim \|T - T_c\|^{-\nu}$		

$R_G \sim (a/\xi_0)^d$; it is the ratio of the two lengths a and ξ_0 which controls the relative width of the fluctuation range,

$$t = |\Delta T / T_c| \simeq \left(\frac{a}{\xi_0}\right)^{2d/(4-d)} \tag{5.56}$$

Clearly, for $d < 4$, corrections to the predictions of mean field theory must be calculated by including fluctuations perturbatively. This may be achieved by using K. Wilson's renormalization group (RG) ideas which grew out of the scaling concepts developed in the 1960s by M.E. Fisher, L.P. Kadanoff, B. Widom and others [2]. In combination with methods from quantum field theory, the RG leads to systematic expressions for the critical exponents in powers of $\epsilon = 4 - d$ (the so-called ϵ expansion). The RG predictions lead to very significant corrections to the mean field critical exponents in three dimensions, which are in good agreement with experimental measurements. Such calculations are well beyond the scope of this book, but the more phenomenological scaling ideas are introduced in the following section.

5.3 Scaling ideas

The need of going beyond the mean field description of second-order phase transitions is clearly demonstrated by the Ginzburg criterion, and became apparent when careful experiments showed systematic deviations of measured critical exponents from their mean field predictions. These exponents pertain to the singular behaviour of thermodynamic properties and of the correlation length in the vicinity of T_c. Some of the key exponents are defined through the scaling laws in table 5.1.

In Ising spin language, the order parameter is the magnetization, and B is the external magnetic field, while in the case of the liquid–gas critical behaviour, the order parameter is $\rho_l - \rho_g$, the susceptibility is the isothermal compressibility, while the ordering field B is $\mu - \mu_c$ (cf. section 4.2). Table 5.2 compares the mean

---

[2] For an in-depth account of the fundamental ideas, see M.E. Fisher, *Rev. Mod. Phys.* **70**, 653 (1998).

Table 5.2. *Numerical values of the critical*
exponents for the liquid–gas transition

Exponent	Mean field	Experimental (3d)	Ising (2d)
α	0	0.11	0 (log)
β	1/2	0.325	1/8
γ	1	1.24	7/4
δ	3	4.82	15
η	0	0.03	1/4
ν	1/2	0.63	1

field values of these critical exponents (as derived, e.g., from Landau theory in section 4.2) to the generally accepted experimental values corresponding to systems belonging to the Ising universality class. The table also lists the exact values for the two-dimensional Ising model, which follow from Onsager's celebrated solution of the model (1944), to illustrate the importance of dimensionality.

The scaling ideas which will be briefly discussed in this section do not provide explicit values of these (and further) exponents, nor indeed of the critical amplitudes (i.e. the prefactors in the scaling relations of table 5.1), but they lead to exact relations between these exponents (or indices).

Focus first on the correlations and susceptibility, governed by the exponents η, ν and γ. The fundamental assumption underlying the contemporary view of critical fluctuations is that near a critical point, the correlation length ξ is the only relevant length scale in the system; all other lengths must be expressed relative to ξ. This simple assumption lies at the heart of the scaling relations between critical exponents characterizing the singular behaviour of thermodynamic properties at the critical point. It is convenient to introduce the dimensionless deviation of the temperature from the critical temperature T_c:

$$t = \frac{T - T_c}{T_c} \tag{5.57}$$

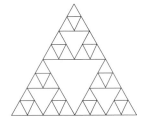

Figure 5.1. The Sierpinski gasket construction, illustrating scale invariance.

At the critical point ($t \to 0$) $\xi \to \infty$ and the system becomes scale invariant (or self-similar). This means that the pattern of local density fluctuations will look similar under any change in scale (or magnification). A simple construct that illustrates scale invariance is the 'Sierpinski gasket', of which the first three iterations are shown in figure 5.1. Iterations may be carried on indefinitely, leading to an obviously self-similar fractal pattern.

The subsequent discussion is restricted to homogeneous fluids where the two-point correlation function $G(\mathbf{r}', \mathbf{r}'')$ defined in (3.32) or (5.31) reduces (within a factor ρ^2) to $h(\mathbf{r}' - \mathbf{r}'') = h(r)$. Scale invariance implies that, within a factor, the correlation function $h(r)$ must be left unchanged under the scale transformation

$r \to r/b$ (where b is any increase of the unit length):

$$h(r/b) = b^p h(r) \tag{5.58}$$

where the exponent p is called the 'dimension' of h. Equation (5.58) expresses the fact that $h(r)$ must be a homogeneous function of r, of degree p, i.e.:

$$h(r) \sim r^{-p} \quad \text{at} \quad t = 0 \tag{5.59}$$

The mean field prediction (5.50) of the previous section is $p = d - 2$, leading to a structure factor $S(k) = 1 + \rho h(k)$ behaving, at small k, like (cf. (5.4) in the limit $t \to 0$):

$$S(k) = k^{-2} \quad \text{at} \quad t = 0 \tag{5.60}$$

Plots of experimental values of $1/S(k)$ versus k^2 show small but significant deviations from linear behaviour at small k. This prompted M.E. Fisher, in 1965, to postulate the following asymptotic behaviour of $h(r)$:

$$h(r) \sim r^{d-2+\eta} \quad t = 0 \tag{5.61}$$

$$h(r) \sim r^{d-2+\eta} H(r/\xi; t) \quad t \neq 0 \tag{5.62}$$

where d is the dimensionality of space ($d = 3$ for bulk liquids) and H is a *scaling function*, such that H goes over to a constant when $t \to 0$. According to table 5.2 the Fisher exponent is small and positive for $d = 3$; $\eta = 0$ in mean field theory. The mean field form (5.50) of h is a special case of (5.61). Since $\xi \sim |t|^{-\nu}$ (cf. table 5.1), the correlation function (5.62) satisfies a homogeneity relation which generalizes (5.58), namely:

$$h(r; t) = b^{-(d-2+\eta)} h(r/b; b^{1/\nu} t) \tag{5.63}$$

which simply expresses the requirement that h is invariant under the change of scale $r \to r/b$ provided the reduced distance from the critical point, t, is scaled by a factor $b^{1/\nu}$.

The corresponding singular part of the structure factor follows from (3.49):

$$S(k; t) \sim \int h(r; t) \exp(i\mathbf{k} \cdot \mathbf{r}) \, d\mathbf{r} \sim b^{-(d-2+\eta)} \int h(r/b; b^{1/\nu} t) \exp(i\mathbf{k} \cdot \mathbf{r}) \, d\mathbf{r}$$
$$\sim b^{2-\eta} S(bk; b^{1/\nu} t) \tag{5.64}$$

valid for any b. When $t = 0$, the choice $b = 1/k$ leads directly to the generalization of (5.60):

$$S(k) \sim k^{-(2-\eta)} \quad t = 0 \tag{5.65}$$

while choosing $b = |t|^{-\nu}$ yields:

$$S(k; t) \sim |t|^{-\gamma} S(k|t|^{-\nu}, \pm 1) \tag{5.66}$$

where the exponent γ is related to ν and η by the scaling relation:

$$\gamma = (2 - \eta)\nu \qquad (5.67)$$

According to (3.52), γ describes the divergence of the susceptibility, or isothermal compressibility:

$$\chi_T(t) \sim S(k = 0, t) \sim |t|^{-\gamma} \qquad (5.68)$$

The relation (5.67) is satisfied as a special case by the mean field results.

Homogeneity arguments can be extended to the thermodynamic potential, as pioneered by B. Widom (1965). A simple way of visualizing the underlying scale invariance assumption is provided by Kadanoff's block spin picture (1966) which is particularly well adapted to the Ising model. Kadanoff's decimation procedure amounts to regrouping the initial N spins of the system into blocks of linear size λ; within each cell of volume λ^d, the individual spin variables are replaced by a single two-state spin variable with a value determined according to some majority rule. The total free energy of the system, F, which is of the order of Nk_BT, is reduced in the decimation process to $F = \overline{N}k_BT$, where $\overline{N} = N/\lambda^d$. However, the reduction of the number of degrees of freedom is accompanied by a renormalization of the interactions between spins, and hence of T_c (cf. equation (4.14)) and finally t. If $F(t, B)$ denotes the free energy as a function of the distance from criticality and of the ordering field B, the Kadanoff block spin transformation may be summarized as follows:

$$N \to \overline{N} = N/\lambda^d \qquad t \to \overline{t} \qquad B \to \overline{B} \qquad (5.69)$$

$$F(t, B) \to \lambda^{-d} F(\overline{t}, \overline{B}) \qquad (5.70)$$

$$\xi \to \overline{\xi} = \xi/\lambda \qquad (5.71)$$

However, since $\xi \sim |t|^{-\nu}$, it follows immediately from (5.71) that $\overline{t} = \lambda^{1/\nu} t$. If it is furthermore assumed that $\overline{B} = \lambda^x B$, where the yet unknown exponent x is the 'dimension' of the ordering field B, then the transformation (5.70) may be cast in the scaling form

$$F(t, B) = \lambda^{-d} F(\lambda^{1/\nu} t, \lambda^x B) \qquad (5.72)$$

The order parameter conjugate to B, in this case the magnetization M, follows from the general relation (4.39):

$$M(t, B) = -\frac{\partial F}{\partial B} = -\frac{1}{\lambda^d} \frac{\partial F(\overline{t}, \overline{B})}{\partial(\overline{B}\lambda^{-x})} = \lambda^{x-d} M(\overline{t}, \overline{B}) \qquad (5.73)$$

Specializing to the case of zero field:

$$M(t, 0) = \lambda^{x-d} M(t\lambda^{1/\nu}, 0) \qquad (5.74)$$

which implies that

$$M(t, 0) = t^{\nu(d-x)} \qquad (5.75)$$

and, from the definition of β (table 5.1), the scaling relation:

$$\beta = v(d - x) \tag{5.76}$$

Similarly, differentiating (5.72) twice with respect to temperature, one arrives at the specific heat in zero field:

$$C(t, 0) \sim t^{dv-2} \tag{5.77}$$

which yields the specific heat exponent

$$\alpha = 2 - vd \tag{5.78}$$

Finally, the susceptibility in zero field may be obtained from (4.49), leading to the following generalization of the mean field result (4.51):

$$\chi(t, 0) \sim t^{dv-2vx} \chi(\pm 1, 0) \tag{5.79}$$

where $\chi(\pm 1, 0)$ are two amplitudes corresponding to $t > 0$ and $t < 0$; the definition of the critical exponent γ then implies a further scaling relation:

$$\gamma = 2vx - dv \tag{5.80}$$

while a combination of (5.77), (5.78) and (5.80) allows elimination of the field dimension x in the following Rushbrooke scaling relation:

$$\alpha + 2\beta + \gamma = 2 \tag{5.81}$$

By specializing (5.73) to the critical isotherm (i.e. $t = 0$), one may derive a further scaling relation involving the exponent δ

$$\alpha + \beta(1 + \delta) = 2 \tag{5.82}$$

It is interesting to note that the scaling relations for exponents of purely thermodynamic quantities, like (5.80) and (5.82) (i.e. which do not contain the exponents associated with correlations), do not involve the dimensionality d. In fact these scaling relations are intimately linked to well known inequalities, expressing thermodynamic stability conditions (e.g. (2.28)).

The scaling arguments of this section allow all critical exponents to be expressed in terms of only two independent exponents, which may be approximately calculated by the renormalization group (RG) technique evoked in the previous section, which amounts, roughly speaking, to a formal iteration of Kadanoff's block spin picture.

5.4 Application 1: Ginzburg criterion for polymer blends

The study of the Ginzburg ratio in section 5.2 has shown the role of dimensionality in assessing the relevance of fluctuations close to a second-order phase transition. For the Ising, or liquid–gas, transition, mean field theory is valid for

$d \geq 4$. Qualitatively, the success of mean field theory in high dimension can be understood from the fact that each atom is, in high dimensions, connected to many more neighbours than in low dimensional systems. Therefore, mean field theory, which effectively amounts to replacing the interaction with neighbours by an average 'molecular field', is more accurate in higher dimensions.

In a polymer blend such as the one considered in section 4.4, this argument leads to the conclusion that mean field theories should be much more accurate than in simple, molecular systems. The key point is that a given chain, because of its fractal structure, spans a large volume and therefore interacts with many neighbouring chains, as in a high dimensional systems. More precisely, a chain of mass M, which has a Gaussian structure, spans a volume $M^{3/2}v$ (v is the average volume per monomer). With a density of chains $\rho_P = 1/(Mv)$, the average number of other chains to be found within this volume is $\rho_P M^{3/2}v \sim M^{1/2}$, which is an increasing function of the molecular weight.

The argument can be made more quantitative by considering the Ginzburg criterion discussed in section 5.2. To this end, the Flory–Huggins free energy should first be recast in the canonical form (5.27). For simplicity, let us consider a symmetric ($\epsilon_{AA} = \epsilon_{BB}$) incompressible blend, with chains of M monomers. If $\phi_A(\mathbf{r})$ and $\phi_B(\mathbf{r})$ are the fractions of sites occupied by A and B monomers, the incompressibility constraint imposes $\phi_A(\mathbf{r}) + \phi_B(\mathbf{r}) = 1$ everywhere. The appropriate order parameter is $\phi_A - \phi_B \equiv \phi$, and the Flory–Huggins free energy density (cf. equation (4.89)) can be written (omitting irrelevant constant terms) as:

$$f(\phi, T) = k_B T v^{-1} \left[\frac{1+\phi}{2M} \ln(1+\phi) + \frac{1-\phi}{2M} \ln(1-\phi) - \frac{\chi(T)\phi^2}{4} \right] \qquad (5.83)$$

Expanding to fourth order in ϕ about $\phi = 0$, one finds

$$f(\phi, T) = k_B T v^{-1} \left[\left(\frac{1}{2M} - \frac{\chi(T)}{4} \right) \phi^2 + \frac{1}{12M} \phi^4 \right] \qquad (5.84)$$

The mean field demixing critical point is therefore located at a temperature T_c such that $\chi(T_c) = \chi_c = 2/M$, as already mentioned in section 4.4. The full free energy, allowing a description of non-uniform fluctuations, is obtained by adding a gradient term to (5.84):

$$F[\phi] = \int \left[f(\phi(\mathbf{r}), T) + \frac{1}{2} c (\nabla \phi(\mathbf{r}))^2 \right] d^3\mathbf{r} \qquad (5.85)$$

As was clear from the derivation of equation (5.10), the coefficient of the gradient term only involves molecular sizes and energies, and is independent of the molecular weight of the chains. According to equation (5.42), the mean field free energy (5.85) can equally well be considered as an effective Hamiltonian of the form (5.27) for the concentration field. The coefficients are obtained as

$$r = \frac{k_B T (\chi_c - \chi(T))}{2v} \qquad u = \frac{k_B T}{3Mv} \qquad (5.86)$$

which, using $\chi(T) = \theta/2T$, implies $\alpha_2 = k_B\theta/4vT_c$. At the mean field level, the correlation function for the concentration fluctuations is given by equation (5.50), and the prefactor for the correlation length (5.49) is $\xi_0 = \sqrt{2vc/k_B\theta}$. Note that this prefactor is independent of molecular weight: the correlation length for the concentration fluctuations does not depend on chain size [3].

The Ginzburg ratio for the polymer blend is easily estimated from equation (5.54). In three dimensions, one finds

$$R_G = (k_B T/3Mv)c^{-3/2}\left(\frac{k_B T(\chi_c - \chi(T))}{2v}\right)^{-1/2} = \frac{1}{\sqrt{M}}\left(\frac{k_B T}{3vc}\right)^{3/2}\sqrt{\frac{\chi_c}{\chi_c - \chi(T)}}$$

$$(5.87)$$

In agreement with the qualitative discussion above, this ratio is, for a given relative distance to the critical point, much smaller than for a molecular liquid with similar interactions. The fluctuation dominated region for a polymer blend is much narrower (reduced by a factor $1/M$) than for a simple mixture, and mean field critical exponents will be observed in an extended temperature range [4].

5.5   Application 2: scaling laws for polymer solutions

A simple statistical description of linear polymer chains in good solvent, based on the freely jointed chain model, was introduced in section 1.5. A linear homopolymer is made up of M identical monomers, and hence of $N = M - 1$ segments of length ℓ, where ℓ may be an effective length (persistence length) taking into account the local stiffness of the chain, and hence regroup several successive chemical monomer units. In general $N \gg 1$ and it is instructive to consider the formal scaling limit $1/N \to 0$, in which end effects become negligible. In a certain sense the scaling limit is reminiscent of the critical limit $t \to 0$ at second-order phase transitions where fluctuations become scale invariant. Similarly, a very long polymer chain is a 'critical' object, characterized by scale invariance which may be exploited to derive simple scaling relations for global polymer properties. In this section a few such relations are derived, starting with single chain properties, corresponding to infinitely dilute polymer solutions, before considering interacting polymer chains in semi-dilute solutions, and finally in concentrated solutions or melts. These concentration regimes are conveniently distinguished by considering the segment concentration, ρ, i.e. the number of

---

[3] These results can also be obtained using the alternative RPA approach outlined in section 3.10. We will make use of this alternative approach when studying the phase behaviour of copolymer melts in section 7.5; we encourage the reader to recover the results for polymer blends along the lines of section 7.5.

[4] The cross-over from mean field to Ising exponents in the vicinity of the critical point has been studied experimentally in F.S. Bates, J.H. Rosedale, Pr. Stepanek, T.P. Lodge, P. Wiltzius, G.H. Fredrickson and R.P. Hjelm, *Phys. Rev. Lett.* **65**, 1893 (1990).

segments per unit volume in a polymer solution. If R_G is the radius of gyration of one polymer, defined in (1.24), and ρ/N is the number of polymers per unit volume, then the volume fraction of polymers may be defined as $\phi_p = \rho R_G^3/N$. Overlap between different polymer coils becomes significant when $\phi_p = 1$, which defines the threshold concentration:

$$\frac{\rho^*}{N} R_G^3 = 1 \tag{5.88}$$

For $\rho \ll \rho^*$, the solution is dilute and polymer chains may be considered as independent. For $\rho \geq \rho^*$, the solution is semi-dilute, and the different polymer chains begin to overlap; it will become clear that $\rho^* \to 0$ in the scaling limit. Concentrated solutions and melts correspond to $\rho \gg \rho^*$.

Single chain

For a freely jointed chain, which is isomorphous to a random walk, and its coarse-grained version, the Gaussian (or bead-spring) model, the mean end-to-end distance, R_0, and the radius of gyration are proportional to the square root of N (cf. equation (1.26)). $R_G \sim N^{1/2}$ is a trivial example of a scaling law, but the exponent holds only for Gaussian chains, just like the exponent $\nu = 1/2$ characterizes the divergence of the correlation length at a critical point only in the mean field approximation (cf. table 5.2). Note that the Gaussian chain statistics implies that the threshold concentration (5.88) scales like $N^{-1/2}$, i.e. $\rho^* \to 0$ as $N \to \infty$.

The Gaussian model does not, however, account for excluded volume effects between distant monomers on the same chain. While neighbouring monomers do not overlap due to the local stiffness of the polymer, nothing prevents distant monomers from overlapping, i.e. the corresponding random walk from intersecting. If excluded volume interactions between distant monomers are taken into account, many conformations of the Gaussian chain become forbidden, and overall the polymer will swell, i.e. its radius of gyration is expected to be larger than predicted by the Gaussian model. In fact light scattering experiments on dilute polymer solutions in 'good' solvent show that R_G scales like:

$$R_G \sim N^\nu \ell \tag{5.89}$$

where the *Flory exponent* ν is universal (i.e. independent of the chemical nature of the monomers) and significantly larger than $1/2$, i.e. $\nu = 0.588$. A rough estimate of ν was obtained by P. Flory by taking into account excluded volume interactions between distant monomers at the second virial coefficient level. This is justified since the monomer density inside a coil, $\rho = N/R_G^3 \sim N^{1-3\nu}$ is very small for large N (since ν is larger than $1/2$). If the monomers were totally independent, the low density approximation for the excess free energy per monomer for a coil of extension $R \simeq R_G$ would be, according to equations (2.79), (2.80) and (2.12),

$$F_{ex} = Nk_B T B_2 \rho = Nk_B T \frac{v\rho}{2} = \frac{1}{2} k_B T v \frac{N^2}{R^3} \qquad (5.90)$$

where v is the excluded volume of a pair of monomers.

The total free energy is obtained by adding to the excess (or excluded volume) part (5.90), the conformational entropy (1.28) associated with the extension R of the chain, i.e.:

$$F(R) = F_0 + F_{ex}(R) + \frac{3k_B T}{2} \frac{R^2}{R_0^2} \qquad (5.91)$$

where $R_0^2 = N\ell^2$

The most probable extension of the chain, identified with the radius of gyration, is then obtained by minimizing the free energy (5.91) with respect to R, which yields

$$R_G = \left(\frac{v\ell^2}{2} \right)^{1/5} N^{3/5} \qquad (5.92)$$

leading to the Flory value $v = 3/5$ for the exponent v in (5.89), remarkably close to measured values, and the best theoretical estimates based on renormalization group calculations[5]. The flattering agreement must be considered as partly accidental, in view of the inconsistencies of the theory, which assumes that the conformational entropy is identical in form to that of a Gaussian chain, and that the monomers are uniformly distributed over the coil volume R^3.

Flory's argument is easily generalized to account for solvent effects. Adopting the cell model for a binary mixture of solvent molecules (species A) and segments (species B) introduced in section 4.1, the number of lattice sites within a volume R^3 is R^3/v (where v, the volume associated with each site, may be identified with the excluded volume of the segments or solvent molecules) and the fraction of cells occupied by a segment is $\phi = Nv/R^3$. The interaction energy (4.9) may hence be cast in the form:

$$U(R) = \frac{R^3}{v} \frac{v\epsilon}{2} \frac{Nv}{R^3} \left(1 - \frac{Nv}{R^3} \right) = U_0 - \frac{v\epsilon}{2} \frac{vN^2}{R^3} \qquad (5.93)$$

which may be added to the free energy (5.91), to yield an expression of identical form, with v replaced by an effective, temperature-dependent excluded volume parameter:

$$v_{eff}(T) = \left(1 - \frac{v\epsilon}{k_B T} \right) v = (1 - 2\chi(T))v \qquad (5.94)$$

where $\chi(T)$ is the Flory–Huggins parameter between solvent and monomers (see equation (4.83)). The radius of gyration is still given by (5.92), with v replaced

---

[5] See e.g. L. Schäfer, *Excluded Volume Effects in Polymer Solutions*, Springer, Berlin, 1999.

by $v_{\text{eff}}(T)$, and the Flory exponent remains unchanged. As a rule, ϵ is a positive quantity, since interactions between molecules of the same species are more favourable than those existing between different species. Hence the coil is seen to contract in order to minimize energetically unfavourable contacts between segments and solvent molecules. The case $\epsilon = 0$ (or $\chi = 0$) corresponds to a so-called *athermal* solvent, with molecules that are chemically identical to the monomers. Note the similarity between equation (5.94) and the virial coefficient (4.87), which was obtained by considering the entropy of mixing between solvent and polymer [6]. At the so-called θ-temperature, defined by $\chi(\theta) = 1/2$, $v_{\text{eff}}(T)$ vanishes, due to a cancellation of the excluded volume and attractive interactions between monomers: the polymer behaves effectively as an ideal (Gaussian) chain. Below $T = \theta$, the result (5.92) with $v = v_{\text{eff}}$ is obviously inapplicable since $v_{\text{eff}} < 0$. The polymer is now in a *poor solvent*, and a more accurate treatment shows that below θ, R_{G} rapidly shrinks to a value $\sim N^{1/3}$, corresponding to a dense globule of segments.

> **Exercise:** Show how the above calculation of R_{G} must be modified for a polymer coil confined to two dimensions, e.g. for a polymer adsorbed on a surface or confined in a narrow slit. Calculate the corresponding value of the Flory exponent v above the θ-temperature.

The scale invariance arguments for critical phenomena presented in section 5.3 may be adapted to polymers in the scaling limit, by considering a transformation reminiscent of Kadanoff's spin decimation. Consider a very long linear polymer $N \gg 1$ and divide the initial N segments of length ℓ into λ groups of successive segments of length ℓ', i.e. consider the coarse-graining transformation $N \rightarrow N' = N/\lambda$, $\ell \rightarrow \ell' = \lambda^{\alpha}\ell$. This transformation should not change the radius of gyration (5.89), which implies that $\alpha = v$.

The scale invariance assumption now states that any physical quantity associated with the polymer (N, ℓ) transforms, under the above decimation, as

$$f(N/\lambda, \lambda^v \ell) = \lambda^p f(N, \ell) \tag{5.95}$$

where the exponent p must be determined by an additional constraint. If the quantity is invariant under the transformation (like R_{G}), then $p = 0$.

As an application, consider the monomer pair distribution function inside an isolated coil, $g_0(r)$, and the corresponding structure factor (or form factor) $S_0(k)$

---

[6] The entropy of mixing is therefore effectively equivalent to the existence of an excluded volume between monomers.

defined in section 3.9. It satisfies the normalization condition:

$$S_0(k = 0) = \int g_0(r) \, d^d \mathbf{r} = N \tag{5.96}$$

The second equality in (5.96) implies that $g_0(r)$ must scale like λ^{-1} in a coarse-graining transformation, and so must its Fourier transform $S_0(k)$. For obvious dimensional reasons, $S_0(k)$ must be a function of the dimensionless argument $k\ell$:

$$S_0(k, N, \ell) = f(N, k\ell) \tag{5.97}$$

Hence, under a coarse-graining transformation $N \to N/\lambda$, $\ell \to \lambda^\nu \ell$, f must satisfy the scaling relation (5.95) with $p = -1$

$$f(N/\lambda, \lambda^\nu k\ell) = \lambda^{-1} f(N, k\ell) \tag{5.98}$$

Since λ is arbitrary, this is possible only provided:

$$f(N, k\ell) = N\psi(kN^\nu \ell) \tag{5.99}$$

so that:

$$S_0(k, N, \ell) = N\psi(kR_G) \tag{5.100}$$

This scaling is indeed observed in measurements of the intensity of light scattered from polymers with different molecular weights. The scaling relation (5.100) is satisfied by the structure factor (3.116) of a Gaussian polymer.

On sufficiently short length scales, $r \ll R_G$, corresponding to the wavenumber regime $kR_G \gg 1$, the form factor $S_0(k)$ is expected to be independent of the degree of polymerization N; since $R_G \sim N^\nu$, equation (5.100) then implies that:

$$S_0(k) \sim N(kR_G)^{-1/\nu} \sim k^{-1/\nu} \qquad \text{for} \qquad kR_G \gg 1 \tag{5.101}$$

Equation (3.116), valid for a Gaussian coil, is recovered with $\nu = 1/2$. Plots of light scattering data for $1/S_0(k)$ versus k^2 show clear deviations from linearity and provide experimental estimates of the exponent ν governing the radius of gyration. Equation (5.101) also implies that the fractal dimension of a polymer chain is $1/\nu$, as can be seen from equation (3.123).

Semi-dilute solutions

Equation (5.88) defines the cross-over segment concentration from the dilute to the semi-dilute regime. Above the θ-point $R_G \sim N^\nu$ and $\rho^* \sim N^{1-3\nu} = N^{-0.8}$; as expected ρ^* becomes very small for large molecular weight polymers. The limitations of the Flory–Huggins theory for the osmotic pressure of dilute polymer solutions were stressed in section 4.4 (cf. equations (4.86) and (4.87)). The Flory–Huggins result is thus of no help in the semi-dilute regime, and a scaling approach will be used to obtain the osmotic equation of state for $\rho \geq \rho^*$. Dimensional analysis requires the osmotic pressure to be of the generic form:

$$\Pi = \rho k_B T f(\rho \ell^3, N) \tag{5.102}$$

where the scaling function f goes to $1/N$ in the limit $\rho \to 0$, to reproduce van't Hoff's law for an ideal solution (cf. equation (4.86)). Under the coarse-graining transformation $N \to N/\lambda$ (and hence $\rho \to \rho/\lambda$) and $\ell \to \lambda^\nu \ell$, Π must remain invariant, so that:

$$\Pi = \rho k_B T f(\rho \ell^3, N) = \frac{\rho}{\lambda} k_B T f(\lambda^{3\nu-1} \rho \ell^3, N/\lambda) \tag{5.103}$$

which can only be satisfied provided:

$$f(\rho \ell^3, N) = \frac{1}{N} \psi \left(\frac{\rho}{N} (N^\nu \ell)^3 \right) \tag{5.104}$$

and hence:

$$\Pi = \frac{\rho}{N} k_B T \psi(\rho/\rho^*) \tag{5.105}$$

where $\rho^* = N/R_G^3 = N/(N^\nu \ell)^3$ is the overlap concentration (5.88). Now, when polymer chains are overlapping (i.e. when $\rho > \rho^*$), the attribution of segments to any particular chain becomes irrelevant, so that Π must be independent of N. In view of (5.105), this is only possible provided $\psi(\rho/\rho^*) \sim (\rho/\rho^*)^{1/(3\nu-1)}$, so that:

$$\Pi \sim \rho^{1/(3\nu-1)} \sim \rho^{9/4} \tag{5.106}$$

The predicted variation of Π with ρ is in good agreement with experimental measurements of the osmotic equation of state of semi-dilute polymer solutions.

Exercise: The correlation length ξ in a semi-dilute or concentrated polymer solution was introduced in section 3.10 (cf. equations (3.151) and (3.152)). Use a scaling argument similar to that invoked for Π, to show that $\xi \sim \rho^{\nu/(1-3\nu)} \sim \rho^{-3/4}$, which differs from the RPA prediction (3.151), $\xi_{RPA} \sim \rho^{-1/2}$. From this result, show that the semi-dilute solution can be pictured as a dense packing of non-overlapping 'correlation blobs' of size ξ each containing g segments, and such that inside a 'blob' the statistics is that of a single excluded volume chain, $\xi \sim g^\nu$. ξ can therefore be seen as the scale above which interactions between chains become dominant over intrachain excluded volume.

Chain extension

In highly concentrated solutions, or in polymer melts, polymer chains overlap more and more, and the swelling of any one chain is counteracted by the presence of the other chains, which tend to screen the excluded volume interactions between segments belonging to the same chain. The screening becomes more and more efficient as the concentration increases, to such an extent that polymer chains in melts behave as ideal Gaussian chains, as was already mentioned in section 3.10. This argument can in fact be extended to the scaling regime above ρ^*. According to the result of the exercise above, the semi-dilute solution can be pictured as a dense packing of correlation blobs of size ξ; taking these blobs as subunits, one is led back to the 'melt' situation of densely packed chains of blobs, and one may argue that

$$R_G \sim (N/g)^{1/2}\xi \qquad (5.107)$$

where $g \sim \xi^{1/\nu}$ is the number of monomers per blob. As a result, one finds a concentration dependence of R_G:

$$R_G \sim N^{1/2}\xi^{1-1/2\nu} \sim N^{1/2}\rho^{\frac{\nu-1/2}{1-3\nu}} \sim \rho^{-1/8} \qquad (5.108)$$

As expected, polymer chains shrink slightly as the concentration increases.

The same result can be derived from a scaling argument which goes as follows. In a dilute solution, $R_G = R_G^{(0)} \sim N^\nu \ell$. At finite concentration, this result must be modified by a scaling function $\psi(\rho/\rho^*)$ similar to that appearing in equation (5.105):

$$R_G = R_G^{(0)}\psi(\rho/\rho^*) = N^\nu \ell\psi(\rho/\rho^*) \qquad (5.109)$$

For $\rho \gg \rho^*$, it was just argued that $R_G \sim N^{1/2}$; since $\rho^* \sim N^{1-3\nu}$, this implies that $\psi(x) = x^\alpha$, where $\alpha = \frac{\nu-1/2}{1-3\nu}$, such that the concentration dependence (5.108) is recovered.

5.6 Application 3: finite size scaling

In the vicinity of a critical point, the correlation length characterizing the extent of correlated fluctuations diverges.This unavoidable feature of criticality would seem to render rather hopeless a simulation approach using the Monte-Carlo or molecular dynamics methods outlined in section 1.6. Indeed, these approaches have a computational cost that scales like the number of microscopic degrees of freedom, and are therefore limited to systems whose linear size L cannot exceed, in three dimensions, a few hundred molecular sizes (which already corresponds to 10^6 particles or lattice sites). If simulations are expected to reproduce reasonably the properties of bulk systems, they should be carried out with systems much larger than any microscopic length scale, including the correlation length. Obviously, this limit, $L \gg \xi$, cannot be reached in the vicinity of T_c.

The finite size scaling method, introduced by M.E. Fisher in the early 1970s, elegantly turns this apparent drawback of numerical approaches into an advantage. By using the flexibility offered by the simulations to vary the system size, together with simple scaling arguments, accurate determinations of the critical exponents are possible. The method has reached a great degree of sophistication, and we will only outline the main ideas below [7].

For concreteness, let us consider the behaviour of the susceptibility, $\chi(t, L)$, in a finite system of size L, as a function of $t = T - T_c$. L is large compared to the atomic dimensions, but of course finite. In a finite system, the susceptibility cannot diverge; however, one expects χ to increase in the vicinity of T_c, and to be close to the values for an infinite system far from T_c, when all microscopic length scales become small compared to L. Therefore $\chi(t, L)$ as a function of t will display a rounded peak in the vicinity of the critical temperature, $t = 0$.

The ratio of $\chi(t, L)$ to its thermodynamic limit $\chi(t, \infty)$ is a dimensionless function of L; as molecular details are expected to be irrelevant in the vicinity of T_c, the only dimensionless ratio that can be formed with L is $L/\xi_\infty(t)$, where $\xi_\infty \sim |t|^{-\nu}$ is the correlation length in the infinite system. Hence one has

$$\frac{\chi(t, L)}{\chi(t, \infty)} = f_\chi \left(\frac{L}{\xi_\infty} \right) \tag{5.110}$$

Using the scaling relation $\chi(t, \infty) \sim |t|^{-\gamma}$, the following scaling behaviour is obtained as a function of L:

$$\chi(t, L) \sim |t|^{-\gamma} f_\chi(L|t|^{-\nu}) = L^{-\gamma/\nu} g_\chi(L|t|^{-\nu}) \tag{5.111}$$

where the second equality defines the scaling function g_χ. Since we argued above that $\chi(t, L)$ must go through a maximum, equation (5.111) implies that the shape of this maximum is entirely determined by the L independent function g_χ. In particular, the maximum is located at a finite value of t, $t_{max} = \text{constant}/L^{1/\nu}$, its height is proportional to $L^{-\gamma/\nu}$ and its width to $L^{-1/\nu}$. Therefore it is possible, by analysing the results obtained for several different values of L, to obtain a precise determination of the exponents γ and ν. Of course the same type of argument can be used for any other quantity that diverges at the transition (e.g. the specific heat) by replacing γ by the appropriate critical exponent.

---

[7] The interested reader may consult the book by V. Privman (ed.), *Finite Size Scaling and Numerical Simulation of Statistical Systems*, World Scientific, Singapore, 1988.

III Interfaces and inhomogeneous fluids

6 Macroscopic description of interfaces

6.1 Interfacial tension and excess quantities

In section 2.3, we defined the surface tension γ as the excess free energy per unit area associated with the presence of a planar interface between two distinct phases. In the thermodynamic, macroscopic representation, an interface is a sharp boundary, entirely characterized by its area. At the microscopic scale, however, the local density (or concentration in the case of mixtures) is continuous. The interface is therefore characterized by an interfacial profile, as illustrated in figure 6.1 for the interface between a pure (single component) liquid and its vapour. The interfacial density profile $\rho(z)$ drops continuously from the bulk density of the liquid phase (phase a) to that of the vapour (phase b) over several molecular diameters. The mathematical boundary that enters the macroscopic description is usually identified with the *Gibbs dividing surface*, a plane located at $z = z_G$, where the position z_G is determined by the condition:

$$\int_{-\infty}^{z_G} \left(\rho(z) - \rho^{(a)} \right) dz + \int_{z_G}^{\infty} \left(\rho(z) - \rho^{(b)} \right) dz = 0 \tag{6.1}$$

which corresponds to the equality of the hatched areas in figure 6.1. Integrating by parts for z_G, equation (6.1) is easily recast in the equivalent, explicit form:

$$z_G = \frac{1}{\rho^{(b)} - \rho^{(a)}} \int_{-\infty}^{\infty} z\rho'(z) \, dz \tag{6.2}$$

Obviously these equalities, presented here for a planar interface, can be extended to describe an interface which is curved on macroscopic scales, much larger than its intrinsic width.

As the presence of an interface modifies the bulk thermodynamic relations, it is tempting to associate specific surface thermodynamic properties with the interface. These include the conjugate variables A and γ, but one may define surface equivalents of other properties, like entropy or partial densities, in the following way. If phases a and b coexist on opposite sides of an interface, like that pictured in figure 6.1, it proves convenient to introduce a reference system made up of two bulk phases a and b, separated by the Gibbs dividing surface. Phase a, as

Figure 6.1. Interfacial
profile and Gibbs dividing
surface for a
liquid–vapour interface.

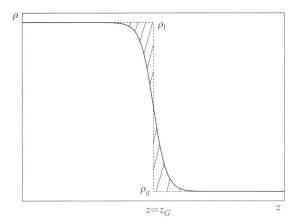

characterized by its bulk properties, is supposed to extend up to $z = z_G$ while
phase b extends beyond $z = z_G$. In this reference system the interface is assumed
to be infinitely sharp, and the density jumps discontinuously from $\rho^{(a)}$ to $\rho^{(b)}$; if
the two phases contain ν chemical species, each one of the partial densities ρ_α
jumps discontinuously at z_G, where z_G must now be defined by some appropriate generalization of the equal area construction in figure 6.1. The total volume
V and the dividing surface being fixed, the two bulk phases occupy volumes
$V^{(a)} + V^{(b)} = V$. If $\{\rho_\alpha^{(a)}\}$ and $\{\rho_\alpha^{(b)}\}$ are the bulk densities of the species in each
phase, the numbers of particles of each species in the two phases of the refer-
ence system are $N_\alpha^{(a)} = \rho_\alpha^{(a)} V^{(a)}$ and $N_\alpha^{(b)} = \rho_\alpha^{(b)} V^{(b)}$. The numbers of particles
associated with the interface are then given by the differences:

$$N_\alpha^{(s)} = N_\alpha - N_\alpha^{(a)} - N_\alpha^{(b)} \qquad 1 \le \alpha \le \nu \tag{6.3}$$

and the adsorption, or surface density, of species α is defined by:

$$\Gamma_\alpha = N_\alpha^{(s)}/A \tag{6.4}$$

The precise values of the Γ_α clearly depend on the position of the Gibbs dividing
surface, since the latter determines the volumes of the bulk phases in the reference
system. However, it is easily verified from the Gibbs–Duhem relations (2.18) for
the bulk phases, that under conditions of constant pressure and temperature, the
sum $\sum_\alpha \Gamma_\alpha \, d\mu_\alpha$ is independent of the choice of dividing surface. The latter
is generally chosen such that one of the adsorptions vanishes, as in the one-
component case (cf. figure 6.1). In a binary solvent–solute system it is convenient
to fix the dividing surface such that the adsorption of the solvent vanishes. The
adsorption of solute may then be positive or negative depending on whether the
solute tends to concentrate preferentially at the interface, or on the contrary is
depleted in the interfacial region.

Other surface thermodynamic properties may be defined in a similar way. Subtracting from (2.67) the corresponding relations for the bulk phases a and b (which lack the interfacial contribution, but are taken at the same pressure and chemical potentials), one arrives at the following expression for the surface free energy:

$$F^{(s)} = \gamma A + \sum_\alpha \mu_\alpha N_\alpha^{(s)} \qquad (6.5)$$

The surface tension may be identified with the surface free energy per unit area only if the dividing surface is chosen such that the second term in (6.5) vanishes. However, for an open system, $\gamma = \Omega^{(s)}/A$, independently of the position of the dividing surface. Starting from the differential relation (2.72), the same subtraction procedure leads to the following differential of the surface free energy:

$$dF^{(s)} = -S^{(s)} \, dT + \gamma \, dA + \sum_\alpha \mu_\alpha \, dN_\alpha^{(s)} \qquad (6.6)$$

where $S^{(s)}$ is the surface entropy, $S^{(s)} = S - S^{(a)} - S^{(b)}$. Subtracting from (6.6) the differential of (6.5), one arrives at the surface equivalent of the bulk Gibbs–Duhem relation (2.18):

$$S^{(s)} \, dT + A \, d\gamma + \sum_\alpha N_\alpha^{(s)} \, d\mu_\alpha = 0 \qquad (6.7)$$

or, dividing through by the surface area A, this can be cast in the form of the *Gibbs adsorption equation*:

$$d\gamma + s^{(s)} \, dT + \sum_\alpha \Gamma_\alpha \, d\mu_\alpha = 0 \qquad (6.8)$$

where $s^{(s)}$ is the surface entropy per unit area. Returning to the binary solvent (species 1)–solute (species 2) situation, and choosing the dividing surface such that $\Gamma_1 = 0$, one immediately derives from equation (6.8) the following useful thermodynamic relation:

$$\left(\frac{\partial \gamma}{\partial \mu_2} \right)_{T, \Gamma_1 = 0} = -\Gamma_2 \qquad (6.9)$$

As already mentioned in section 2.6, the adsorption of surfactant molecules at the water–oil interface strongly reduces the interfacial tension between these immiscible liquids. This should be qualitatively clear from equation (6.5), since strong adsorption will increase the surface entropy, and hence reduce the surface free energy, at least as long as interactions between adsorbed surfactant molecules can be neglected. The effect can equivalently be understood from the constant temperature version of the Gibbs adsorption equation: for preferential (i.e. positive) adsorption of surfactant molecules at the water–oil interface, any increase of surfactant concentration, and hence of chemical potential in the bulk, will lead to a lowering of γ. In fact for non-interacting surfactant molecules, considered in section 2.6, equation (6.9), with $\mu_2 = k_B T \ln \Gamma_2$ yields a linear reduction in the surface tension with the adsorption, consistent with equation (2.102).

A concept closely related to surface tension is that of *adhesion*. Consider two immiscible liquids, forming bulk phases a and b separated by an interface. Each of these liquid phases has a surface tension (labelled here $\gamma^{(a)}$ and $\gamma^{(b)}$) characteristic of the interface with its own vapour. The interface between a and b is similarly characterized by an interfacial tension $\gamma^{(a,b)}$. The reversible work per unit area required to separate the two liquid phases, i.e. to destroy the interface, and to create the two liquid–vapour interfaces is:

$$w^{(a,b)} = \gamma^{(a)} + \gamma^{(b)} - \gamma^{(a,b)} \tag{6.10}$$

$w^{(a,b)}$ is referred to as the work of adhesion, which depends on the state variables characterizing the two phases in mutual thermodynamic equilibrium. Adhesion can also be defined for a liquid–solid interface along similar lines. If one considers a hypothetical surface separating two subvolumes of the same liquid, then the reversible work needed to pull the two subvolumes apart is the work of *cohesion* $w^{(a,a)} = 2\gamma^{(a)}$, since $\gamma^{(a,a)}$ is obviously zero.

Surfactant molecules can also be adsorbed at the water–air interface, with their hydrophobic tails sticking out of the water into the air. Such *Langmuir monolayers* provide widely studied two-dimensional systems with rich phase behaviour as the area occupied by the surfactant molecules at the surface is varied in a *Langmuir trough*, which involves a sliding two-dimensional piston to control the area. The monolayer exerts a *surface pressure*, i.e. a force per unit length, which has the same dimension as the surface tension, but should not be confused with the latter.

6.2   Geometry of curved surfaces

Interfacial tension and other excess quantities are usually defined in reference to a planar interface. Curvature effects, however, are often essential. In order to minimize its interfacial energy at constant volume, a liquid drop takes a spherical shape. Fluctuations about a planar interface are also governed by curvature effects. Generally, interfacial quantities such as the surface tension γ are assumed to be unaffected by curvature, as long as the radius of curvature is large compared to molecular sizes. Curvature effects are thus essentially geometric in nature, so that a brief reminder of the geometric description of a curved surface is in order here.

Locally, any smooth surface can be described in a system of coordinates formed by the plane tangent to the surface and the direction normal to it. If M is the current point on the surface, we will denote this system of coordinates by $MXYZ$, where the Z direction is the normal to the surface at point M. In this system of coordinates, the equation of the surface in the vicinity of point M is quadratic, of the form

$$Z(X, Y) = AX^2 + BY^2 + 2CXY + \mathcal{O}(X^3, Y^3) \tag{6.11}$$

The three coefficients A, B, C form the 2×2 matrix of second derivatives. This curvature matrix is symmetric, and can be diagonalized, with real eigenvalues and orthogonal eigenvectors. Calling X' and Y' the new orthogonal directions in the MXY planes, such that the matrix is diagonal, one has the new equation for the surface

$$Z(X', Y') = \frac{X'^2}{2R_1} + \frac{Y'^2}{2R_2} + \mathcal{O}(X'^3, Y'^3) \tag{6.12}$$

which defines the two principal curvature directions X' and Y' and the associated principal radii of curvature, R_1 and R_2. Note that the signs of R_1 and R_2 depend on the arbitrary choice that has been made for orienting the normal to the surface.

An alternative description of the local curvature is often given in terms of the two invariants of the curvature matrix. The trace $R_1^{-1} + R_2^{-1} = 2H$ defines the *mean* curvature H, while the determinant $(R_1 R_2)^{-1} = K$ is the *Gaussian* curvature K.

A surface that deviates only slightly from a plane, say the xOy plane, can be described conveniently by a function $h(x, y)$ of the two projected coordinates. A point M of the surface has coordinates $(x, y, h(x, y))$. Note that this description excludes consideration of overhangs. Two vectors tangent to the surface are $\mathbf{u}_x = (1, 0, h_x)$ and $\mathbf{u}_y = (0, 1, h_y)$[1]. The element of area underpinned by an element of *projected* area $dx \times dy$ is given by the vector product $\mathbf{u}_x \wedge \mathbf{u}_y$, and is therefore $dA = \sqrt{1 + h_x^2 + h_y^2}\, dx\, dy$. The unit normal $\mathbf{N}$ to the surface is proportional to $\mathbf{u}_x \wedge \mathbf{u}_y$, and has coordinates $N_x = -h_x$, $N_y = -h_y$, $N_z = 1$. In order to obtain the expression of the curvature matrix, it is necessary to express the equation of the surface in the coordinates spanned by the tangent and normal vectors. The calculation is rather lengthy[2], and we will merely quote the final results for the mean and Gaussian curvature, which are

$$H = \frac{1}{2}\left(1 + h_x^2 + h_y^2\right)^{-3/2}\left(\left(1 + h_y^2\right)h_{xx} + \left(1 + h_x^2\right)h_{yy} - 2h_x h_y h_{xy}\right) \simeq \frac{1}{2}(h_{xx} + h_{yy}) \tag{6.13}$$

$$K = \left(1 + h_x^2 + h_y^2\right)^{-2}\left(h_{xx}h_{yy} - h_{xy}^2\right) \simeq h_{xx}h_{yy} - h_{xy}^2 \tag{6.14}$$

The second equality in equations (6.13) and (6.14) corresponds to the case of weakly undulated surfaces, $h_x \ll 1$ and $h_y \ll 1$.

---

[1] The subscript in h_x denotes a partial derivative, $h_x = \frac{\partial h}{\partial x}$, $h_{xx} = \frac{\partial^2 h}{\partial x^2}\dots$
[2] For details see e.g. the book by S. Safran (further reading in this chapter).

Figure 6.2. (a) Vertical cut through a drop of water (phase 1) at the oil (phase 2) – air (phase 3) interface. (b) Illustration of the force balance at the contact line. (c) Drops of liquid on a solid substrate. θ is the contact angle. $\theta < \pi/2$ (left) corresponds to partial wetting, while $\theta > \pi/2$ corresponds to a non-wetting situation.

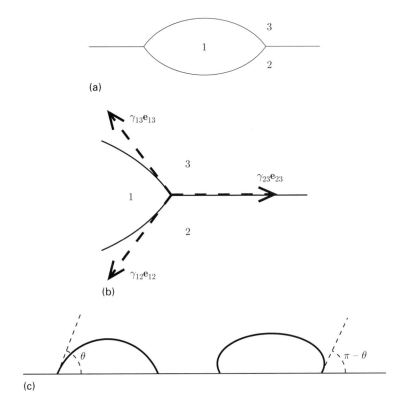

6.3 Wetting phenomena

Three-phase contact

Liquid drops in contact with a vapour are often supported either by a solid substrate or by another fluid phase (see figure 6.2). The three different phases meet on a line, the contact line, which is the intersection of the three interfaces. Each interface exerts a force proportional to the surface tension and tangent to the interface on this contact line, and the balance of forces can be written in the vector form (see figure 6.2 for the definition of the notation)

$$\gamma_{12}\mathbf{e}_{12} + \gamma_{23}\mathbf{e}_{23} + \gamma_{31}\mathbf{e}_{31} = 0 \qquad (6.15)$$

This vector relation determines, up to a global rotation, the geometry of the three interfaces near the contact line. The angles between the interfaces are fixed by the values of the surface tensions.

A slightly simpler situation is encountered when one of the phases is a flat solid substrate. The contact angle, i.e. the angle between the flat substrate and the liquid–vapour (or liquid–liquid) interface, is determined by projecting the vector equality on the tangent to the surface. The result, known as Young's equality, reads

$$\gamma_{SL} + \gamma_{LV} \cos\theta = \gamma_{SV} \qquad (6.16)$$

In the normal direction, the force exerted by the contact line must be compensated by a deformation of the solid. On a dimensional basis, this deformation can be estimated as γ_{LV}/E, where E is the Young modulus. Using reasonable values for E ($E \sim 10^{10}$ Pa) and γ_{LV}, this deformation is indeed found to be completely negligible for bulk substrates[3].

Wetting

Equation (6.15) can be fulfilled only if the three surface tensions can form the three sides of a triangle. If this is not the case, e.g. if $\gamma_{31} > \gamma_{12} + \gamma_{23}$, the surface energy of the 1–3 interface is so high that it is thermodynamically more favourable to form a film of phase 2 between the two phases 1 and 3. At equilibrium, there is no contact line, and phase 2 is said to wet the 1–3 interface completely.

A similar reasoning holds for the case of a contact line with a solid substrate. Equation (6.16) can be satisfied only if $\gamma_{SL} - \gamma_{LV} < \gamma_{SV} < \gamma_{SL} + \gamma_{LV}$. If $\gamma_{SV} > \gamma_{SL} + \gamma_{LV}$, a liquid film will form on the substrate ($\theta = 0$). The liquid is said to wet the solid completely. Such a situation is often encountered when the solid surface is a high energy one (i.e. the cohesion energy of the solid is high), as with surfaces of metals or covalent materials.

If $\gamma_{SV} < \gamma_{SL} + \gamma_{LV}$, the contact angle is finite. It is usual to distinguish the partial wetting case, $\theta < \pi/2$, and the non-wetting case, $\theta > \pi/2$[4]. The extreme non-wetting case corresponds to $\gamma_{SL} > \gamma_{LV} + \gamma_{SV}, \theta = \pi$, in which case a vapour film forms at the liquid–solid interface. Non-wetting substrates are obtained either when the liquid is highly cohesive (e.g. mercury), or when the substrate has received a special chemical treatment so as to make γ_{SL} particularly high (e.g. grafting of aliphatic chains on a surface makes it water repellent).

6.4 Capillary pressure and capillary condensation

Laplace pressure formula

Laplace's law expresses the mechanical equilibrium condition across a curved interface. Consider a surface element $d^2 S = dx_1 \times dx_2$, as illustrated in figure 6.3, with radii of curvature R_1 and R_2 in the two orthogonal directions that define its sides. Setting $dx_1 = R_1 d\theta_1$ and $dx_2 = R_2 d\theta_2$, and orienting the surface in the

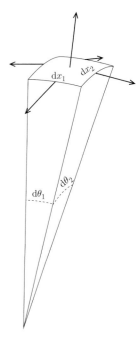

Figure 6.3. Element of surface dS, illustrating the balance of forces between the surface tensions and the pressure forces.

[3] Very thin solid layers, however, can deform under the influence of capillary forces.
[4] The reason for this distinction will become clear when we study the capillary condensation problem, section 6.4.

direction of its concavity, the balance of forces in the normal direction yields

$$(P_{int} - P_{ext})\, dx_1\, dx_2 - \gamma\, dx_1\, d\theta_2 - \gamma\, dx_2\, d\theta_1 = 0 \qquad (6.17)$$

The difference in pressure between the interior (convex) side and the exterior is thus given by the Young–Laplace equation,

$$P_{int} - P_{ext} = \gamma \left(\frac{1}{R_1} + \frac{1}{R_2} \right) = 2\gamma H \qquad (6.18)$$

The above proof of the Laplace formula is purely mechanical. It is interesting to consider an alternative derivation, based on thermodynamics. We will consider an interface between a liquid and a vapour phase. The vapour is constrained by a piston to be at pressure P_0. For simplicity, we restrict the calculation to the case of an interface defined by a surface $z = h(x)$, invariant by translation along the y axis. The liquid occupies the volume $z < h(x)$. At equilibrium, the Gibbs function $G = F + P_0 V$ must be extremal with respect to variations of the interface position, $\delta h(x)$. δG can be expressed as

$$\delta G = \gamma\, \delta A - P_{int}\, \delta V + P_0\, \delta V \qquad (6.19)$$

Using $dA = \sqrt{1 + h_x^2}\, dx\, dy$, and $\delta V = \delta h\, dx\, dy$, equation (6.19) yields

$$\delta G = (P_0 - P_{int}) \int dx \int dy\, \delta h(x) + \gamma \int dx \int dy \frac{h_{xx}\delta h_x}{\sqrt{1 + h_x^2}}$$

$$= \int dx \int dy \left[P_0 - P_{int} - \gamma \frac{h_{xx}}{\left(1 + h_x^2\right)^{3/2}} \right] \delta h(x) + \gamma \int dy \left[\frac{h_x}{\sqrt{1 + h_x^2}} \delta h(x) \right]_0^{L_x} \qquad (6.20)$$

Specializing to variations δh that vanish at the boundary, one cancels the last term in equation (6.20). The term between square brackets must also vanish, since G is stationary. This yields

$$P_0 - P_{int} = \gamma \frac{h_{xx}}{\left(1 + h_x^2\right)^{3/2}} \qquad (6.21)$$

which, using equation (6.13), is equivalent to equation (6.18).

Exercise: Generalize this proof of the Laplace formula to a general surface $z = h(x, y)$, and recover equation (6.13).

Capillary condensation

Surface effects have a strong influence on the location of phase transitions. Consider for example a liquid–vapour transition, that takes place at a temperature T at the saturation pressure $P_{sat}(T)$. P_{sat} is defined by the equality of chemical potentials, $\mu_L(P_{sat}, T) = \mu_V(P_{sat}, T)$. Consider now the case of a vapour pressure P slightly smaller than P_{sat}. If a droplet of liquid of radius R forms in this vapour, its internal pressure will be $P + 2\gamma/R$. The droplet will coexist with the vapour provided $\mu_L(P + 2\gamma/R, T) = \mu_V(P, T)$. Using $\frac{\partial \mu_L}{\partial P} = 1/\rho_L = \text{constant}$ (incompressible fluid), and $\mu_V(P, T) = k_B T \ln P + f(T)$, one obtains the condition for coexistence of a droplet of radius R with the vapour

$$k_B T \rho_L \ln \frac{P}{P_{sat}} = P - P_{sat} + \frac{2\gamma}{R} \qquad (6.22)$$

known as the Kelvin relation.

Solid surfaces will modify phase coexistence in a very similar way. Consider a thin slit of width h between two parallel solid walls. This slit is in contact with a vapour reservoir, at fixed chemical potential μ (and pressure P). If $\gamma_{SL} < \gamma_{SV}$, forming a liquid bridge between the two surfaces may be thermodynamically favourable, even if $P < P_{sat}$. This is the *capillary condensation* phenomenon, which explains the accumulation of moisture in porous media such as sand or concrete. Quantitatively, the coexistence between a liquid confined between the two plates and the vapour at fixed chemical potential is determined by minimizing the grand potential of the confined system. For the liquid, the grand potential per unit area is $\Omega_L/A = hf_L - \mu h \rho_L + 2\gamma_{SL}$. For the vapour, $\Omega_V/A = hf_V - \mu h \rho_V + 2\gamma_{SV}$. Minimizing the grand potential with respect to the liquid density, one obtains the condition that the chemical potential μ_L of the liquid must be equal to μ. This is possible only if the pressure P_L in the liquid is different from P, namely

$$\mu_L(P_L, T) = \mu = \mu_V(P, T) \Rightarrow P_L - P_{sat} = k_B T \rho_L \ln \frac{P}{P_{sat}} \qquad (6.23)$$

With this condition, the grand potentials in the liquid and vapour phases are given by (cf. equation (2.72))

$$\Omega_L/A = -P_L h + 2\gamma_{SL}$$
$$\Omega_V/A = -P h + 2\gamma_{SV} \qquad (6.24)$$

Condensation will take place when $\Omega_L < \Omega_V$, which translates into

$$P - P_{sat} - k_B T \rho_L \ln \frac{P}{P_{sat}} < \frac{2}{h}(\gamma_{SV} - \gamma_{SL}) \qquad (6.25)$$

The term $P - P_{sat}$ in equation (6.25), which is at most of order $k_B T \rho_V$, can be neglected compared to $k_B T \rho_L$. The condition for condensation between two walls separated by h, at pressure P, is therefore

$$h < 2 \frac{(\gamma_{SV} - \gamma_{SL})}{k_B T \rho_L \ln(P_{sat}/P)} \qquad (6.26)$$

Obviously this condensation is possible only if $\gamma_{SV} - \gamma_{SL} > 0$, which is to say that the contact angle is smaller than $\pi/2$. The possibility of capillary condensation is

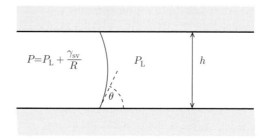

Figure 6.4. Schematic view of the meniscus separating a confined liquid from the vapour phase, at coexistence. The radius of curvature R of the meniscus is such that $h = 2R\cos\theta$. The pressure drop across this meniscus is $\gamma_{LV}/R = (\gamma_{SV} - \gamma_{SL})/h = P - P_L$.

the qualitative difference between the partial wetting and non-wetting situations. Quantitatively, the typical distance at which capillary condensation takes place is $(\gamma_{SV} - \gamma_{SL})/(k_B T \rho_L)$. This distance is rather small, typically a few nanometers, which means that solid surfaces have to be in very close contact in order for the condensation to take place.

Finally, it is interesting to note that, at coexistence, the pressure difference between the liquid and the vapour is exactly the Laplace pressure difference across the meniscus that separates the liquid and the vapour phases, as indicated in figure 6.4.

Exercise: Generalize the above study of capillary condensation to a cylindrical pore of radius R.

6.5 Disjoining pressure and film stability

Up to now, we have considered interfaces between bulk phases, that extend to infinity in the direction normal to the interface. A different description is needed for very thin films. For such films, the thickness can be so small that the finite range of intermolecular forces has to be taken into account. Imagine for example a thin film of liquid separating a solid substrate from its vapour, in the wetting case $\gamma_{SL} + \gamma_{LV} < \gamma_{SV}$. When the film becomes very thin, the interaction of the solid and the vapour across the liquid film has to be taken into account. Hence the grand potential for a film area A and film thickness h, at fixed chemical potential μ, will be written in the form

$$\Omega = \Omega_S + \Omega_V + \Omega_L + A(\gamma_{SL} + \gamma_{LV} + W(h)) \tag{6.27}$$

where $W(h)$ is the term that describes the interaction between the vapour and the solid, and depends on the thickness h of the film. The liquid grand potential is

$-P_{\mathrm{L}} A h$, and the grand potential of the vapour is $-P_{\mathrm{V}}(H - h)A$, where H is the total (macroscopic) thickness of the system. A minimization with respect to the film thickness yields

$$P_{\mathrm{L}} = P_{\mathrm{V}} + \frac{\mathrm{d}W}{\mathrm{d}h} = P_{\mathrm{V}} - \Pi_{\mathrm{d}}(h) \tag{6.28}$$

where the second equality defines the disjoining pressure $\Pi_{\mathrm{d}}(h) = -\frac{\mathrm{d}W}{\mathrm{d}h}$.

A positive disjoining pressure means that the two interfaces repel each other. If this is the case, a liquid film with a thickness h determined by equation (6.28) will cover the surface, even if the chemical potential μ is smaller than μ_{sat}. At coexistence ($\mu = \mu_{\mathrm{sat}}$), the liquid film will grow and occupy the whole volume.

The dependence of $W(h)$ on h is therefore crucial for determining the thickness of thin films. For $h \rightarrow 0$, $W(h)$ must reduce to $\gamma_{\mathrm{SV}} - \gamma_{\mathrm{SL}} - \gamma_{\mathrm{LV}}$. For a film thickness large compared to molecular sizes, $W(h)$ will be determined by the van der Waals r^{-6} interactions. The resulting interaction, when integrated over a half space, is of the form

$$W(h) = \frac{A_{\mathrm{SLV}}}{12\pi h^2} \tag{6.29}$$

The *Hamaker constant* A_{SLV} depends on the interactions between the substrate, the liquid and the vapour. A_{SLV} is positive, since the liquid, with a higher density, has a stronger interaction with the substrate. Typical values for the Hamaker constant are in the range 10^{-21} J (for interactions between organic substances) to 10^{-18} J (for interactions involving high energy, metallic surfaces). The shape of $W(h)$ is illustrated in figure 6.5.

In the case where the amount of liquid is limited, $W(h)$ will determine the minimal thickness reached by the wetting film. For a liquid 'puddle' with fixed total volume $V = Ah$, the free energy is $f_{\mathrm{L}} Ah + A(\gamma_{\mathrm{SL}} + \gamma_{\mathrm{LV}} - \gamma_{\mathrm{SV}} + W(h))$.

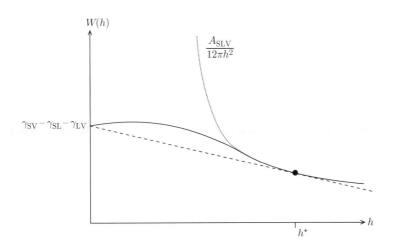

Figure 6.5. Typical shape of the $W(h)$ function for a wetting liquid. The minimal film thickness h^* corresponding to equation (6.30) is also indicated.

Minimizing with respect to h at fixed V, one obtains the equilibrium condition

$$W(h) + \gamma_{SL} + \gamma_{LV} - \gamma_{SV} - h\frac{dW}{dh} = 0 \tag{6.30}$$

For van der Waals interactions of the form (6.29), this leads to a minimal thickness $h^* = (A_{SLV}/12\pi(\gamma_{SV} - \gamma_{SL} - \gamma_{LV}))^{1/2}$, typically of the order of a few molecular sizes.

It is interesting to note that the thickness h^* defined by equation (6.30) corresponds to a double tangent, or convex envelope, construction on the curve $W(h)$. A thin film with a thickness between 0 and h^* will therefore lower its grand potential by decomposing into regions of thickness h^* and 'dry' regions ($h = 0$).

Further reading

Thermodynamics of interfaces is reviewed in most thermodynamics books, including Tabor's *Gases,Liquids and Solids* (see Further reading, chapter 1) or Atkins' *Physical Chemistry* (see chapter 2).

More specialized books on this topic include the monograph by S.A. Safran, *Statistical Thermodynamics of Surfaces, Interfaces, and Membranes*, Addison-Wesley, Reading, MA, 1994, and the book by B.V. Derjaguin, N.V. Churaev, V.M. Muller, *Surface Forces*, Consultants Bureau, New York, 1987, written by the pioneers of the Russian school in the field, who in particular introduced the concept of disjoining pressure.

7 The density functional approach

7.1 Variational principle

A homogeneous fluid is invariant under translations, and its local density $\rho(\mathbf{r})$, defined in equation (3.31), reduces to the bulk density $\rho = N/V$. For an open system, ρ is uniquely determined by the chemical potential μ of the molecules, fixed by a reservoir. Translational invariance can, however, be broken by the presence of confining surfaces, like the walls of a container, by the presence of interfaces separating different phases, or by some external potential $V_{\text{ext}}(\mathbf{r})$ acting on the particles. Conceptually, the latter situation is the simplest way of studying inhomogeneities in liquids. The inhomogeneity induced by $V_{\text{ext}}(\mathbf{r})$ will be characterized by a non-uniform, local density $\rho(\mathbf{r})$. If we consider once more an open system, it is clear from the statistical average in equation (3.31) that $\rho(\mathbf{r})$ is uniquely determined by the 'local' chemical potential $\psi(\mathbf{r}) = \mu - V_{\text{ext}}(\mathbf{r})$, for fixed values of the temperature T and volume V (for simplicity the present discussion is restricted to fluids containing a single chemical species). While under these conditions the grand potential Ω is a function of μ in the absence of an external field (i.e. for $V_{\text{ext}}(\mathbf{r}) = 0$), it is now a functional[1] of $\psi(\mathbf{r})$ in the presence of this potential, $\Omega = \Omega[\psi(\mathbf{r})]$.

In the presence of the external field, the total potential energy of N interacting particles is

$$V_N(\{\mathbf{r}_i\}) + \sum_i V_{\text{ext}}(\mathbf{r}_i) \tag{7.1}$$

Consider an infinitesimal change $\delta V_{\text{ext}}(\mathbf{r})$ of the external potential, which results (for fixed μ) in an infinitesimal change $\delta\psi(\mathbf{r})$ of the 'local' chemical potential. A straightforward statistical mechanics calculation, along the lines of the linear response calculation of section 3.6, yields the following result for the change in grand potential:

$$\delta\Omega = \Omega[\psi(\mathbf{r}) + \delta\psi(\mathbf{r})] - \Omega[\psi(\mathbf{r})] = -\int d\mathbf{r}\, \rho(\mathbf{r})\, \delta\psi(\mathbf{r}) \tag{7.2}$$

---

[1] A functional associates a number with a function. A simple example is the integral functional, $f \to I[f] = \int dx\, f(x)$. A functional is conventionally denoted by square brackets.

Equation (7.2) shows that the local density $\rho(\mathbf{r})$ is given by the functional derivative of Ω with respect to $\psi(\mathbf{r})$,

$$\rho(\mathbf{r}) = -\frac{\delta \Omega[\psi]}{\delta V_{ext}(\mathbf{r})} \tag{7.3}$$

Just as the macroscopic density ρ is generally a more practical thermodynamic variable than μ in a homogeneous system, so the local density $\rho(\mathbf{r})$ is in many cases a more convenient local variable than $\psi(\mathbf{r})$. This is particularly true in the context of phase transitions, in which inhomogeneities arise spontaneously in the absence of an external potential. Switching from μ to ρ is achieved by a generalized Legendre transformation. The standard transformation (2.19), which applies to homogeneous systems, generalizes to

$$\Omega[\psi] = F[\rho] - \int \rho(\mathbf{r})\psi(\mathbf{r})\, d\mathbf{r} \tag{7.4}$$

where $F[\rho]$ is a functional of the local density $\rho(\mathbf{r})$, which is the local variable conjugate to $\psi(\mathbf{r})$ (just as N is conjugate to μ). $F[\rho]$ is a (Helmholtz) free energy functional, with no explicit dependence on the external potential $V_{ext}(\mathbf{r})$. F is therefore, and this is the important point, an intrinsic property of the fluid, which can be discussed without any reference to an external perturbation. For a fixed 'local' chemical potential $\psi(\mathbf{r})$, $\rho(\mathbf{r})$ plays the role of an order parameter, and the equilibrium local density is that which minimizes the right-hand side of equation (7.4), i.e. the free energy obeys the following variational principle

$$\frac{\delta}{\delta \rho(\mathbf{r})}\left\{ F[\rho] - \int \rho(\mathbf{r})\psi(\mathbf{r})\, d\mathbf{r} \right\} = 0 \tag{7.5}$$

which expresses the fact that the equilibrium density must satisfy the equation

$$\frac{\delta F[\rho]}{\delta \rho(\mathbf{r})} + V_{ext}(\mathbf{r}) = \mu \tag{7.6}$$

For a fixed number of particles, the chemical potential is in fact the Lagrange multiplier associated with the constraint $\int \rho(\mathbf{r})\, d\mathbf{r} = N$.

This key variational principle was first established in the 1960s by Hohenberg and Kohn[2], in the context of the inhomogeneous electron gas at zero temperature ($F[\rho]$ reduces then to the ground state energy functional $E[\rho]$), and later generalized to finite temperature by Mermin[3].

Equation (7.6) allows, in principle, a determination of the density profile $\rho(\mathbf{r})$ for any external potential. However, the functional $F[\rho]$, being a property of an interacting many-body system, is in general a highly non-trivial object, so that the apparent simplicity of equation (7.6) is rather illusory. $F[\rho]$ is exactly known

[2] P.C. Hohenberg and W. Kohn, *Phys. Rev. B* **136**, 864 (1964).
[3] N. Mermin, *Phys. Rev. A* **137**, 1441 (1965).

only for an ideal gas of non-interacting atoms, where it generalizes the result (2.40), valid for a homogeneous gas, in the form

$$F_{id}[\rho] = k_B T \int d^3 \mathbf{r} \, \rho(\mathbf{r})(\ln(\rho(\mathbf{r})\Lambda^3) - 1) \tag{7.7}$$

If equation (7.7) is inserted into equation (7.6), the usual Boltzmann distribution of non-interacting particles in an external potential, $\rho(\mathbf{r}) \sim \exp(-V_{ext}(\mathbf{r})/k_B T)$, is recovered. In more realistic situations, only approximations to $F[\rho]$ are known, some of which will be discussed in the next section.

7.2 Some approximate functionals

The quadratic or Gaussian approximation

Practical approximations for the free energy functionals $F[\rho]$ can be obtained only in certain limiting cases. We first consider the case of a density which is only weakly inhomogeneous, i.e. $\rho(\mathbf{r}) = \rho_0 + \delta\rho(\mathbf{r})$, with $\rho_0 = N/V$ the average uniform density and $\delta\rho$ a weak modulation, $\delta\rho(\mathbf{r})/\rho \ll 1$ everywhere. In this case one can expand the functional to quadratic order in the form

$$F[\rho_0 + \delta\rho(\mathbf{r})] = V f_0(\rho_0) + \frac{1}{2} \int \int d\mathbf{r} \, d\mathbf{r}' \, A(\mathbf{r}, \mathbf{r}') \, \delta\rho(\mathbf{r}) \, \delta\rho(\mathbf{r}') + \mathcal{O}(\delta\rho^3) \tag{7.8}$$

$f_0(\rho)$ is the free energy density of a uniform system. The absence of a linear term in the expansion stems from the fact that, in the absence of any external potential, a uniform density ($\delta\rho = 0$) must be a minimum of the free energy functional. The function $A(\mathbf{r}, \mathbf{r}')$ is a property of the uniform system, and invariance under translation implies that it only depends on the difference $\mathbf{r} - \mathbf{r}'$. It is convenient to rewrite equation (7.8) in terms of the Fourier components $\delta\rho(\mathbf{k})$ of the density modulation, as

$$F[\rho_0 + \delta\rho(\mathbf{r})] = V f_0(\rho_0) + \frac{1}{2} \sum_k A(\mathbf{k}) \, \delta\rho(\mathbf{k}) \, \delta\rho(-\mathbf{k}) + \mathcal{O}(\delta\rho^3) \tag{7.9}$$

In the presence of a weak external potential, equations (7.6) and (7.9) can be combined to show that the density response is proportional to the external potential and given by

$$A(\mathbf{k}) \, \delta\rho(\mathbf{k}) = -V_{ext}(\mathbf{k}) \tag{7.10}$$

Comparing with equation (3.65), one can relate $A(\mathbf{k})$ to the structure factor and rewrite equation (7.9) in the form

$$F[\rho_0 + \delta\rho(\mathbf{r})] = V f_0(\rho_0) + \frac{k_B T}{2\rho_0 V} \sum_k \frac{\delta\rho(\mathbf{k}) \, \delta\rho(-\mathbf{k})}{S(k)} + \mathcal{O}(\delta\rho^3) \tag{7.11}$$

The free energy cost in creating a density modulation with wavevector $\mathbf{k}$ is thus proportional to $1/S(k)$. For small fluctuations, the statistical weight of a fluctuation is Gaussian, proportional to $\exp - [\delta\rho(\mathbf{k}) \, \delta\rho(-\mathbf{k})/2\rho_0 V S(k)]$.

The local density and square gradient approximations

Another case in which the free energy functional can be expressed in a reasonably simple form is that of long wavelength inhomogeneities, such that $\nabla \rho / \rho \ll 1/\xi$ where ξ is the typical range of density correlations in the bulk fluid (typically ξ is the range of the oscillations in the pair correlation function for a simple fluid). The procedure in that case is to write the functional F in the form

$$F[\rho] = \int d\mathbf{r} \, f(\rho(\mathbf{r}), \nabla \rho, \nabla \nabla \rho ...) \tag{7.12}$$

where f is now a *function* of ρ and all its spatial derivatives. Assuming that the gradient is small, one can expand in the form

$$F[\rho] = \int d\mathbf{r} \left[f_0(\rho(\mathbf{r})) + \frac{k_B T \xi_0(\rho(\mathbf{r}))^2}{2\rho} (\nabla \rho(\mathbf{r}))^2 + \cdots \right] \tag{7.13}$$

Truncating the expansion after the first term $f_0(\rho(\mathbf{r}))$ in the expansion, one obtains the so-called local density approximation (LDA), which simply treats the inhomogeneous system by adding up the contributions from different, nearly homogeneous, regions. The first correction to this approximation is *quadratic* in the gradient [4]. For convenience, the coefficient of this quadratic term has been written in terms of a density dependent length $\xi_0(\rho)$, which may in fact be related to the direct correlation function of the homogeneous fluid. Equation (7.13) is known as the 'square gradient' approximation.

Miscellaneous

Ornstein–Zernike form of the structure factor
The square gradient and quadratic approximations can be combined in the case of small *and* long wavelength inhomogeneities. For small inhomogeneities, equation (7.13) can be expanded in the form

$$F[\rho] = V f_0(\rho_0) + \frac{k_B T}{2V} \sum_k \left(\frac{d^2 f_0}{d\rho_0^2} + \frac{k_B T k^2 \xi_0(\rho_0)^2}{\rho_0} \right) \delta\rho(\mathbf{k}) \, \delta\rho(-\mathbf{k}) + \mathcal{O}(\delta\rho^3) \tag{7.14}$$

Comparing with equation (7.11), one is led to the conclusion that the structure factor can be expressed, for small enough wavevectors, in the so called *Ornstein–Zernike* form

$$S(k) = \frac{S(0)}{1 + k^2 \xi^2} \quad \text{with} \quad S(0)^{-1} = \frac{\rho_0}{k_B T} \frac{d^2 f_0}{d\rho^2} \quad \text{and} \quad \xi^2 = S(0)\xi_0(\rho_0)^2 \tag{7.15}$$

[4] A linear correction would not be invariant under the $\nabla \rho \to -\nabla \rho$ transformation (mirror reflection symmetry). A correction of the form $\nabla^2 \rho$ is of the same order as the $(\nabla \rho)^2$ term in equation (7.13), but an integration by parts shows that the two terms can be transformed into each other.

This form is especially useful for studying critical phenomena, as discussed earlier in section 5.1.

Long-range forces

The square gradient approximation, equation (7.13), was derived under the assumption that any functional could be written as a local functional of the density and its derivatives. Although this might seem at first a natural assumption, the simple example of an electrostatic energy functional,

$$\frac{1}{2} \int \int d\mathbf{r} \, d\mathbf{r}' \, \frac{\rho(\mathbf{r})\rho(\mathbf{r}')}{|\mathbf{r} - \mathbf{r}'|} \tag{7.16}$$

shows that this assumption is incorrect in the case of long-range interactions. If one tries to reduce equation (7.16) to a local form by expanding $\rho(\mathbf{r}')$ in the form $\rho(\mathbf{r}') = \rho(\mathbf{r}) + (\mathbf{r}' - \mathbf{r}) \cdot \nabla\rho(\mathbf{r})$, it is easily seen that the coefficients in the resulting functional are all divergent. Hence a square gradient approximation – or any other kind of approximation involving only a local density and its derivatives – cannot be valid for systems with slowly decaying interactions. The strategy for dealing with such systems will be to isolate the long-range interaction between molecules, $v_{lr}(r)$, and to write the free energy functional in the form

$$F[\rho] = F_{sr}[\rho] + \frac{1}{2} \int \int d^3\mathbf{r} \, d^3\mathbf{r}' \, \rho(\mathbf{r}) v_{lr}(\mathbf{r} - \mathbf{r}')\rho(\mathbf{r}') \tag{7.17}$$

where F_{sr} is a functional that correctly describes the short-range part of the correlations. An example of such an approach is the functional

$$F[\rho] = k_B T \int d^3\mathbf{r} \, \rho(\mathbf{r})(\ln(\rho(\mathbf{r})\Lambda^3) - 1) + \frac{1}{2} \int \int d^3\mathbf{r} \, d^3\mathbf{r}' \, \rho(\mathbf{r}) v_{lr}(\mathbf{r} - \mathbf{r}')\rho(\mathbf{r}') \tag{7.18}$$

in which the short-range part has been replaced by the ideal gas free energy. This functional contains the physical ingredients of the Poisson–Boltzmann equation for charged fluids (section 7.6), namely an ideal gas approximation for the entropic part and a mean field treatment of the long-range interaction. If excluded volume effects are properly included on a local level, the functional (7.18) also forms the basis of the van der Waals approximation for neutral fluids (section 4.3).

Density functionals and mean field approximations

The basic theorem that underlies the use of density functionals, as expressed in equation (7.5), does not rely on any mean field approximation. Nevertheless, the *approximate* functionals such as those defined in equations (7.13) and (7.18), generally involve some kind of mean field approximation. As an example, the square gradient approximation, with the choice of a mean field expression for the free energy $f_0(\rho)$, corresponds exactly to the so-called Landau–Ginzburg approximation in the theory of phase transitions.

The obvious implication is that systems in which long wavelength fluctuations play an important role will not be properly described in this context. A reasonable approach in such cases will be to consider the approximate free energy functionals as providing a good description of short wavelength fluctuations, and to use them as *effective Hamiltonians* for the description of long wavelength fluctuations. Examples of this type will be encountered when dealing with interfacial fluctuations (section 8.1).

7.3 Application 1: the fluid–fluid interface

If a binary liquid mixture is taken below the phase transition point, the two phases are separated by interfacial regions in which the composition varies between those of the coexisting equilibrium phases. In this section, we consider the detailed structure of such a planar interfacial region, and show how the density functional approach allows a determination of the interfacial profile and the interfacial tension.

For simplicity, we consider a symmetric, incompressible mixture of two components, whose phase diagram was discussed in section 4.1. We denote by $\rho_A(\mathbf{r})$ the local number density of the A component. Thanks to the incompressibility condition, $\rho_A(\mathbf{r}) + \rho_B(\mathbf{r}) = \rho$, one density field is sufficient to describe the system. For a planar interface parallel to the xOy plane, the density profile $\rho_A(z)$ is a function of z only. We can therefore write the square gradient approximation free energy in the form

$$F[\rho_A] = L^2 \int dz \left[\frac{1}{2} m(\rho_A) \left(\frac{d\rho_A}{dz} \right)^2 + f(\rho_A) \right] \qquad (7.19)$$

Here f is the double well free energy that results in the phase diagram depicted in figure 4.2, $m(\rho)$ is a density and temperature dependent coefficient (in the notation of section 7.2, $m = k_B T \xi_0^2 / \rho_A$, see equation (7.13)), and L^2 is the interfacial area. In the vicinity of the critical point, we will use an approximate quartic form of this free energy (from which we substract the bulk contribution) in the form

$$f(\rho_A) = \frac{C}{2} \left(\rho_A - \rho_A^{(1)} \right)^2 \left(\rho_A - \rho_A^{(2)} \right)^2 \qquad (7.20)$$

where $\rho_A^{(1)}$ and $\rho_A^{(2)}$ denote the densities in the two coexisting phases, and C is a constant. Although this expression may not be quantitatively accurate, it captures the essential features of the free energy and allows tractable calculations.

The interfacial profile is obtained by minimizing the free energy (7.19) with respect to variations of $\rho_A(z)$, with fixed boundary conditions $\rho_A(z) \to \rho_A^{(1)}$ when $z \to -\infty$, and $\rho_A(z) \to \rho_A^{(2)}$ when $z \to +\infty$. The minimization yields the following differential equation

$$m(\rho_A)\frac{d^2\rho_A}{dz^2} + \frac{1}{2}\frac{dm}{d\rho_A}\left(\frac{d\rho_A}{dz}\right)^2 = \frac{df(\rho_A)}{d\rho_A} \tag{7.21}$$

Equation (7.21) can be integrated[5], yielding

$$\frac{1}{2}m(\rho_A)\left(\frac{d\rho_A}{dz}\right)^2 - f(\rho_A) = 0 \tag{7.22}$$

Inserting the result of the minimization, equation (7.22), in the free energy, and using ρ rather than z as the integration variable, one obtains a simple expression for the excess interfacial free energy of the equilibrium interfacial profile

$$\min_{\rho} F[\rho] = A\int_{\rho_A^{(1)}}^{\rho_A^{(2)}} d\rho\sqrt{2m(\rho)f(\rho)} \tag{7.23}$$

Hence the surface tension can be obtained without a detailed computation of the interfacial profile[6]. Using the quartic free energy (7.20), and assuming for simplicity that m is independent of ρ, one has

$$\gamma = \int_{\rho_A^{(1)}}^{\rho_A^{(2)}} d\rho\sqrt{2m(\rho)f(\rho)} = \sqrt{mC}\left(\rho_A^{(2)} - \rho_A^{(1)}\right)^3 \tag{7.24}$$

The interfacial tension behaves therefore as the cube of the density difference between the two phases. Close to the critical point, mean field theory predicts that $(\rho_A^{(2)} - \rho_A^{(1)}) \sim (T_c - T)^{1/2}$, and therefore $\gamma \sim (T_c - T)^{3/2}$. Although this mean field prediction correctly predicts the vanishing of γ near the critical point, the exponent is as usual slightly incorrect, the actual value being closer to 2 in three dimensions.

We now turn to the explicit determination of the interfacial profile, by solving the differential equation (7.22), for constant $m(\rho_A)$. The resulting profile is

$$\rho_A(z) = \frac{\rho_A^{(1)}}{1 + \exp(-z/\xi)} + \frac{\rho_A^{(2)}}{1 + \exp(z/\xi)}$$

$$= \frac{1}{2}\left(\rho_A^{(2)} + \rho_A^{(1)}\right) + \frac{1}{2}\left(\rho_A^{(2)} - \rho_A^{(1)}\right)\tanh(z/2\xi) \tag{7.25}$$

where $\xi = \sqrt{m/C}|\rho_A^{(2)} - \rho_A^{(1)}|^{-1}$. ξ defines the *interfacial width*, and diverges close to the critical point.

The square gradient approximation used in this section is valid only for density variations that take place over scales which are large compared to molecular sizes.

---

[5] By multiplying both sides by $\frac{d\rho_A}{dz}$, in analogy with the integration of Newton's equation to derive energy conservation in mechanical systems.

[6] It is often the case that the value of the density functional *at the minimum* can be expressed much more simply than the original functional.

Hence, equation (7.26) shows that the approach is consistent only for large ξ, i.e. near the critical point. Nevertheless, the 'hyperbolic tangent' profile described by (7.26) will be a reasonable approximation even relatively far from T_c.

Finally, it is interesting to note that the interfacial width ξ is identical to the correlation length of the density fluctuations, that can be obtained from the Ornstein–Zernike form of the structure factor. Transient fluctuations in the density involve the same length scale as interface formation.

7.4 Application 2: the adsorbed polymer layer

We have seen in section 6.1 that a solute will in general have a preferential adsorption at the interface between two fluid phases. The same is of course true at a fluid–solid interface. For small solute molecules, the adsorption depends very strongly on details of the molecular interactions, as was already discussed in the case of surfactants, and the adsorbed molecules are confined to a thin interfacial layer. The situation is completely different with long polymers. In this case the large scale introduced by the molecules allows a coarse-grained description, and the adsorbed layer has interesting universal features. Adsorption of polymers is important in colloidal science, since the adsorbed layer will strongly modify the interactions between colloids and ensure colloid stabilization (see section 7.7)

We consider a semi-dilute polymer solution in the vicinity of a solid wall, located in the plane $z = 0$. Far from the wall, the monomer concentration is ρ_0, and the chemical potential of the monomer is μ_0. The wall attracts the monomer with a short-range potential. As usual, we can use the incompressibility condition to ignore the solvent and write the excess grand potential associated with the interface in the form

$$\Omega_{ex} = \Omega_s + A \int_0^\infty dz \left(\frac{m(\rho)}{2} \left(\frac{d\rho}{dz} \right)^2 + f(\rho) - \mu_0 \rho(z) + \Pi_0 \right) \qquad (7.26)$$

Using the notation of section 3.10, we take $m(\rho) = k_B T b^2 / 24\rho$ and $f(\rho)/k_B T = \frac{1}{2} v \rho^2$. Equation (7.26) then reproduces the RPA structure factor of the solution, equation (3.150)[7]. A is the interfacial area, and Π_0 the osmotic pressure in the homogeneous solution. Ω_s is the contribution to the grand potential from the surface, which we write in the form

$$\Omega_s = \gamma_0 A - k_B T A \gamma_1 b \rho_s \qquad (7.27)$$

γ_0 is the interfacial tension of the interface in the absence of polymer, and the second term, in which γ_1 is a dimensionless constant and $\rho_s = \rho(z = 0)$, is an

---

[7] This means that an analysis of this square gradient free energy along the lines of section 5.1 will yield the RPA structure factor.

estimate of the energy gain from polymer adsorption at the solid wall. The calculation proceeds along lines very similar to those of section 7.3. Minimization of equation (7.26) at fixed ρ_s yields the differential equation, equivalent to (7.22)

$$\frac{1}{2}m(\rho)\left(\frac{d\rho}{dz}\right)^2 + u(\rho) = 0 \quad \text{with} \quad u(\rho) = -\frac{k_B T}{2}v\rho^2 + \mu_0\rho - \Pi_0 = -\frac{k_B T}{2}v(\rho - \rho_0)^2 \tag{7.28}$$

One can then express the integral in (7.26) at the minimum in a form similar to that in equation (7.24), and minimize the total expression (including the Ω_s term) with respect to ρ_s. This yields

$$\gamma_1 b k_B T = \sqrt{2m(\rho_s)u(\rho_s)} = m(\rho_s)\left.\frac{d\rho}{dz}\right|_{z=0} \tag{7.29}$$

where the second equality is obtained using (7.28). Solving equation (7.28) with the initial condition (7.29) at the wall and the condition $\rho(z) \to \rho_0$ far into the liquid, one obtains

$$\rho(z) = \rho_0 \coth^2\left(\frac{z + z_0}{\xi}\right) \tag{7.30}$$

where $\xi = b/\sqrt{3v\rho_0}$ is the correlation length in the solution (see section 3.10), and z_0 is fixed by the condition

$$b = 6\gamma_1\xi \cosh\frac{z_0}{\xi} \sinh\frac{z_0}{\xi} \tag{7.31}$$

The case of strong adsorption, $\frac{b}{\gamma_1} \ll \xi$, is particularly interesting. In this case $z_0 \simeq b/(6\gamma_1)$ and the concentration profile can be separated into three different regions, as illustrated in figure 7.1:

- a proximal region $z < z_0$, where $\rho(z)$ is approximately constant, equal to γ_1/v;
- a scaling region $z_0 < z < \xi$, in which the profile behaves as a power law, $\rho(z) \sim \frac{a}{3v(z+z_0)^2}$; in this region, the concentration profile is independent of ρ_0 or γ_1;
- a cross-over region, $\xi < z$, in which the concentration crosses over to the bulk value ρ_0.

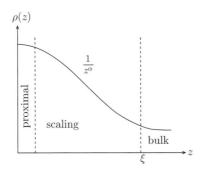

$\rho(z)$

proximal

scaling

$\frac{1}{z^\alpha}$

bulk

ξ

z

Figure 7.1. Schematic density profile of a polymer layer adsorbed to a planar solid substrate, exhibiting the proximal, scaling and bulk regions. ξ is the bulk correlation length.

Remarkably, the adsorbed quantity, $\Gamma = \int_0^\infty (\rho(z) - \rho_0)\,dz$, is independent of ρ_0. The main contribution comes from the scaling region and is approximately $\Gamma = \gamma_1/b^2$.

Although the approach presented in this section is purely mean field, the main features are recovered in non-mean-field treatments of the problem. In particular, the existence of a scaling region, extending over a distance ξ (the bulk correlation length), in which the concentration has a power law decay, is a robust feature. The fact that the adsorbed quantity is independent of ρ_0 is also recovered in non-mean-field theories. The simplest non-mean-field treatment of the scaling region is due to de Gennes. His argument, essentially a dimensional one, is that the concentration ρ at a distance z from the wall is such that the local correlation length $\xi(\rho(z))$ is equal to z, which is the only relevant length in the problem. The local correlation length is defined as the correlation length in a homogeneous solution with concentration $c(z)$. Using the mean field expression for $\xi(\rho)$, $\xi \sim \rho^{-1/2}$, the $1/z^2$ scaling obtained above is recovered. If, however, the semi-dilute correlation length $\xi \sim \rho^{-3/4}$ is used, this yields a scaling $\rho(z) \sim z^{-4/3}$, which is in fact the correct one, as checked by experiments using neutron or X-ray reflectivity.

7.5　Application 3: self-assembly of copolymers

Block copolymers are macromolecules composed of sequences, or blocks, of chemically distinct monomers. Such polymers are prepared by anionic polymerization, with sequential addition of chemically different monomers to a reactive chain ('living' polymer). In this short section, we restrict our considerations to the case of diblock molecules, with a sequence of type A monomers followed by a sequence of type B monomers. For the sake of simplicity, each block will be assumed to have similar characteristics (monomer size, persistence length). The chemical difference between the two types of monomer is accounted for by a non-zero χ parameter in the language of the Flory–Huggins theory (see section 4.4). For chains of M monomers, the length of the A block will be denoted by fM $(0 < f < 1)$ and the length of the B block is $(1 - f)M$.

Although a block copolymer system is made of two types of monomer, it is important to realize that a block copolymer melt is, nevertheless, a one-component system from a thermodynamic point of view, since it contains only one species (the copolymer chain). While f will be referred to as the composition, it is not a thermodynamic variable, but a fixed parameter, characterizing the molecular architecture.

In a binary mixture of A and B chains with lengths fM and $(1 - f)M$, a phase separation would be observed when lowering the temperature, when χM reaches some critical value. In the one-component system considered here, macroscopic separation is impossible (the two blocks being chemically linked) and will be replaced by a so-called *microphase separation transition* (MST), in which the segregation between the two types of monomer takes place on the molecular scale. Such a segregation is achieved through the formation of ordered *mesophases* with various morphologies. The simplest case, which is realized for values of f close to $1/2$, is that of a lamellar phase (figure 7.2), in which A-rich lamellae alternate with B-rich lamellae. Much more complex morphologies can arise for less symmetric cases, as illustrated in figure 7.3.

A detailed consideration of the various morphologies that are found in a di-block copolymer system is beyond the scope of this book. We concentrate here on the case of nearly symmetric diblocks, that give rise to a microphase separation towards a lamellar phase. Our aim is to understand the nature of the transition

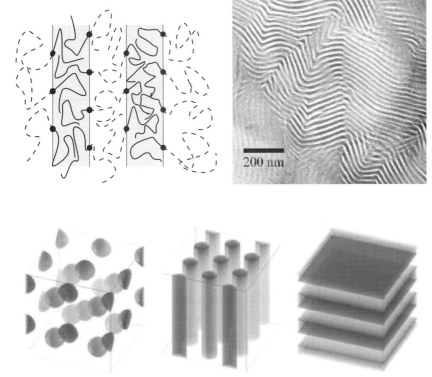

200 nm

Figure 7.2. Left: molecular organization in the lamellar phase. The incompatible A and B blocks of polymer chains are represented by full and dashed curves respectively. The dots correspond to the linkage points of A and B chains. Right: electron micrograph of a lamellar phase in a polystyrene-polybutadiene copolymer. (Courtesy of Professor E.L. Thomas, MIT.)

Figure 7.3. Several morphologies observed in diblock copolymers. Left: cubic phase, observed at small f; the A parts of the chains form ordered micelles, ordered on an FCC lattice and immersed in a matrix formed by the B parts. Middle: hexagonal phase, observed for intermediate values of f; in this phase, A monomers form cylinders organized on a triangular lattice. Right: lamellar phase, observed for f close to $1/2$.

from the disordered melt at high temperature towards the lamellar phase, and the evolution of the lamellar spacing with temperature. Experimentally, the transition is characterized by an increase of the scattering at a wavevector corresponding to the periodicity of the lamellae, similar to the Bragg peaks of crystalline solids. The transition can also be monitored by measuring the elastic properties; while the melt behaves like a viscous liquid (equation (1.7)), the ordered phase has a finite (albeit small) elastic modulus and behaves like a solid. Note that the transition does not spontaneously give rise to well ordered samples like the one shown in figure 7.2. Such samples are obtained by subjecting the lamellar phase to a shear stress, inducing plastic flow that aligns the lamellae.

The simplest theoretical description of the transition is obtained by combining the Flory–Huggins (section 4.4) and RPA (section 3.10) approaches. The RPA equations are somewhat more complex than for polymer solutions, since even in the non-interacting systems correlations between the two types of species are present (i.e. the partial structure factor $S_{AB}^{(0)}$ does not vanish). These correlations are obviously related to the chemical link that forces the two subchains to remain within one chain size of each other. For external potentials V_A and V_B acting, respectively, on the A and B monomers, application of the RPA approach leads to

$$\rho_A(\mathbf{k}) = -\frac{\rho S_{AA}^{(0)}(\mathbf{k})}{k_B T}(V_A(\mathbf{k}) - \epsilon_{AA}b^3\rho_A(\mathbf{k}) - \epsilon_{AB}b^3\rho_B(\mathbf{k}) + \Phi(\mathbf{k}))$$

$$-\frac{\rho S_{AB}^{(0)}(\mathbf{k})}{k_B T}(V_B(\mathbf{k}) - \epsilon_{AB}b^3\rho_A(\mathbf{k}) - \epsilon_{BB}b^3\rho_B(\mathbf{k}) + \Phi(\mathbf{k})) \quad (7.32)$$

$$\rho_B(\mathbf{k}) = -\frac{\rho S_{BA}^{(0)}(\mathbf{k})}{k_B T}(V_A(\mathbf{k}) - \epsilon_{AA}b^3\rho_A(\mathbf{k}) - \epsilon_{AB}b^3\rho_B(\mathbf{k}) + \Phi(\mathbf{k}))$$

$$-\frac{\rho S_{BB}^{(0)}(\mathbf{k})}{k_B T}(V_B(\mathbf{k}) - \epsilon_{AB}b^3\rho_A(\mathbf{k}) - \epsilon_{BB}b^3\rho_B(\mathbf{k}) + \Phi(\mathbf{k})) \quad (7.33)$$

Here b^3 is the volume per monomer. The notation in (7.32) and (7.33) is similar to that used in section 4.4. In deriving equations (7.32) and (7.33), the intermolecular interaction has been divided into two parts. The part related to the attraction between monomers (involving the parameters ϵ_{ij}) has been expressed explicitly, in the spirit of section 3.10. The part related to repulsive interactions (the excluded volume), which is assumed to be identical for both species, has been kept in the form of an unspecified potential Φ. Instead of trying to make these excluded volume interactions explicit, we will simply assume that their role is to ensure the incompressibility of the system. Therefore Φ will be obtained by adding to (7.32) and (7.33) a third equation that enforces the incompressibility constraint

$$\rho_A(\mathbf{k}) + \rho_B(\mathbf{k}) = 0 \quad (7.34)$$

The inversion of equations (7.32) and (7.33), supplemented by (7.34), is straightforward, although somewhat lengthy. The final result for the RPA structure factor of the incompressible copolymer melt can be cast in the form

$$\frac{1}{S(\mathbf{k})} = \frac{S_{AA}^{(0)}(\mathbf{k}) + S_{BB}^{(0)}(\mathbf{k}) + 2S_{AB}^{(0)}(\mathbf{k})}{S_{AA}^{(0)}(\mathbf{k})S_{BB}^{(0)}(\mathbf{k}) - (S_{AB}^{(0)}(\mathbf{k}))^2} - 2\chi(T) \tag{7.35}$$

where $S(\mathbf{k}) = S_{AA}(\mathbf{k}) = S_{BB}(\mathbf{k}) = -S_{AB}(\mathbf{k})$ is, as in a blend, the concentration–concentration structure factor (that gives rise to the contrast in a small-angle scattering experiment). $\chi(T) = (\epsilon_{AA} + \epsilon_{BB} - 2\epsilon_{AB})/2k_B T$ is the usual Flory–Huggins parameter. In practice, this parameter can be obtained for a given copolymer from a knowledge of the phase diagram of the corresponding polymer blend. In order to use equation (7.35), it is necessary to compute the structure factors in the non-interacting system. These are obtained within the Gaussian description of the non-interacting chain, in its continuous version, e.g.

$$S_{AB}^{(0)}(k) = M \int_0^f ds \int_f^1 ds' \exp(-k^2 b^2 |s - s'|/6) \tag{7.36}$$

which is simply a generalization of equation (3.116) for a Gaussian chain. A detailed calculation shows that the concentration–concentration structure factor of the non-interacting, incompressible system ($\chi = 0$), i.e. the inverse of the first term on the right-hand side of equation (7.35), has a maximum of order M at a finite wavevector $k^* = x(f)/(bM^{1/2})$, where $x(f)$ is a dimensionless function such that $x(1/2) = 4.6$. This maximum is of the form $S_{\max}(k^*) = Ms(f)$, where $s(f)$ is again a dimensionless function, with $s(1/2) \simeq 10$. Again, this maximum is related to the existence of correlations between the two halves of the chain. At large wavelength the system is a one-component system, perfectly homogeneous with respect to concentration fluctuations, so that $S(0) = 0$. For a large wavevector, $S(k)$ also goes to zero, as in a polymer melt[8]. Between these two limits a maximum must be observed, at a wavevector characteristic of the inverse of the chain size.

The obvious implication is that the inverse of the total structure factor will have a minimum equal to $1/Ms(f) - 2\chi(T)$ at $k = k^*$. When $\chi(T)$ is increased, $1/Ms(f) - 2\chi(T)$ eventually vanishes, indicating that the concentration fluctuations at $k = k^*$ are divergent. When this happens, the linear response analysis of section 3.6 implies that the system is unstable with respect to concentration fluctuations at wavevector k^*, and will spontaneously undergo a transition towards a mesophase with a non-uniform concentration. The condition $1/Ms(f) - 2\chi(T) = 0$ therefore defines a spinodal line in the (f, χ) plane.

In order actually to predict the phase diagram of diblock copolymers, it is necessary to go beyond this simple analysis of the structure factor in the isotropic melt, and to compute the free energy of modulated phases with various symmetries. The simplest analysis[9] involves a Landau expansion of the free energy

---

[8] More precisely to unity, if the discreteness of the beads is taken into account.

[9] L. Leibler, *Macromolecules* 13, 602 (1980).

in terms of the Fourier components of the concentration. According to equation (7.9), the quadratic term in this expansion is determined by the inverse of the structure factor, and a diverging $S(k^*)$ will signal a spinodal line, i.e. the instability of the disordered phase. The phase diagram is calculated by comparing the free energies of ordered phases with different symmetries, which requires a computation of the higher order (cubic and quartic) terms in the free energy.

In the simplest case of symmetric diblocks ($f = 1/2$), and of a lamellar structure, the Landau expansion in terms of the first Fourier component of the density has no cubic term for symmetry reasons, and the quartic term turns out to be positive. The theory therefore predicts a second-order transition from the isotropic melt to the lamellar phase. A more detailed study [10] shows that when fluctuations beyond mean field are taken into account, a weakly first-order transition is actually predicted, in agreement with experiments.

The approach based on the Landau expansion, as described above, is appropriate in the so-called weak segregation limit, in which the concentration modulation is nearly sinusoidal – only wavevectors near the peak of $S(k)$ contribute – and has a small amplitude. This is in general true near the transition lines (crystallization of copolymers is often described as 'weak crystallization'). However, as the temperature decreases and the incompatibility parameter χ increases, the interfaces between the segregated regions tend to sharpen, and a correct description of the concentration profile would involve more and more harmonics in the Fourier expansion, and higher order terms in the Landau expansion, making the approach impractical. The determination of morphologies in this 'strong segregation' limit is therefore based on a different approach, in which the interfaces are treated as infinitely sharp, with a mass-independent surface free energy γ. The total free energy is the sum of the stretching free energy of the two halves of the chain, which are of course tethered to the interfaces, and of the interfacial free energy. As an example of this approach, we can estimate the evolution of the lamellar spacing in the lamellar phase as a function of molecular weight and temperature. If we denote by d the lamellar spacing and by A the area of an interface, the total interfacial energy is

$$E_{\text{interf}} = \gamma A \left(\frac{V}{dA} \right) \tag{7.37}$$

where the term in parentheses is the total number of lamellae in a system of volume V. The stretching energy is of the order of

$$E_{\text{stretching}}/k_B T = N_{\text{chains}} \frac{d^2}{Mb^2} = \left(\frac{\rho V}{M} \right) \frac{d^2}{Mb^2} \tag{7.38}$$

Minimizing the sum $E_{\text{interf}} + E_{\text{stretching}}$ with respect to the spacing d, one obtains

$$d \sim M^{2/3} (\gamma b^2 / \rho k_B T)^{1/3} \sim M^{2/3} \chi^{1/6} b \tag{7.39}$$

[10] G. Fredrickson, E. Helfand, *J. Chem. Phys.* **87**, 697 (1987).

where we have used $\rho = 1/b^3$ and $\gamma \simeq \chi^{1/2} k_B T/b^2$. The latter expression for the surface tension is appropriate for a blend at low enough temperatures, so that the entropic term in the free energy can be neglected. When this is the case, use of equation (7.24) with a bulk free energy density proportional to $\chi(T) k_B T \rho_A (\rho - \rho_A) V$ immediately yields the above scaling for the surface tension. The scaling of d with M and χ is found to be consistent with experimental and numerical results.

7.6 Application 4: electric double-layers

Macromolecules, lipid bilayer membranes or colloidal particles dissolved or suspended in water will almost invariably acquire an overall electric charge by releasing anions or cations through ionization or dissociation of polar groups. For example, carboxylic groups will release protons through the dissociation reaction:

$$COOH \rightarrow COO^- + H^+$$

leaving behind a negative charge on the macromolecule or on the surface of membranes or colloids. The released ions in solution are called *counterions*, since they carry an electric charge of opposite sign to that of the macromolecule (generally referred to as a polyelectrolyte) or surface. These counterions are electrostatically attracted to the charged macromolecules or surfaces, but in order to increase their entropy, they will also tend to move thermally away from the latter into the bulk of the aqueous solution. The balance between the two opposing effects leads to the formation of an electric *double-layer*. Moreover, the aqueous solvent generally contains anions and cations from dissolved electrolytes (e.g. salts), if only H_3O^+ and OH^- ions resulting from the dissociation of pure water, at pH 7. Ions carrying a charge of the same sign as the polyelectrolyte or charged surface are called *coions*, and are repelled by the latter; ions of opposite charge are again counterions, and both coions and counterions contribute to the electric double-layer.

In this section we shall restrict our attention to two simple geometries, namely infinite planar surfaces, as a crude description of rigid membranes or lamellar colloids (e.g. clays), and infinitely long cylinders, appropriate for the description of stiff polyelectrolytes (like DNA). The key problem will be to determine the thickness λ of the electric double-layer, as a function of surface charge and electrolyte concentration or, in a more detailed description, the density profiles $\rho_\alpha(\mathbf{r})$ of the various ionic species, and the resulting electrostatic potential $\psi(\mathbf{r})$ near the charged surface. Due to the symmetry of the planar and cylindrical geometries, the electric double-layers reduce to a diffuse slab (planar case) or sheath (cylindrical case) with density profiles depending on a single variable.

On the scale of the thickness λ, which will turn out to be much larger than molecular dimensions, the discrete nature of the solvent may be ignored, and the

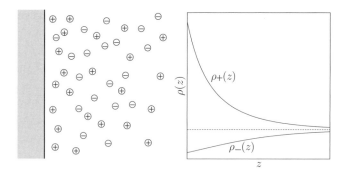

Figure 7.4. Schematic representation of an electric double-layer near a negatively charged planar surface. Counterions and coions are represented as small circles. The right-hand side shows typical counterion and coion density profiles, $\rho_+(z)$ and $\rho_-(z)$, as functions of distance z from the plane.

latter will act as a mere dielectric continuum, characterized by its macroscopic permittivity ϵ ($\epsilon = 78$ for water at room temperature); this simplification is akin to that used in the 'primitive model' of simple electrolytes (section 3.10). We will successively consider electric double-layers near a uniformly charged plane, between two parallel planes, and around an infinite, uniformly charged cylinder. Double-layers around spherical colloidal particles will be considered in the next section.

A single charged plane

Consider an infinite planar surface separating a dielectric medium (e.g. glass, or polystyrene) extending to the left, from an ionic solution on the right, as shown in figure 7.4. The surface carries a uniform charge density σ (charge per unit area), which will be assumed to be negative. For the sake of simplicity, the positive counterions and negative coions will be assumed to be monovalent (e.g. Na^+ and Cl^-). The corresponding density profiles $\rho_+(z)$ and $\rho_-(z)$ depend only on z, the coordinate orthogonal to the plane; the microions cannot penetrate to the left of the planar surface, so that $\rho_\pm(z < 0) = 0$.

To compute the density profiles approximately, we adopt a mean field free energy functional of the form (7.18) which neglects correlations between microions. The latter behave as an ideal gas of independent point charges moving in a self-consistent electric potential $\psi(z)$. The corresponding free energy functional (per unit area) is the sum of ideal and Coulomb terms:

$$F[\rho_+(z), \rho_-(z)] = F_{id}[\rho_+(z)] + F_{id}[\rho_-(z)] + F_c[\rho_c(z)]$$

$$= \sum_{\alpha=+,-} k_B T \int_0^\infty \rho_\alpha(z) \left(\ln(\rho_\alpha(z)\Lambda_\alpha^3) - 1\right) dz + \frac{e}{2} \int_0^\infty \rho_c(z)\psi(z) \, dz$$

$$(7.40)$$

where $e\rho_c(z) = e(\rho_+(z) - \rho_-(z))$ is the charge density, which is related to the local electrostatic potential $\psi(z)$ by Poisson's equation:

$$\frac{d^2\psi(z)}{dz^2} = -\frac{e}{\epsilon_0\epsilon}\rho_c(z) \qquad (7.41)$$

The variational principle (7.5) leads now to the following set of two coupled Euler–Lagrange equations:

$$k_B T \log(\rho_\alpha(z)\Lambda_\alpha^3) \pm e\psi(z) = \mu_\alpha \qquad \alpha = +, - \tag{7.42}$$

In taking the functional derivatives of (7.40), it must be remembered that ψ is related to ρ_c by the linear Poisson equation (7.41), so that the Coulomb term in the free energy functional is in fact quadratic in $\rho_c(z)$, as in equation (7.18), with the potential v_{lr} replaced by the Green's function associated with the differential equations (7.41) for given boundary conditions on the charged surface [11]. Exponentiating both sides of (7.42), one arrives at the Boltzmann distributions:

$$\rho_\pm(z) = \rho_\pm \exp(\mp\beta e\psi(z)) \tag{7.43}$$

ρ_+ and ρ_- are the uniform (macroscopic) counterion and coion number concentrations far from the charged surface, where the potential goes to zero; due to charge neutrality in the bulk, $\rho_+ = \rho_- = \rho_0$. Overall charge neutrality requires:

$$e \int_0^\infty \rho_c(z)\, dz = -\sigma \tag{7.44}$$

Substitution of (7.43) into (7.41) leads to the Poisson–Boltzmann (PB) equation:

$$\frac{d^2\psi(z)}{dz^2} = \frac{2e\rho_0}{\epsilon_0\epsilon} \sinh(\beta e\psi(z)) \tag{7.45}$$

This non-linear second-order differential equation for ψ must be solved subject to the two boundary conditions:

$$\lim_{z\to\infty} \frac{d\psi}{dz} = 0 \qquad -\frac{d\psi}{dz}\bigg|_{z=0} = \frac{\sigma}{\epsilon_0\epsilon} \tag{7.46}$$

The 'constant charge' boundary condition (7.46) is physically reasonable for most colloidal systems. Under some circumstances a 'constant potential' boundary condition, or an intermediate 'self-regulating' boundary condition may be more appropriate [12].

Consider now the local number concentration of microions, $\rho_N(z) = \rho_+(z) + \rho_-(z)$. Its gradient is easily calculated from (7.43) and (7.45):

$$\frac{d\rho_N(z)}{dz} = \rho_0\beta e \frac{d\psi(z)}{dz} [\exp(\beta e\psi(z)) - \exp(-\beta e\psi(z))]$$

$$= \epsilon_0\epsilon\beta \frac{d\psi(z)}{dz} \frac{d^2\psi(z)}{dz^2} = \frac{\epsilon_0\epsilon\beta}{2} \frac{d}{dz} \left(\frac{d\psi(z)}{dz}\right)^2 \tag{7.47}$$

[11] A detailed introduction to electrostatics and Green's functions may be found, e.g. in J. D. Jackson, *Classical Electrodynamics*, 3rd edn., Wiley, New York, 1999.
[12] B.W. Ninham and V. A. Parsegian, *J. Theor. Biol.* **31**, 405 (1971).

Integration of both sides of (7.47) from z to ∞ leads to the following relation between the local microion concentration and the local electric field $E(z) = -\frac{d\psi}{dz}$:

$$k_B T(\rho_N(z) - 2\rho_0) = \frac{\epsilon_0 \epsilon}{2} E(z)^2 \qquad (7.48)$$

Since the microions are non-interacting, the left-hand side is the difference in local osmotic pressure $\Pi(z) = k_B T \rho_N(z)$, between a point at z, and in the bulk (where $\rho_N(z) \to 2\rho_0$); the right-hand side is the electrostatic pressure at z, which vanishes in the bulk. Equation (7.48) thus expresses that a difference in osmotic pressure must be balanced by a difference in electrostatic pressure to ensure mechanical equilibrium.

Exercise: Show that the second boundary condition in (7.46) follows directly from (7.44) and (7.45). By differentiating both sides of equation (7.48), recover the hydrostatic equilibrium equation:

$$\frac{d\Pi}{dz} = F(z) \qquad (7.49)$$

where $F(z)$ is the local force per unit volume acting on the ionic solution; express $F(z)$ in terms of the local electric field.

Applying (7.48) at $z = 0$ and using the second boundary condition (7.46), one obtains an expression for the enhancement of the microion concentration at contact over the bulk concentration:

$$\rho_N(0) = 2\rho_0 + \frac{\sigma^2}{2\epsilon_0 \epsilon k_B T} \qquad (7.50)$$

The enhancement grows quadratically with σ, and becomes quite substantial for high surface charges. In fact at sufficiently high values of σ, the local microion density becomes unphysically large, such that the corresponding packing fraction (calculated with the ionic radii R_α) would exceed close-packing. This breakdown of PB theory, which systematically overestimates contact densities, can be traced back to the neglect of correlations between microions. This deficiency may be partly remedied by introducing a layer of counterions tightly bound to the highly charged surface, the so-called *Stern layer* of 'adsorbed' counterions; this reduces the surface charge from its bare value to some effective value $|\sigma_{\text{eff}}| < |\sigma|$, which may be considered as a phenomenological, adjustable parameter. The PB theory sketched above applies then only to the 'diffuse' electric double-layer, beyond the Stern layer. This concept has a more clearly defined validity in the case of cylindrical geometry to be considered later.

The PB equation (7.45) may in fact be solved analytically [13]. With the dimensionless potential $\Phi(z) = e\psi(z)/k_B T$ equation (7.45) reduces to

$$\frac{d^2\Phi(z)}{dz^2} = \kappa_D^2 \sinh(\Phi(z)) \tag{7.51}$$

where κ_D is the inverse of the Debye screening length, equation (3.144). The solution of the differential equation (7.51) is

$$\psi(z) = \frac{k_B T}{e} \Phi(z) = \frac{4k_B T}{e} \operatorname{arctanh}(g \exp(-\kappa_D z)) \tag{7.52}$$

where g is related to the surface potential $\psi_0 = \psi(z=0)$ by

$$g = \tanh\left(\frac{e\psi_0}{4k_B T}\right) \tag{7.53}$$

The resulting concentration profiles follow from equation (7.43):

$$\rho_+(z) = \rho_0 \left(\frac{1 - g\exp(-\kappa_D z)}{1 + g\exp(-\kappa_D z)}\right)^2 \tag{7.54}$$

and $\rho_-(z)$ is obtained by replacing g with $-g$. The surface potential ψ_0 and surface charge σ are related by the boundary condition (7.46), which yields Grahame's equation:

$$\sigma = (8\rho_0 \epsilon_0 \epsilon k_B T)^{1/2} \sinh\left(\frac{e\psi_0}{2k_B T}\right) \tag{7.55}$$

If the surface potential is less than about 25 mV at room temperature, the right-hand side may be linearized to yield $\sigma = \epsilon_0 \epsilon \kappa_D \psi_0$, showing that for low surface potentials, the electric double-layer behaves as a condenser of width equal to the Debye length. Since $\psi(z) < \psi_0$ for $z > 0$, the right-hand side of the PB equation (7.51) may be linearized with respect to $\Phi < 1$. The resulting linear differential equation is easily solved with the result

$$\psi(z) = \psi_0 \exp(-\kappa_D z) = \frac{\sigma}{\epsilon_0 \epsilon \kappa_D} \exp(-\kappa_D z) \tag{7.56}$$

while the corresponding density profiles reduce to:

$$\rho_\pm(z) = \rho_0 \pm \kappa_D \frac{|\sigma|}{2e} \exp(-\kappa_D z) \tag{7.57}$$

showing that the coion and counterion density profiles decay exponentially towards their bulk value; the width of the electric double-layer is of the order of λ_D, which varies as $1/\sqrt{\rho_0}$. The exponential decay is reminiscent of the exponential screening of the correlation functions in bulk electrolytes (cf. equation (3.142)), and mirrors the close relationship between the RPA, or Debye–Hückel theory for ionic solutions, and linearized PB theory of electric double-layers.

---

[13] G. Gouy, *J. Phys. (Paris)* **9**, 457 (1910). D.L. Chapman, *Philos. Mag.* **25**, 475 (1913).

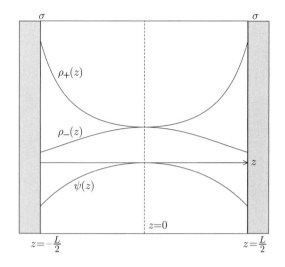

Two parallel charged planes

Now consider the case of two infinite planar surfaces carrying identical charge densities σ, and placed at $z = \pm L/2$, as shown in figure 7.5. The solvent confined between the two planes contains monovalent coions and counterions, which are in equilibrium with a solution in an infinite reservoir fixing the chemical potentials of non-interacting ions, $\mu_\alpha = k_B T \ln(\rho_\alpha \Lambda_\alpha^3)$ (ρ_α being the bulk concentrations). The midplane $z = 0$ is a symmetry plane for the potential $\psi(z)$ and for the ion density profiles. In particular, $E(z = 0) = -\frac{d\psi(z)}{dz}|_{z=0} = 0$, while the electroneutrality constraint within the slab now reads:

$$e \int_0^{L/2} \rho_c(z) \, dz = e \int_{-L/2}^0 \rho_c(z) \, dz = -\sigma \tag{7.58}$$

The PB equation (7.45) must now be solved in the interval $-L/2 \leq z \leq 0$, subject to the boundary conditions:

$$-\frac{d\psi(z)}{dz}\bigg|_{z=-L/2} = \frac{\sigma}{\epsilon_0 \epsilon} \qquad \frac{d\psi(z)}{dz}\bigg|_{z=0} = 0 \tag{7.59}$$

The disjoining pressure is the force per unit area to be applied to the mutually repelling charged surfaces to maintain them at a distance L from each other. Let $P(z)$ be the total local pressure exerted on a test surface placed at z. Clearly, for equilibrium to be achieved, $P(z)$ must be constant throughout the slab between the planes, i.e. $dP/dz = 0$. $P(z)$ is the sum of the osmotic pressure $\Pi(z)$ exerted by the ions, and of an electrostatic contribution related to Maxwell's electrostatic stress tensor [14]. The hydrostatic equilibrium condition (7.49) is satisfied provided

$$P = \Pi(z) - \frac{\epsilon_0\epsilon}{2}\left(\frac{d\psi(z)}{dz}\right)^2 = k_B T\rho_N(z) - \frac{\epsilon_0\epsilon}{2}E(z)^2 \qquad (7.60)$$

where the second term is to be identified with the electrostatic contribution. The right-hand side in equation (7.60) is most conveniently evaluated at $z = 0$, where $E = 0$. The disjoining pressure then reads:

$$\Delta P = P(L) - P(\infty) = k_B T(\rho_N(0) - \rho) \qquad (7.61)$$

and is hence determined by the total ion density in mid-plane from which the bulk ion density is subtracted. If applied at $z = \pm L/2$, equation (7.60), together with the boundary condition (7.59), yields the contact theorem:

$$P = k_B T\sum_\alpha \rho_\alpha(z = \pm L/2) - \frac{\sigma^2}{2\epsilon_0\epsilon} \qquad (7.62)$$

When $L \to \infty$, P coincides with the bulk pressure of the electrolyte, and equation (7.62) provides in fact an exact relationship between the latter, the charge density on a planar surface and the densities of the ionic species at the plane of closest approach, provided the exact density profiles are used, rather than their PB approximation [15]. The PB equation (7.45), with the appropriate boundary conditions (7.59), cannot be solved analytically in closed form, except in the somewhat academic case of zero salt concentration (i.e. when only counterions are present between the plates).

For sufficiently low surface charge, the right-hand side of the PB equation (7.45) or (7.51) may again be linearized; the solution of the resulting linear differential equation $d^2\Phi/dz^2 = \kappa_D^2\Phi$, satisfying the boundary conditions (7.59), is [16]

$$\psi(z) = \frac{\psi_0}{\sinh(\kappa_D L/2)}\cosh(\kappa_D z) \qquad (7.63)$$

where $\psi_0 = \sigma/\epsilon_0\epsilon\kappa_D$; the resulting concentration profiles follow from the linearized version of (7.43). The potential in mid-plane is $\psi_0/\sinh(\kappa_D L/2) \simeq 2\psi_0\exp(-\kappa_D L/2)$ valid for $L \gg \lambda_D$, while the disjoining pressure calculated from (7.61) is, to lowest non-vanishing order in $\psi(0)$

$$\Delta P(L) = \frac{\rho}{2}(\beta e\psi(0))^2 = \frac{2\sigma^2}{\epsilon_0\epsilon}\exp(-\kappa_D L) \qquad (7.64)$$

[14] See e.g. L. Landau and E. Lifshitz, *Electrodynamics of Continuous Media*, Pergamon Press, Oxford, 1959.

[15] D. Henderson, L. Blum and J. L. Lebowitz, *J. Electroanal. Chem.* **102**, 315 (1979). S. L. Carnie and D. Y. C. Chan, *J. Chem. Phys.* **74**, 1293 (1981).

[16] See e.g. D. Andelman, in *Structure and Dynamics of Membranes*, ed. R. Lipowsky and F. Sackmann, Elsevier, Amsterdam, 1995.

ΔP decays exponentially with the distance L between the two charged surfaces. However, an analysis of the full, non-linear PB equation in the limit where $L \ll \lambda_D$ shows that $\Delta P(L) \sim 1/L$ as $L \to 0$. This result may be understood by noting that, when $L \to 0$, only counterions are left within the slab. Charge neutrality requires that their mean density be $2e/\Sigma L$, where $\Sigma = e/|\sigma|$ is the area per unit charge on the plates; this density far exceeds the bulk density of electrolyte for large surface charges, so that the disjoining pressure (7.61) is of the order of $2k_B T/\Sigma L$.

The concentration of salt between the charged surfaces is less than the concentration in the reservoir. The salt concentration in the slab is equal to the mean concentration ρ'_- of coions:

$$\rho'_- = \frac{2}{L} \int_0^{L/2} \rho_-(z)\, dz \qquad (7.65)$$

Using the linearized concentration profile derived from (7.63) and (7.43), one arrives at the result:

$$\rho'_- = \rho_- \left(1 - \frac{1}{L\Sigma\rho_-} \right) \qquad (7.66)$$

The reduction from the reservoir concentration ρ_0 is very significant in a situation of strong confinement (i.e. small L); this expulsion of salt is the Donnan effect.

Exercise: Calculate the disjoining pressure between two uniformly charged planes in the absence of salt. To that purpose write down the PB equation in the presence of counterions only, and show that its solution, subject to the boundary conditions (7.59), is:

$$\psi(z) = \frac{k_B T}{e} \ln[\cos^2(\kappa z)] \qquad (7.67)$$

Specify the expression for the inverse screening length κ.

Derive from this the counterion density profile, keeping in mind that the total number of counterions per unit area is fixed by charge neutrality, and finally the disjoining pressure $\Delta P(L)$.

An infinite charged cylinder

Consider now the sheath-like electric double-layer around a uniformly charged cylindrical rod of radius a and length L. This is a model for rod-like colloids, like boehmite, or viruses, like the tobacco mosaic virus (TMV). The aspect ratio L/a of the latter is about 20, and if the main interest is in the immediate vicinity of the rod, end effects are unimportant, and one may usefully consider the simpler case of an infinitely long rod ($L \to \infty$). Moreover, a charged rod may also serve as a model for stiff polyelectrolytes, i.e. water-soluble polymers made up of ionizable monomers, carrying a line charge, or charge per unit length, ξ. A simple generalization of Flory's argument for the size of a non-intersecting polymer shows that the electrostatic interactions between charged monomers tend to swell the polyelectrolyte chain. Indeed, if M is the total number of segments of length b, and f is the fraction of monovalently charged segments, then the Flory free energy may be written as the sum of the entropic elastic energy and of a Coulombic energy; within numerical factors of order 1:

$$F_{\text{Flory}} = k_{\text{B}}T \left[\frac{R^2}{Mb^2} + \frac{(Mf)^2 \ell_{\text{B}}}{R} \right] \tag{7.68}$$

The equilibrium size R is obtained by minimizing (7.68) with respect to R, with the result:

$$R \sim M f^{2/3} (\ell_{\text{B}} b^2)^{1/3} \tag{7.69}$$

showing that the polyelectrolyte chain is fully stretched, since R is directly proportional to the number of segments.

An important length scale for charged rods or polyelectrolytes is the distance $\ell = e/|\xi|$ between successive monovalent charges; for polyelectrolytes it is related to the fraction of charged monomers by $\ell = b/f$.

The electrostatic field around an infinite cylindrical rod is radial by symmetry, and its amplitude $E(r)$ follows directly from Gauss' theorem by calculating the flux of the radial field through a coaxial cylinder of radius r; the result is $E(r) = \xi/(2\pi\epsilon_0\epsilon r)$, and the corresponding electrostatic potential is:

$$\psi_0(r) = -\frac{\xi}{2\pi\epsilon_0\epsilon} \ln(r/\lambda) \tag{7.70}$$

where λ is an arbitrary length determined by the boundary condition imposed on $\psi(r)$.

The radial structure of the double-layer sheath may again be studied within PB theory. The cylindrical Poisson equation is:

$$\frac{1}{r} \frac{d}{dr} \left[r \frac{d\psi}{dr} \right] = -\frac{e}{\epsilon_0\epsilon} \rho_c(r) \tag{7.71}$$

where $e\rho_c(r)$ denotes the radial charge density. Consider first a single charged rod in a monovalent ionic solution. The reduced potential $\Phi(r) = \beta e\psi(r)$ satisfies the PB equation:

$$\frac{1}{r} \frac{d}{dr} \left[r \frac{d\Phi}{dr} \right] = \kappa_{\text{D}}^2 \sinh \Phi(r) \tag{7.72}$$

to be solved subject to boundary conditions similar to (7.59), but adapted to the cylindrical geometry. For sufficiently low line charge $|\xi|$ the reduced potential $\Phi(r)$ is small everywhere, so that equation (7.72) may be linearized by replacing sinh Φ by Φ. The solution for the electrostatic potential is:

$$\psi(r) = \frac{k_B T}{e} \Phi(r) = \frac{\xi}{2\pi\epsilon_0\epsilon} K_0(\kappa_D r) \tag{7.73}$$

where K_0 is a modified Bessel function of the second kind; $\psi(r)$ goes over to $\psi_0(r)$ at short distances, and decays exponentially for $r > \lambda_D$, due to screening.

For sufficiently large line charge ξ, non-linear effects become important and counterion condensation sets in according to a scenario first proposed by Manning [17]. Close to the charged rod, the total electrostatic potential $\psi(r)$ goes over to $\psi_0(r)$, within a constant, so that the distribution of counterions is $\rho(r) = \rho_0 \exp(-\Phi(r)) \sim r^{-2\Gamma}$, where $\Gamma = \ell_B/\ell$ is a dimensionless coupling parameter, equal to the ratio of the Coulomb energy between two monovalent nearest-neighbour charges along the rod, $e^2/(4\pi\epsilon_0\epsilon\ell)$, over the thermal energy $k_B T$. The total number of counterions, per unit length of the rod, within a distance r from the axis of the rod is:

$$n(r) = \int_a^r \rho(r')2\pi r' \, dr' = 2\pi\rho_0 \int_a^r r'^{(1-2\Gamma)} \, dr' \tag{7.74}$$

The integral clearly diverges at its lower bound as $a \to 0$ when $\Gamma > 1$, which corresponds to a high line charge, or equivalently to a large fraction f of charged segments on a polyelectrolyte ($f > b/\ell_B$). This divergence signals a strong accumulation or 'condensation' of the counterions onto the rod or polyelectrolyte chain. The condensed counterions will partially compensate for the line charge and reduce its effective value until the radio Γ drops again below 1, and counterion condensation ceases.

The polyelectrolyte and the sheath of condensed counterions is then equivalent to a polyelectrolyte with an average distance between its charged segments equal to ℓ_B ($\Gamma = 1$ and $f = b/\ell_B$). The remaining counterions form the 'diffuse' part of the electric double-layer which, in the presence of salt, extends radially over a distance of the order of λ_D. Counterion condensation has a number of measurable consequences; in polyelectrolyte solutions it leads to a reduction of the osmotic pressure, which is mainly due to the counterions, since the fraction of the latter which are condensed does not contribute to the pressure. The condensed counterions are mobile along the axis of the rod or polyelectrolyte and form a kind of one-dimensional Coulomb gas. The correlated fluctuations of the counterions condensed on two neighbouring parallel rods lead to an effective attraction between equally charged rods, via a purely classical mechanism reminiscent of the quantum fluctuations of bound electrons which give rise to the van der Waals attraction between molecules [18].

[17] G. S. Manning, *J. Chem. Phys.* **51**, 954 (1969).
[18] For a review on polyelectrolytes, see J.L. Barrat and J.F. Joanny, *Adv. Chem. Phys.* **XCIV**, 1 (1996).

Concentrated suspensions of parallel charged rods or polyelectrolyte chains, like DNA 'bundles', are best studied within a Wigner–Seitz cell model. Each rod together with its associated counterions is enclosed on average by a surface of zero electric field, which must be parallel to the rod. The Wigner–Seitz model assumes that this surface is a cylinder coaxial with the rod, of radius R determined by the number n of rods per unit area, through $n\pi R^2 = 1$. The cylindrical PB equation to be solved is:

$$\frac{1}{r}\frac{d}{dr}\left[r\frac{d\Phi}{dr}\right] = -\kappa_D^2 \exp(-\Phi(r)) \tag{7.75}$$

where the inverse screening length is determined by the macroscopic counterion density $1/(\pi R^2 \ell)$. The PB equation (7.75) was solved analytically by Fuoss and coworkers [19]; the two integration constants are determined by the boundary condition $d\Phi/dr = 0$ at $r = R$ and the charge neutrality constraint:

$$2\pi \int_a^R \rho_c(r) r\, dr = 1/\ell \tag{7.76}$$

It is interesting to note that Fuoss' solution has a different analytic form below and above the Manning threshold $\Gamma = 1$. For $\Gamma < 1$, the solution is:

$$\psi(r) = \frac{k_B T}{e} \ln\left[\frac{\kappa_D^2 r^2}{2\alpha^2} \sinh^2\left(\alpha \ln(r/R) - \operatorname{arctanh}\alpha\right)\right] \tag{7.77}$$

where α is related to Γ by $\Gamma = (1 - \alpha^2)(1 - \alpha/\tanh(\alpha \ln(b/R)))$, while for $\Gamma > 1$, the corresponding solution is obtained by replacing all hyperbolic functions by their trigonometric counterparts. As in the planar case examined earlier in this section, the total pressure has ideal osmotic and electrostatic contributions; the latter vanishes at the Wigner–Seitz boundary surface, so that the pressure is simply given by

$$P = k_B T \rho_c(r = R) = k_B T \rho_c \exp(-\Phi(r = R)) \tag{7.78}$$

The total pressure calculated by substituting Fuoss' solution into (7.78) indeed shows the reduction predicted by Manning's counterion condensation mechanism.

7.7 Application 5: colloid stability

The stability of suspensions of natural or synthetic, mineral or polymeric colloidal particles is of prime importance for a broad range of industrial applications. Depending on the solvent and various physical conditions, like ionic strength or the concentration of added polymer, the suspension may be homogeneous and stable over macroscopically long times (corresponding to a dispersed phase) or, on

[19] R. M. Fuoss, A. Katchalsky and S. Lifson, *Proc. Natl. Acad. Sci. USA* **37**, 579 (1951).

the contrary, *flocculate* (or coagulate) rapidly and precipitate. The flocculation of colloidal suspensions occurs when strong attractive forces between the colloidal particles lead to their aggregation into dense (precipitate) or fractal (gel) clusters. The long-range attraction which is in most cases responsible for flocculation is the ubiquitous van der Waals interaction between colloidal particles, which results from the sum of all van der Waals attractions (or dispersion forces) between the very large number (typically 10^{10} in micrometre-sized particles) of molecules belonging to neighbouring colloidal particles.

To grasp the importance of these attractions, it is worthwhile to calculate the van der Waals interaction between two spherical colloidal particles as a function of their distance. The van der Waals–London dispersion interactions between two molecules are of the general form:

$$w(r) = -\frac{C}{r^n} \qquad (7.79)$$

and the leading contribution is expected to be the non-retarded dipole-induced dipole energy, for which $n = 6$ (cf. section 1.2). To arrive at the desired result between two spherical colloids, it proves convenient to consider first some simpler geometries.

The interaction energy between a single atom placed at a distance d from an infinite planar surface, and all atoms placed in the infinite half-space beyond that plane, with a uniform number density ρ, is easily calculated in cylindrical coordinates, with the z-axis along the normal to the plane:

$$w(d) = -2\pi\rho C \int_d^\infty dz \int_0^\infty \frac{r\,dr}{(r^2 + z^2)^{n/2}} = -\frac{2\pi\rho C}{(n-2)(n-3)}\frac{1}{d^{n-3}} \qquad (7.80)$$

Note that for the leading dispersion interaction ($n = 6$), $w(d) \sim d^{-3}$ is of much longer range than the interaction potential (7.79) between two atoms or molecules.

Next consider the interaction energy between an infinite planar surface, and a surface element ΔS placed at a distance d and bounding an infinite cylindrical sample orthogonal to the planar surface, as shown in figure 7.6. If the molecular number densities to the left of the plane and in the cylinder are ρ_1 and ρ_2, the total interaction energy is easily calculated by integrating the result (7.80) over the volume of the cylinder:

$$W(d) = -\frac{2\pi C \rho_1 \rho_2 \Delta S}{(n-2)(n-3)} \int_0^\infty \frac{dz}{(z+d)^{n-3}} = -\frac{2\pi C \rho_1 \rho_2 \Delta S}{(n-2)(n-3)(n-4)}\frac{1}{d^{n-4}} \qquad (7.81)$$

The interaction energy per unit area between two parallel planar surfaces separated by d is hence:

$$w_{pp}(d) = \frac{W(d)}{\Delta S} = -\frac{2\pi C \rho_1 \rho_2}{(n-2)(n-3)(n-4)}\frac{1}{d^{n-4}} \qquad (7.82)$$

This result may immediately be put to use to calculate the interaction energy between a planar surface (bounding a semi-infinite medium with molecular

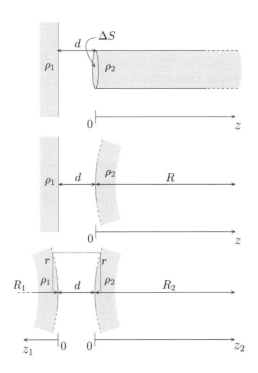

Figure 7.6. Geometries for the calculation of the total interaction energy between a semi-infinite slab of density ρ_1 on the left and a cylindrical volume element of density ρ_2 on the right (upper frame), between a semi-infinite slab and a sphere of radius R (middle frame), and between two spheres of radii R_1 and R_2.

density ρ_1) and a spherical colloidal particle of radius R, and molecular density ρ_2, with its centre placed at $d + R$ from the planar surface (cf. figure 7.6). For distances $d \ll R$, the total interaction energy is obtained by integrating the result (7.81) over coaxial annular cylindrical regions. The surface element $dS = 2\pi r\, dr$ is placed at a distance $Z = d + z$ from the plane, where r and z are related by $R^2 = r^2 + (R - z)^2$, i.e. $r^2 = 2Rz - z^2 \simeq 2Rz$, such that $2r\, dr = 2R\, dz = 2R\, dZ$, and:

$$
w_{\text{sp}}(d) = 2\pi \int_d^\infty w_{\text{pp}}(Z) r\, dr = 2\pi R \int_d^\infty w_{\text{pp}}(Z)\, dZ
$$

$$
= -\frac{4\pi^2 C R \rho_1 \rho_2}{(n-2)(n-3)(n-4)(n-5)} \frac{1}{d^{n-5}}
$$

(7.83)

Note that this expression is valid only for $d \ll R$, and that it is this assumption which allows the upper integration limit to be taken to ∞.

The case of two spheres of radii R_1 and R_2, and densities ρ_1 and ρ_2, may be treated in a similar fashion. Consider two parallel planar surfaces orthogonal to the line of centres, such that their intersections with the two spheres are circles of the same radius r; the corresponding z coordinates are then given by $r^2 = R_1^2 - (R_1 - z_1)^2 \simeq 2R_1 z_1 = R_2^2 - (R_2 - z_2)^2 \simeq 2R_2 z_2$ (cf. figure 7.6), and the distance between the two planes is $Z = d + z_1 + z_2$. Derjaguin's approximation [20] amounts to assuming that the annular area $2\pi r \, dr$ to the left interacts with a locally flat surface (rather than the curved spherical surface) on the right (or vice versa); this approximation is valid only if $d \ll R_1, R_2$. The result (7.81) may then be used again, to yield:

$$w_{ss}(d) = 2\pi \int_{Z=d}^{Z=\infty} w_{pp}(Z) r \, dr = -\frac{4\pi^2 C \rho_1 \rho_2}{(n-2)(n-3)(n-4)(n-5)} \frac{R_1 R_2}{(R_1 + R_2)} \frac{1}{d^{n-5}}$$

(7.84)

In practice, the dominant term is the lowest order dispersion interaction, $n = 6$. Defining the Hamaker constant $A = \pi^2 C \rho_1 \rho_2$ (which has the dimension of energy), equation (7.84) reduces, in the case of spheres of equal diameter $R = R_1 = R_2 = \sigma/2$, to:

$$w_{ss}(r) = -\frac{A}{24} \frac{\sigma}{(r - \sigma)} \qquad |r - \sigma| \ll \sigma$$

(7.85)

where r denotes now, as usual, the centre-to-centre distance. Note that the interaction (7.85) is singular at contact. In fact when the closest distance $d = r - \sigma$ between the surfaces of the two spheres becomes of the order of molecular size, the continuum picture breaks down and the singularity is smoothed out. The attraction remains, however, very strong in general, and leads to flocculation if no precautions are taken to stabilize the suspension, as discussed below.

Note that in the opposite limit $r \gg \sigma$, the two spheres behave, in first approximation, like point particles, and the attractive potential $w_{ss}(r)$ recovers its familiar r^{-6} form; explicitly

$$w_{ss}(r) = -\frac{A}{36} \left(\frac{\sigma}{r}\right)^6$$

(7.86)

An analytic expression for $w_{ss}(r)$ valid for all distances r, was derived by Hamaker [21]. Stabilization of colloidal suspensions against flocculation requires a strongly repulsive interaction between two colloidal particles to keep them apart and shield the strong van der Waals attraction at contact. In most practical situations, this repulsive shielding may be achieved by one of two mechanisms. In aqueous solutions, radicals on the surface of the particles (e.g. polystyrene 'balls') exposed to water dissociate and release counterions into the solution, leaving a substantial surface charge on the particles. This leads to the formation of electric double-layers around the particles, similar to those examined in the previous section, which have been shown to repel for distances of the order of the Debye screening length.

In the case of colloidal particles (like polymethylmethacrylate (PMMA) latex 'balls') suspended in organic solvents, the particles may be sterically stabilized

[20] B.V. Derjaguin, *Kolloid Z.* **69**, 155 (1934).
[21] H.C. Hamaker, *Physica* **4**, 1058 (1937).

against flocculation by grafting or adsorbing polymer chains on their surface. The polymer 'brushes' on two approaching colloids will strongly resist interpenetration (which would reduce the number of available conformations of the grafted chains). This repulsion of entropic origin will lead to an effective diameter of the colloids larger than their bare diameter σ, and hence shield the strong van der Waals attraction.

Turning the attention first to charge stabilization, one could proceed as for the case of two charged plates studied in section 7.6, and solve the Poisson–Boltzmann equation for the microion density profiles around two spheres, as a function of their mutual distance, to calculate the effective interaction between the two overlapping electric double-layers [22]. A somewhat different point of view is adopted here, which is valid at finite (but low) colloid concentration, and for low structural or effective surface charges. For the sake of simplicity, the salt-free case will be considered here, i.e. the system under consideration contains N spherical colloidal particles of radius R (henceforth referred to as 'polyions'), carrying a total charge uniformly distributed on their surface, and NZ monovalent point counterions, in a volume V. The solvent is treated as a dielectric continuum with permittivity ϵ. At this stage only excluded volume and Coulomb interactions between polyions and counterions are considered. In view of the considerable asymmetry in size between the two species, it is natural to seek a formal reduction of the initial two-component system to an effective one-component system of 'dressed' polyions. The objective is to define rigorously the effective interactions between the 'dressed' polyions. The Hamiltonian of the initial two-component system splits into three terms:

$$H = H_p + H_c + V_{pc} \tag{7.87}$$

where the polyion H_p and counterion H_c terms contain kinetic and potential energy contributions involving only interactions between particles of one species, while V_{pc} is the interaction energy between the two species:

$$V_{pc} = \sum_{i=1}^{N} \sum_{j=1}^{NZ} v_{pc}(\mathbf{R}_i - \mathbf{r}_j) \tag{7.88}$$

$$v_{pc}(r) = \infty \qquad r < R$$
$$= -\frac{Ze^2}{4\pi\epsilon r} \qquad r > R \tag{7.89}$$

In these equations the $\mathbf{R}_i$ and the $\mathbf{r}_j$ refer to positions of polyions and counterions respectively.

---

[22] The classic presentation of this approach is in E.J.W. Verwey and J.Th.G. Overbeek, *Theory of Stability of Lyophobic Colloids*, Elsevier, Amsterdam, 1948.

At a fixed inverse temperature, $\beta = 1/k_B T$, the Helmholtz free energy of the two-component system may be formally expressed as [23]:

$$
\begin{aligned}
\exp(-\beta F) = \text{Tr}_p \, \text{Tr}_c \, \exp(-\beta H) &= \text{Tr}_p \, \exp-\beta(H_p + F_c(\{\mathbf{R}_i\})) \\
&= \text{Tr}_p \, \exp(-\beta H_p^{\text{eff}})
\end{aligned}
\tag{7.90}
$$

where the effective polyion Hamiltonian is the sum of H_p and of the free energy $F_c(\{\mathbf{R}_i\})$ of an inhomogeneous gas of counterions in the 'external field' (7.88) of the polyions, which depends parametrically on the positions $\{\mathbf{R}_i\}$ of the latter.

Subtracting the trivial kinetic energy of the polyions, one is led to the key result that the effective interaction energy between the polyions is the sum of their direct interactions (steric and Coulombic) and of an indirect interaction, induced by, and averaged over, the counterions, which is given exactly by their free energy for any configuration of the polyions.

$$
V_{pp}^{\text{eff}}(\{\mathbf{R}_i\}) = V_{pp}(\{\mathbf{R}_i\}) + F_c(\{\mathbf{R}_i\})
\tag{7.91}
$$

The result (7.91) is very general, and applies whenever a reduction from a multi-component to an effective one-component system is made by tracing out the degrees of freedom of the species of 'small' particles in the initial mixture. An example in hand is the depletion interaction introduced in section 2.7. As stressed already there, the effective interaction has an entropic component (through F), is state dependent, and is not necessarily pair-wise additive, even if the direct interaction energy V_{pp} is.

For the formally exact result (7.91) to be useful, one must be able to evaluate the free energy F_c of the inhomogeneous fluid of counterions. DFT provides the obvious tool. The required free energy will correspond to the minimum of the free energy functional, introduced in equation (7.5), with respect to variations of the local counterion density. In a mean field perspective, where counterion correlations are neglected, the 'intrinsic' free energy functional $F[\rho(\mathbf{r})]$ splits into ideal and Coulombic parts (see also section 7.2):

$$
\begin{aligned}
F[\rho(\mathbf{r})] &= F_{\text{id}}[\rho(\mathbf{r})] + F_{\text{Coul}}[\rho(\mathbf{r})] \\
&= k_B T \int \rho(\mathbf{r}) \left(\ln(\rho(\mathbf{r})\Lambda^3) - 1 \right) d\mathbf{r} + \frac{1}{2} \int \rho(\mathbf{r})\psi(\mathbf{r}) \, d\mathbf{r}
\end{aligned}
\tag{7.92}
$$

where, outside the colloidal particles, the local electrostatic potential is related to the charge density $e\rho(\mathbf{r})$ (assuming monovalent counterions) by the three-dimensional form of Poisson's equation (7.41).

$$
\nabla^2 \psi(\mathbf{r}) = -\frac{e}{\epsilon_0 \epsilon} \rho(\mathbf{r})
\tag{7.93}
$$

If the polyions are weakly charged, or if their structural charge has been reduced by a tightly bound Stern layer to a low effective charge, they will induce a weak

---

[23] In equation (7.90) we introduce the short-hand 'trace' notation for integration over the phase space of polyions or counterions, e.g. $\text{Tr}_p = \frac{1}{N! h^{3N}} \int d\Gamma_N$.

inhomogeneity of the counterion density, which will be rewritten in the form $\rho(\mathbf{r}) = \rho_c + \Delta\rho(\mathbf{r})$, where ρ_c is the mean counterion density NZ/V. The integrand of $F_{id}[\rho(\mathbf{r})]$ may be expanded to quadratic order in $\Delta\rho(\mathbf{r})$ so that, in view of the linear relation between $\rho(\mathbf{r})$ and $\psi(\mathbf{r})$, the intrinsic free energy reduces to a quadratic functional of $\Delta\rho(\mathbf{r})$:

$$F[\rho(\mathbf{r})] = F_{id}(\rho_c) + k_B T \ln(\lambda^3 \rho_c) \int \Delta\rho(\mathbf{r})\,d\mathbf{r} + \frac{k_B T}{2\rho_c} \int (\Delta\rho(\mathbf{r}))^2\,d\mathbf{r} + \frac{e}{2} \int \rho(\mathbf{r})\psi(\mathbf{r})\,d\mathbf{r}$$

(7.94)

The first term is the free energy of a homogeneous ideal gas of density ρ_c. Substitution of (7.94) into the variational principle (7.5), leads to the Euler–Lagrange equation:

$$k_B T \ln(\Lambda^3 \rho_c) + \frac{k_B T}{\rho_c} \Delta\rho(\mathbf{r}) + e\psi(\mathbf{r}) = \mu - V_{ext}(\mathbf{r})$$

(7.95)

For non-interacting counterions, $\mu = k_B T \ln(\Lambda^3 \rho_c)$, while $V_{ext}(\mathbf{r}) = e\psi_{ext}(\mathbf{r})$, where $\psi_{ext}(\mathbf{r})$ is the external electrostatic potential acting on the counterions, due to the polyions. The Euler–Lagrange equation hence simplifies to:

$$\Delta\rho(\mathbf{r}) = -\frac{e\rho_c}{k_B T} (\psi(\mathbf{r}) + \psi_{ext}(\mathbf{r})) = -\frac{e\rho_c}{k_B T} \int \frac{e\rho(\mathbf{r}') - Ze\rho_{ext}(\mathbf{r}')}{4\pi\epsilon_0 |\mathbf{r} - \mathbf{r}'|}\,d\mathbf{r}'$$

(7.96)

where $\rho_{ext}(\mathbf{r}) = \sum_i \delta(\mathbf{r} - \mathbf{R}_i)$ is the local density of polyions, momentarily considered to be point particles. Taking Fourier transforms of both sides of equation (7.96), invoking the convolution theorem and remembering that the Fourier transform of $1/4\pi r$ is $1/k^2$, one arrives at an expression for the Fourier components of the local counterion density, namely:

$$\rho(\mathbf{k}) = \frac{Z\kappa_D^2}{k^2 + \kappa_D^2} \sum_{i=1}^{N} \exp(i\mathbf{k} \cdot \mathbf{R}_i)$$

(7.97)

Inverse Fourier transformation yields the counterion density profile:

$$\rho(\mathbf{r}) = \sum_{i=1}^{N} \rho_i(\mathbf{r}) = \sum_{i=1}^{N} \frac{Z\kappa_D^2}{4\pi} \frac{\exp(-\kappa_D |\mathbf{r} - \mathbf{R}_i|)}{|\mathbf{r} - \mathbf{R}_i|}$$

(7.98)

The local density thus appears as a superposition of counterion density profiles associated with each of the N polyions. It is easily verified that each of the N profiles integrates up to Z, thus ensuring overall charge neutrality.

Exercise: Show that equation (7.98) follows directly from the general result (7.11), valid for quadratic free energy functionals, and from the RPA (or Debye–Hückel) expression (3.141) for the structure factor of a homogeneous gas of counterions.

Since counterions cannot penetrate the polyion core, $\rho_c(\mathbf{r}) = 0$ whenever $|\mathbf{r} - \mathbf{R}_i| < R$; the corresponding normalization condition is:

$$\int_{|\mathbf{r}-\mathbf{R}_i|>R} \rho_i(\mathbf{r})\,d\mathbf{r} = Z \tag{7.99}$$

which is satisfied by (7.98), provided the polyion valence Z is renormalized to

$$Z' = Z\frac{\exp(\kappa_D R)}{1 + \kappa_D R} \tag{7.100}$$

The total electrostatic potential may likewise be written as a superposition of N screened potentials due to the 'dressed' polyions:

$$\psi_t(\mathbf{r}) = \sum_{i=1}^{N} \frac{Z'e}{4\pi\epsilon_0\epsilon} \frac{\exp(-\kappa_D|\mathbf{r} - \mathbf{R}_i|)}{|\mathbf{r} - \mathbf{R}_i|} \tag{7.101}$$

If the optimum density profiles (7.98) and potentials (7.101) are substituted into the free energy functional (7.94) (including the 'external' field contribution of the polyions), and the resulting free energy of the counterions is added to the direct polyion interaction energy, as in (7.91), the following expression for the effective interaction energy of the 'dressed' polyions results:

$$V_{pp}^{eff}(\{\mathbf{R}_i\}) = V_0 + \sum_i \sum_{j<i} v_{eff}(\mathbf{R}_i - \mathbf{R}_j) \tag{7.102}$$

where the effective pair potential is:

$$v_{eff}(\mathbf{R}_i - \mathbf{R}_j) = \int \psi_i(\mathbf{r})\rho_j(\mathbf{r})\,d\mathbf{r} = \frac{Z'^2 e^2}{4\pi\epsilon_0\epsilon} \frac{\exp(-\kappa_D|\mathbf{R}_i - \mathbf{R}_j|)}{|\mathbf{R}_i - \mathbf{R}_j|} \tag{7.103}$$

The pair-wise additivity of the polyion interaction energy is a consequence of the quadratic approximation to the free energy functional; many-body contributions would arise if the full functional (7.94) were used. The 'zero-body' or 'volume' term V_0 in (7.102) is state dependent and includes the 'self-energy' of the 'dressed' polyions [24].

The effective polyion interaction (7.103) remains valid in the presence of salt, provided the corresponding inverse Debye screening length is used; the 'volume' term then contains additional contributions depending on the salt concentration, which may lead to separation into dilute and concentrated colloidal phases in highly dialysed solutions.

Adding the direct van der Waals attraction between the spherical polyions (the limiting forms of which are given by (7.85) and (7.86)) to the effective double-layer repulsion (7.103) results in the DLVO potential, named after Derjaguin, Landau, Verwey and Overbeek. The balance between the repulsive electrostatic contribution and the van der Waals attraction is controlled by the screening length and hence ultimately by the salt concentration, as illustrated in figure 7.7. At low

[24] R. van Roij, M. Dijkstra and J.P. Hansen, *Phys. Rev. E* **59**, 2010 (1999).

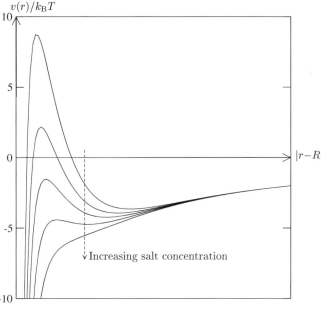

Figure 7.7. DLVO potential between two charged spherical colloidal particles of radius R, as a function of the distance r, where r is the centre-to-centre distance. The potential $v(r)$ is the sum of the screened Coulomb interaction (7.103), and of the dispersion interaction (cf. equations (7.85) and (7.86)). The total potential is shown for five different salt concentrations in increasing order from top to bottom. At low salt concentration the high repulsive barrier due to the electrostatic part prevents flocculation.

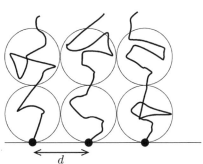

Figure 7.8. Schematic polymer brush; the mean distance between grafting points to the substrate is d. The circles surround sub-units (blobs) of the same size.

salt concentration, the repulsion decays sufficiently slowly for a potential barrier of several $k_B T$ to prevent particles from coming sufficiently close to experience the strong attraction of the 'primary' minimum induced by the divergent van der Waals attraction (7.85) at contact (electrostatic stabilization). When salt is added, λ_D decreases, and the electrostatic potential barrier decreases in amplitude, eventually disappearing, thus leading to flocculation of the colloidal dispersion.

An alternative to electrostatic stabilization is steric stabilization by grafted polymer 'brushes', which may be broadly understood as follows. Consider first a brush of polymers end-grafted to a planar surface in good solvent. If the mean distance d between grafting sites is less than the radius of gyration R_G of a single polymer, the polymers will stretch perpendicularly to the surface, to avoid inter-penetration. Each elongated coil can be regarded as a linear sequence of subunits of diameter d (which coincides then with the correlation length ξ) (see figure 7.8).

Each subunit behaves like an isolated chain, such that the number m of monomers or segments of length b in one subunit is $m = (d/b)^{1/\nu}$, where $\nu \simeq 3/5$ is the Flory exponent appropriate for good solvent. If M is the total number of monomers, the thickness H of the brush will be [25] $H = Mb(b/d)^{2/3}$. When two plates coated with polymer brushes are pushed together, the brushes will touch when the distance between plates $L = 2H$. Since the brushes resist interpenetration, the coils in each brush are compressed, and the monomer concentration c will locally increase by a factor $2H/L$. This leads to an enhanced osmotic pressure, and hence to a repulsive force $(2H/L)^{9/4}$ between the plates, where the exponent 9/4 is that appropriate for semi-dilute solutions of polymers. For polymer-coated colloids, this steric repulsion will efficiently screen the van der Waals attraction at short distances, provided H is sufficiently large.

Further reading

A very complete overview of interfacial physics, including polymer interfaces, can be found in the set of lecture notes edited by J. Charvolin, J-F. Joanny and J. Zinn-Justin, *Liquids at Interfaces*, North Holland, Amsterdam, 1990.

Simple density functionals and the fluid fluid interface are reviewed in the book by Safran (see chapter 6).

F.S. Bates and G.H. Frederickson, *Annu. Rev. Phys. Chem.* **41**, 525 (1990), give a very accessible review of block copolymer thermodynamics, considering both theoretical and experimental aspects. More recent references are also found in the book by Larson (see chapter 1).

Recent progress on interactions between electric double layers is reviewed by J-P. Hansen and H. Löwen, *Annu. Rev. Phys. Chem.* **51**, 209 (2000).

---

[25] S. Alexander, *J. Phys. (Paris)* **38**, 989 (1977).

8 Curvature and fluctuations

8.1 Fluctuations of interfaces and capillary waves

The description of the fluid–fluid interface in section 7.3 was restricted to planar interfaces, with a density profile depending only on one coordinate. This restriction to a special family of density profiles is characteristic of a mean field approach. An actual interface, however, is a fluctuating object. In fact, in the absence of any external field, the position of the interface can fluctuate throughout the accessible volume, so that the *average* density is, in principle, uniform! The density functional approach, being essentially a mean field one, is not appropriate for treating these fluctuations. A complete treatment, starting from a microscopic description of the system, is equally hopeless. The physical solution consists in using a coarse grained approach, in which the phase space of the system is reduced to the large wavelength coordinates that define the position of a weakly curved interface. The energy of this weakly curved interface is of the form

$$H_{\text{cap}} = \gamma \int \int \mathrm{d}^2 A \tag{8.1}$$

where $\mathrm{d}^2 A$ is the element of area of the interface, and γ is the surface tension for a planar interface, calculated e.g. using the theory of section 7.3. The 'energy' in equation (8.1), also known as a 'capillary Hamiltonian' is actually a free energy, as indicated by the presence of the surface tension γ, which depends on thermodynamic parameters like the temperature. This effective Hamiltonian therefore includes fluctuations of short (molecular) wavelength, which are embedded in the definition of γ. The total free energy of the system is obtained by integrating the Boltzmann weight $\exp(-H_{\text{cap}}/k_B T)$ over a phase space defined by the possible positions of the interface.

The calculation can be carried out if one restricts the phase space integration to those configurations that deviate only weakly from the planar configuration. As discussed in section 6.2, such interfaces can be described by a function $h(x, y)$, and the element of area is $\mathrm{d}^2 A = \mathrm{d}x\mathrm{d}y\sqrt{1 + h_x^2 + h_y^2} \simeq \mathrm{d}x\mathrm{d}y\,(1 + h_x^2/2 + h_y^2/2)$. The capillary Hamiltonian is most easily expressed by introducing the Fourier representation for a system of size $L \times L$

$$h(\mathbf{q}) = \int d^2 \mathbf{R} h(\mathbf{R}) \exp(i\mathbf{q} \cdot \mathbf{R}) \qquad h(\mathbf{R}) = \frac{1}{L^2} \sum_q h(\mathbf{q}) \exp(-i\mathbf{q} \cdot \mathbf{R}) \qquad (8.2)$$

where $\mathbf{R} = (x, y)$ and $\mathbf{q} = (q_x, q_y)$ are two-dimensional vectors. With this notation, the capillary Hamiltonian can be rewritten (up to an extensive constant γA) as

$$H_{\text{cap}} = \frac{\gamma}{2L^2} \sum_q \mathbf{q}^2 |h(\mathbf{q})|^2 \qquad (8.3)$$

and the partition function is expressed as an integral over the complex Fourier components $h(\mathbf{q})$

$$Q = \int \prod_q d^2 h(\mathbf{q}) \exp \left(-\frac{\gamma}{2k_B T L^2} \sum_q \mathbf{q}^2 |h(\mathbf{q})|^2 \right) \qquad (8.4)$$

This partition function is a simple Gaussian integral, so that averages can be easily expressed. One finds

$$\langle |h(\mathbf{q})|^2 \rangle = \frac{2k_B T L^2}{\gamma q^2} \qquad (8.5)$$

and

$$\langle |h(\mathbf{R})|^2 \rangle = \frac{k_B T}{\gamma L^2} \sum_q q^{-2} \qquad (8.6)$$

The sum over wavevectors can be transformed into an integral, using the identity, valid for large systems,

$$\frac{1}{L^2} \sum_q = \frac{1}{(2\pi)^2} \int d^2 \mathbf{q} \qquad (8.7)$$

The lower bound in the integral is $q_{\text{min}} = 2\pi/L$, determined by the system size. The upper bound can be taken to be $q_{\text{max}} = 2\pi/\xi$, where ξ is the interfacial width (see section 7.3) below which the coarse-grained description does not hold any more. Equation (8.6) then yields

$$\langle |h(\mathbf{R})|^2 \rangle = \frac{k_B T}{4\pi^2 \gamma} \int_{q_{\text{min}}}^{q_{\text{max}}} \frac{2\pi q \, dq}{q^2} = \frac{k_B T}{2\pi \gamma} \ln \left(\frac{L}{\xi} \right) \qquad (8.8)$$

Equation (8.8) indicates that the effective width of the interface, when the fluctuations are included, is actually a divergent quantity. The average width, as measured by equation (8.8), increases logarithmically with system size. Although such a divergence may seem alarming at first sight, there are several reasons why it is not very important in practice. The first one is related to the weakness of the logarithmic divergence. A simple numerical application shows that even for an interface of one square kilometer, the width defined by (8.8) would not exceed a few nanometres. The second reason is that gravity will actually introduce a cut-off at a wavevector q_g much larger than $2\pi/L$, as shown in the following exercise.

Exercise: The liquid and the vapour have different densities. If $\Delta\rho$ is the mass density difference, show that for a gravity field g along the z axis, the gravitational energy of the curved interface can be expressed as

$$\frac{g\Delta\rho}{2} \int dx \int dy \, h(x, y)^2 \tag{8.9}$$

Show next that

$$\langle |h(\mathbf{q})|^2 \rangle = \frac{2k_{\mathrm{B}} T L^2}{\gamma(q^2 + \Lambda^{-2})} \tag{8.10}$$

where $\Lambda = \sqrt{\gamma/g\Delta\rho}$ is the capillary length, i.e. the length scale above which gravity effects become dominant over capillary ones.
Deduce from (8.10) that for $L \gg \Lambda$

$$\langle |h(\mathbf{R})|^2 \rangle = \frac{k_{\mathrm{B}} T}{2\pi\gamma} \ln\left(\frac{\Lambda}{\xi}\right) \tag{8.11}$$

so that the interfacial width due to fluctuations is not divergent.

The surface fluctuations described in this section are usually termed *capillary waves*, in reference to their dynamical behaviour. A fluctuation of the interface with wavevector $\mathbf{q}$ is a collective mode in the sense discussed in section 11.5, with a dispersion relation that can be computed from hydrodynamics[1]. These collective modes have been studied in light-scattering experiments[2] similar to the neutron-scattering experiments described in section 11.6.

8.2   Membranes and curvature moduli

Fluid membranes are two-dimensional objects made of surfactant, or amphiphilic molecules, which tend, for high enough concentrations, to self-assemble in such a way as to minimize the amount of molecules exposed to an unfavourable solvent (see section 2.6). Membranes may be of the monolayer type, separating two different solvents, or bilayers separating similar solvents (figure 2.4). We will consider in the following the case of amphiphiles that are highly insoluble, so that the fraction of solvated molecules is essentially negligible. In this case, adding more surfactant molecules to the system results in an increase of the area

---

[1] See e.g. L. Landau and E. Lifshitz, *Hydrodynamics*, Pergamon Press, Oxford, 1959.
[2] J. Charvolin, J-F. Joanny and J. Zinn-Justin (eds.), *Liquids at Interfaces*, North Holland, Amsterdam, 1990.

of the membranes contained in the system. Under these conditions, it turns out that the interfacial tension associated with the membranes is effectively zero. This somewhat surprising result can be understood as follows. The number of surfactant molecules per unit area in a membrane, Σ, is a free parameter, which will adjust in such a way as to minimize the free energy of the system. If $f(\Sigma, T)$ is the free energy per unit area of a membrane with surface density Σ, one has at equilibrium,

$$\frac{\partial f(\Sigma, T)}{\partial \Sigma} = 0 \qquad (8.12)$$

Increasing the membrane area at fixed number of molecules is equivalent to modifying Σ. Equation (8.12) indicates that the free energy cost of this operation, which defines the surface tension of the membrane, actually vanishes.

When the surface tension vanishes, the surface free energy is dominated by terms that involve higher order contributions in terms of the deviation from a planar shape. These deviations are characterized by the curvatures of the surface, defined in section 6.2. The lowest order contributions from these curvatures to the free energy can be written in the generic form

$$F_{\text{curv}} = \frac{1}{2} \int d^2 A \left(\kappa (H - H_0)^2 + \bar{\kappa} K \right) \qquad (8.13)$$

where H and K are the curvatures introduced in section 6.2. This free energy depends on three parameters characteristic of the membrane, κ, H_0 and $\bar{\kappa}$. H_0 describes the *spontaneous curvature* of the membrane. In a monolayer, the tail and head of the amphiphilic molecule generally have different volumes, so that the membrane will spontaneously bend in such a way as to optimize the global packing (figure 2.4). H_0, however, vanishes in bilayers for symmetry reasons. The *curvature moduli* κ and $\bar{\kappa}$ are known as the bending and splay moduli, respectively. κ quantifies the energy penalty associated with a deviation of the mean curvature from its 'natural' value H_0. $\bar{\kappa}$, in contrast, describes the penalty associated with the formation of saddle-points in the surface. Typical values of κ and $\bar{\kappa}$ for phospholipidic membranes are in the range $(1-100)k_B T$ at room temperature, so that thermal fluctuations are often essential.

The nature of the fluctuations in systems governed by the free energy (8.13) will be considered in the next section. Before we do so, let us mention the special behaviour of the splay term in equation (8.13). The so-called Gauss–Bonnet theorem specifies that for a surface of fixed topology, this term is a constant that depends only on the topology and not on the shape of the surface. Specifically, one has

$$\int K \, d^2 A = 4\pi(1 - g) \qquad (8.14)$$

where g is the genus of the surface, i.e. the number of holes in the surface: $g = 0$ for a sphere, $g = 1$ for a torus. The equality (8.14) stems from the fact that the integrand, $K \, d^2 A$, is invariant under a transformation that maps a surface onto a topologically equivalent one (e.g. a sphere onto an ellipsoid). The important implication is that, unless one wants to compare the free energies of topologically different surfaces, the saddle-splay term can be ignored.

8.3 Fluctuations of membranes

For a weakly curved bilayer ($H_0 = 0$), described by a function $h(x, y)$, the curvature free energy (8.13) can be written in the form

$$F_{\text{curv}} = \int dx \int dy \frac{\kappa}{2} (\nabla^2 h)^2 = \frac{\kappa}{2L^2} \sum_q q^4 |h(\mathbf{q})|^2 \qquad (8.15)$$

where, as in section 8.1, we have used equation (6.13) for H, and the Fourier representation of the function $h(x, y)$ for a system of size $L \times L$. The term involving the Gaussian curvature has been ignored, since we are interested in fluctuations that affect only the shape of the membrane, leaving its topology unchanged.

As in section 8.1, the free energy (8.15) will be considered as an effective Hamiltonian for the long wavelength fluctuations, assuming all contributions from short-range correlations have been lumped into the definition of the curvature moduli. As for the area fluctuations described by the capillary Hamiltonian (8.3), the statistics of the membrane fluctuations described by the Hamiltonian (8.15) is Gaussian and one may write, in analogy with equations (8.4), (8.5), and (8.6)

$$\langle |h(\mathbf{q})|^2 \rangle = \frac{2 k_{\text{B}} T L^2}{\kappa q^4} \qquad (8.16)$$

and

$$\langle |h(\mathbf{R})|^2 \rangle = \frac{k_{\text{B}} T}{\kappa L^2} \sum_q q^{-4} \simeq \frac{k_{\text{B}} T}{\pi \kappa} \left(\frac{L}{2\pi} \right)^2 \qquad (8.17)$$

where the lower cut-off $q = 2\pi/L$ has once again been used. Equation (8.17) shows that the fluctuations have a very strong effect on the position of the membrane. The uncertainty in the vertical coordinate is proportional to the in-plane dimension. A membrane is a much 'softer' and strongly fluctuating object than an interface with non-zero surface tension. Clearly this makes the description as a weakly undulated interface questionable for large membranes. In order to estimate the length scale at which fluctuations become dominant, we compute the average deviation of the normal to the membrane from the z direction. The normal $\mathbf{N}(x, y)$ has coordinates $(h_x, h_y, 1)$ (see section 6.2) so that the mean squared deviation is given by

$$\langle (\mathbf{N}(x, y) - \mathbf{e}_z)^2 \rangle = L^{-4} \sum_q q^2 \langle |h(\mathbf{q})|^2 \rangle \simeq \frac{k_B T}{2\pi\kappa} \ln\left(\frac{L}{\xi}\right) \qquad (8.18)$$

where the large q cut-off, determined by the membrane width, is equal to $2\pi/\xi$. From equation (8.18) we deduce that the membrane will appreciably deviate from the xy plane as soon as its dimension exceeds the persistence length

$$\xi_p = \xi \exp\left(\frac{2\pi\kappa}{k_B T}\right) \qquad (8.19)$$

Gravity, which acted as a cut-off mechanism for fluctuations of interfaces, is generally irrelevant in membrane systems, which are often symmetric (bilayers). Therefore systems containing many membranes will be dominated by fluctuation effects. In the next section we discuss, as an example, the steric interactions between membranes generated by these fluctuations.

8.4   Steric interactions between membranes

Systems containing a large amount of membranes tend to display a large scale (mesoscopic) organization. In a large fraction of the phase diagram, this organization takes the form of *lamellar* phases, i.e. a periodic stacking of membranes perpendicularly to a fixed direction. The spatial period of this arrangement is determined by the various interactions between membranes, which may be of the van der Waals or electrostatic type, but also include a confinement interaction known as the Helfrich interaction. Physically, this interaction comes from the entropy reduction of a membrane confined by its neighbours in the stack. This entropy penalty can be calculated as follows. Let d be the distance between membranes in the stack. According to (8.17), the confinement will be felt by a membrane when its size reaches L_d such that

$$\frac{k_B T}{\pi\kappa} \left(\frac{L_d}{2\pi}\right)^2 = d^2 \qquad (8.20)$$

In order to compute the free energy of a given membrane, we will assume that it can be divided into blocks of lateral size $L_d \times L_d$, each block being unaffected by the interactions between membranes. The free energy of a block can be computed from the curvature free energy (8.15). For simplicity, we satisfy ourselves with an order of magnitude estimate, replacing $(\nabla^2 h)$ in equation (8.15) by d/L_d^2. Within this rough approximation, the free energy of a block is

$$F(L_d, d) \simeq \kappa L_d^2 \left(\frac{d}{L_d^2}\right)^2 \qquad (8.21)$$

and the total free energy for a membrane of size $L \times L$ is obtained by adding the contributions from different blocks of size $L_d \times L_d$ as

$$F(L,d) \simeq L^2 \frac{(k_B T)^2}{\kappa d^2} \tag{8.22}$$

This free energy corresponds to a repulsive interaction between the membranes, that varies as the inverse square of their distance. When combined with the other types of interactions listed above, this interaction will determine the periodicity of the lamellar stack.

Further reading

Again the best references are Safran's *Statistical Thermodynamic of Interfaces* (Chapter 6) and the proceedings of the Les Houches summer school, *Liquids at Interfaces* (see chapter 7). In the latter see in particular the lectures by W. Helfrich and by J. Meunier.

P.M. Chaikin and T.C. Lubensky, *Principles of Condensed Matter Physics*, Cambridge University Press, Cambridge, 1995, has a long discussion of membranes and generally speaking of fluctuating interfaces.

The book edited by R. Lipowsky and E. Sackmann, *Structure and Dynamics of Membranes*, volume 1, *Handbook of Biological Physics*, Elsevier, Amsterdam, 1995, contains many interesting and recent experimental contributions.

IV Dynamics

9 Phenomenological description of transport processes

9.1 Fluxes, affinities and transport coefficients

The first important concept that underlies any treatment of transport processes is that of *local thermodynamic equilibrium* (LTE). LTE amounts to the assumption that a fluid under non-equilibrium conditions can still be divided into small, mesoscopic regions within which the thermodynamic relations valid for the bulk material retain their validity. In the continuum limit, the variations of the thermodynamic quantities between these regions will be described by thermodynamic 'fields', which are functions of space and time, and whose spatial variation is slow compared to molecular dimensions. Examples include the density field $\rho(\mathbf{r}, t)$, or the temperature field $T(\mathbf{r}, t)$ in an inhomogeneous fluid. The validity of the LTE assumption is a non-obvious result, which relies on a separation of time scales between a microscopic time (typically a few picoseconds) during which molecules equilibrate with respect to their local environment, and a macroscopic evolution time which is determined, as we shall see below in some examples, by the phenomenological evolution equations.

For concreteness, let us consider a solution that can be characterized by two *conserved quantities*, its energy U and the number of solute particles N_s. The total number of molecules (solvent + solute) is assumed to be constant, which essentially amounts to treating the solution as incompressible and allows a simplified description in terms of two thermodynamic parameters only. The bulk thermodynamics of the system is entirely determined by the entropy function $S(U, N_s)$. In particular, this function determines the so-called *affinities* relative to the conserved quantities, defined as

$$(\partial S/\partial E)_{N_s} = \gamma_E = 1/T \tag{9.1}$$

$$(\partial S/\partial N_s)_E = \gamma_{N_s} = -\mu_s/T \tag{9.2}$$

where μ_s is the chemical potential of the solute molecules. A non-equilibrium system will be described by energy and solute density fields, $u(\mathbf{r}, t)$ and $\rho_s(\mathbf{r}, t)$. The local thermodynamic equilibrium assumption allows one to define local values of the affinities, $\gamma_E(\mathbf{r}, t)$ and $\gamma_{N_s}(\mathbf{r}, t)$. Very generally, the second law of thermodynamics implies that an imbalance in the affinities will result in a flow of the corresponding quantity between different parts of the system. It is then natural to assume that the corresponding *fluxes* (particle or energy currents) can be expressed as linear combinations of the *gradients* of the local affinities, a 'linear response' hypothesis expressed as

$$\mathbf{j}_E(\mathbf{r}, t) = L_{EE}\nabla\gamma_E(\mathbf{r}, t) + L_{EN}\nabla\gamma_N(\mathbf{r}, t) \tag{9.3}$$

$$\mathbf{j}_N(\mathbf{r}, t) = L_{NE}\nabla\gamma_E(\mathbf{r}, t) + L_{NN}\nabla\gamma_N(\mathbf{r}, t) \tag{9.4}$$

where for simplicity the usual notation N has been used instead of N_s. The coefficients L_{ij} in these equations are called Onsager coefficients. The diagonal coefficients are obviously related to the usual diffusion constant and thermal conductivity. For example, if the chemical potential is uniform, the first equation is just Fourier's law for heat conduction, $\mathbf{j}_E = -\lambda\nabla T$ with the thermal conductivity λ given by $\lambda = L_{EE}/T^2$. Similarly, if T is uniform and the solution is dilute, $\mu = k_B T \ln \rho$ and the second equation reduces to Fick's law $\mathbf{j}_N = -D\nabla\rho$ with the diffusion constant $D = L_{NN}k_B/\rho$. The significance of the off-diagonal coefficients L_{EN} and L_{NE} is less intuitive. Their existence means, in particular, that a temperature gradient can induce a particle flux. That such an effect is possible is demonstrated by the well known Soret experiment, in which a temperature gradient is established in a fluid slab bounded by two solid walls. In the resulting stationary non-equilibrium state, the current of solute particles is of course zero. Equation (9.4) implies the existence of a gradient in the chemical potential proportional to the temperature gradient. The corresponding gradient in the solute concentration can be detected, for example, by measuring the index of refraction as a function of position.

A very important property of the Onsager coefficients is that, when the linear relations between the fluxes and the gradients are expressed in terms of the proper affinities (as we have done above), the matrix formed by the coefficients is symmetric. In our particular case, this means that $L_{EN} = L_{NE}$. At this point, we simply mention that this intriguing relationship can be traced back to the fact that at a microscopic level, any system is governed by time reversible equations. A simplified proof of these *Onsager reciprocity relations* will be given in section 11.5.

In the following applications, we discuss in some detail the properties of the diffusion equation, which is a particularly important phenomenological transport equation. We then show how this equation can be used to describe the phenomenon of spinodal decomposition.

9.2   Application 1: the diffusion equation

The diffusion equation describes the evolution of solute concentration in a solvent with uniform temperature. It is easily obtained by combining the transport equation (9.3) with the mass conservation equation $\partial\rho/\partial t = -\nabla\mathbf{j}_N$, yielding

$$\frac{\partial\rho(\mathbf{r}, t)}{\partial t} = -\nabla L_{NN}\nabla\gamma_N(\mathbf{r}, t) \tag{9.5}$$

If we assume a weak concentration of solute molecules, then $\gamma_N = k_B \ln(\rho)$. Hence if L_{NN} behaves as $D\rho/k_B$, (9.5) reduces to

$$\frac{\partial\rho(\mathbf{r}, t)}{\partial t} = D\nabla^2\rho(\mathbf{r}, t) \tag{9.6}$$

the usual diffusion equation. D is the diffusion coefficient, expressed in units of cm^2/s, and must be positive, as will become evident below. It is quite instructive before we study this equation to understand why L_{NN} should be proportional to the density, since this will also provide a first encounter with a so-called 'fluctuation–dissipation' relation, first established by Einstein in 1905. Let us consider the case in which the solute particles are subjected to an external force $\mathbf{F}$ (this could be an electric field for charged particles, or the gravitational field for large solutes such as colloids). The current of particles will then be the sum of two contributions. The first one is the diffusion current that results from the inhomogeneity in the particle density, and the second one is the *drift* current related to the external force. At the phenomenological level, one may write

$$\begin{aligned}\mathbf{j}_N(\mathbf{r}, t) &= \mathbf{j}_{\text{diffusion}} + \mathbf{j}_{\text{drift}}\\ &= L_{NN}\nabla\gamma_N(\mathbf{r}, t) + \rho\lambda\mathbf{F}\end{aligned} \tag{9.7}$$

where we have introduced the *mobility* λ of the particles. The drift current takes the form of a particle density times an average velocity $\lambda\mathbf{F}$ of a solute particle drifting under the influence of the external force $\mathbf{F}$. If we now consider the case where $\mathbf{F}$ is the derivative of some external potential $V_{\text{ext}}(\mathbf{r}, t)$, we know that the system can reach an equilibrium state in which the (dilute) solute molecules will be distributed according to the Boltzmann weight, i.e.

$$\rho(\mathbf{r}, t) = \rho_{\text{eq}}(\mathbf{r}) \sim \exp\left(-\beta V_{\text{ext}}(\mathbf{r})\right) \tag{9.8}$$

In such an equilibrium state, the total current (diffusion + drift) must obviously vanish. If we insert the density given by equation (9.8) into equation (9.7), and use $\mu = k_B T \log\rho$ and $\mathbf{F} = -\nabla V_{\text{ext}}$, we find that for any external potential $V_{\text{ext}}(\mathbf{r})$,

$$L_{NN}\beta\nabla V_{\text{ext}}(\mathbf{r}) - \lambda\rho(\mathbf{r})\nabla V_{\text{ext}}(\mathbf{r}) = 0 \tag{9.9}$$

Since this holds for any $V_{\text{ext}}(\mathbf{r})$, L_{NN} must be of the form $k_B D\rho$. Furthermore, a new relationship, known as the Einstein relation, must hold between the mobility λ and the diffusion constant D, namely

$$D = \lambda k_B T \tag{9.10}$$

Hence the *equilibrium* state associated with the Boltzmann distribution (9.8) is not a purely static equilibrium, but rather a fluctuating situation in which a drift current is continuously compensated by a diffusion current.

When associated with the expression for the mobility of a sphere of diameter σ in a liquid of viscosity η derived by Stokes, $\lambda = (6\pi\eta R)^{-1}$ [1], equation (9.10) becomes a very useful relationship between the diffusion constant of a molecule of 'size' (diameter) σ and the viscosity η of the solvent. This Stokes–Einstein relation

$$D = k_B T / 3\pi\eta\sigma \qquad (9.11)$$

holds remarkably well [2], even when the solute molecule is not much larger than the solvent molecules, as would seem to be required by the use of the Stokes expression for the mobility, which follows from a macroscopic description.

The diffusion equation (9.6) is a rather simple partial differential equation, whose solution, however, depends crucially on the boundary conditions. In an infinite volume the equation is best dealt with by introducing the Fourier components $\rho(\mathbf{k}, t)$ of the density, each of which satisfies the simple differential equation

$$\frac{\partial\rho(\mathbf{k}, t)}{\partial t} = -D\mathbf{k}^2\rho(\mathbf{k}, t) \qquad (9.12)$$

Hence the relaxation time of a Fourier component with wavevector k is $\tau(k) = -1/Dk^2$, which implies that the diffusion process rapidly smoothes out an initially rough concentration profile. The divergence in $\tau(k)$ at small wavevector is a natural consequence of the conservation law. Relaxing a long wavelength inhomogeneity requires transport of matter over large length scales, which is a slow process.

For any given initial condition, the solution of the diffusion equation is obtained by an inverse Fourier transform. The most instructive situation corresponds to an initial condition $\rho(\mathbf{r}, t = 0) = N_s\delta(\mathbf{r} - \mathbf{r}_0)$, i.e. a localized injection of N_s solute particles at $\mathbf{r}_0$. In that case one easily obtains [3]

$$\rho(\mathbf{k}, t) = N_s \exp(-\mathbf{k}^2 Dt) \qquad (9.13)$$

---

[1] The derivation of this relation can be found in many hydrodynamic textbooks, see e.g. L. Landau and E.M. Lifshitz, *Hydrodynamics*, Pergamon Press, Oxford, 1959.
[2] Of course, this equation can also be understood as defining a 'hydrodynamic radius' of the solute. It turns out that this determination of the size will be rather close to that obtained from other, more 'chemical', definitions.
[3] One may be surprised here that for any positive t, the probability of finding a particle arbitrarily far from the origin is non-zero. This obviously unphysical feature can be traced back to the breakdown of the local thermodynamic equilibrium assumption in the region where $r \gg Dt$.

and

$$\rho(\mathbf{r}, t) = N_s(4\pi Dt)^{-3/2} \exp(-(\mathbf{r} - \mathbf{r}_0)^2/2Dt) \tag{9.14}$$

This Gaussian function (the *Green's function* of the diffusion equation) shows that the mean squared displacement of the particles from their original positions, $\int d\mathbf{r}\, \rho(\mathbf{r}, t)(\mathbf{r} - \mathbf{r}_0)^2 = 6Dt$, increases linearly with time, again a characteristic feature of diffusion processes.

9.3 Application 2: spinodal decomposition

Let us now consider the case where the 'solute' molecules have a finite concentration. The diffusion equation derived in the above section is now replaced by the slightly more complicated form obtained from equations (9.2) and (9.4) with a uniform temperature

$$\frac{\partial \rho(\mathbf{r}, t)}{\partial t} = \nabla \cdot L_{NN} \nabla \frac{\mu(\mathbf{r}, t)}{T} \tag{9.15}$$

In the *local density approximation* (see section 7.2)), the chemical potential is simply a function of the local density. More generally, it can be defined as $\delta F[\rho]/\delta\rho(\mathbf{r}, t)$, where F is the free energy functional introduced in chapter 7.

For weak concentration gradients, the local density approximation can be used and the equation can be linearized around the uniform state. This results in an equation which takes the same form as in the dilute case (equation (9.6)), with the single particle diffusion constant replaced by a *collective* diffusion constant[4],

$$D_c = \frac{L_{NN}}{T} \left(\frac{\partial \mu}{\partial \rho} \right)_T \tag{9.16}$$

$$= \frac{L_{NN}}{T} \left(\frac{\partial^2 f}{\partial \rho^2} \right)_T \tag{9.17}$$

where $f(\rho, T)$ is the free energy per unit volume. Let us now consider the particularly interesting case of a binary mixture that is quenched down to a temperature below its spinodal line. The free energy versus composition curve is then concave, i.e. the second derivative $d^2 f/d\rho^2$ is negative. This means that $D_c < 0$, and an initial fluctuation in the concentration will grow exponentially as time passes. This growth is the dynamical expression of the thermodynamic instability of the mixture, and the diffusion equation can then be used to study quantitatively the time evolution of the *spinodal decomposition* phenomenon.

Obviously, if the initial density fluctuations grow exponentially, the local density approximation, as well as the linearization of equation (9.15), rapidly fail. A more correct description of the evolution of density fluctuations is then given by the so called Cahn–Hilliard equation, which takes into account both the square

---

[4] This collective diffusion constant will differ from the self diffusion coefficient defined in the previous section when interactions between tracers are not negligible.

gradient term in the free energy functional and the full double well free energy $f(\rho, T)$ (see chapter 7, sections 7.2 and 7.3)

$$\frac{\partial \rho(\mathbf{r}, t)}{\partial t} = -\frac{L_{NN}}{T} \nabla^2 \left[\frac{k_B T \xi_0^2}{\rho} \nabla^2 \rho(\mathbf{r}, t) - \frac{\partial^2 f}{\partial \rho^2} \right] \qquad (9.18)$$

The Laplacian term, which originates from taking the functional derivative of the square-gradient in equation 7.13, will limit the growth rate of the large wavevector Fourier components. At the linear level, this growth rate is $|D_c|k^2 - (L_{NN}k_B\xi_0^2/\rho)k^4$. Since the system has been quenched from a high temperature state, the initial value of the density fluctuation is random, and its mean squared value can be characterized by the structure factor $S(k, t = 0) = \langle|\rho(\mathbf{k}, t = 0)|^2\rangle$. Assuming that at high T this structure factor is essentially flat, the k dependence of the growth rate, which has a maximum at $k^2 \sim |D_c|/L_{NN}\xi_0^2$, will imply the appearance of a peak at finite wavevector in $S(\mathbf{k}, t)$. This corresponds to the emergence of a characteristic size in the growth process, the *domain size*.

Very soon however the linear approximation becomes inappropriate, and the full equation (9.18) must be solved numerically, with the random initial conditions that correspond to the high temperature, frozen-in concentration fluctuations. After some time, the system can be considered as being composed of domains with the concentration of the pure phases separated by interfaces (with an interfacial structure that has locally the sigmoid shape described in section 7.3). The growth of the domains is *self-similar*, i.e. configurations obtained at different times differ only by a scale factor, as illustrated in figure 9.1 Formally, this results in a structure factor having a universal scaling form, $S(k, t) = L^2(t)\Phi(kL(t))$, where $L(t)$ is the typical size of the domains (or correlation length) at time t. The scaling function $\Phi(x)$ has a maximum around $x = 1$, so that, like in the early (linear) phase, $S(k, t)$ has a maximum for finite wavevector, which drifts toward smaller k as the spinodal (or *coarsening*) process evolves. Although very little is known analytically on the solution of equation (9.18), heuristic scaling arguments can be used to predict the growth of $L(t)$ with time. The scaling argument, which can be considerably refined, is roughly the following. The change in the free energy per unit volume across an interface of radius of curvature L is of order γ/L (see chapter 6), where γ is the interfacial energy. If $L(t)$ is the only relevant length scale in the problem (the *scaling assumption*) such changes take place over the scale L. The gradient in the chemical potential, and hence the current, will therefore be proportional to γ/L^2. Since the growth velocity of a domain, dL/dt, is proportional to the current, one may write

$$\frac{dL}{dt} \sim \frac{\gamma}{L^2} \qquad (9.19)$$

which implies that for large t, $L \sim (\gamma t)^{1/3}$.

The approach to spinodal decomposition outlined in this section is, in any case, oversimplified. For long times, the dominant mechanism for mass transport

Figure 9.1. Numerical
solution of equation
(9.18), starting from a
random distribution at
$t = 0$. The light and dark
regions correspond to the
different phases. The two
snapshots are taken at
times that differ by a
factor of ten. The typical
domain size has then
increased by a factor of
approximately 2.15.

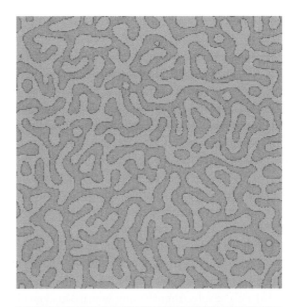

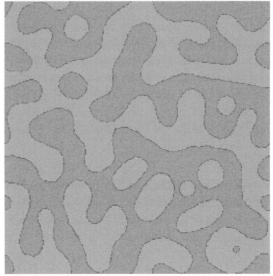

is convection rather than diffusion. In a fluid undergoing spinodal decomposition, internal convective motions will be driven by the presence of interfacial stresses. This feature can modify the long time scaling of $\xi(t)$ with time, in a way which is not yet completely understood[5]. In crystalline solids, even though diffusion

---

[5] A dimensional analysis, balancing the viscous stress at scale L $\eta \partial v / \partial L$ (η is the viscosity) against the interfacial stress γ / L implies (using $v \sim L/t$) the scaling $L(t) \sim \gamma t / \eta$. The cross-over between the diffusive growth and this hydrodynamic regime takes place for $t \sim \eta^{3/2} / \gamma$.

is indeed the main transport mechanism, elastic effects have to be taken into account, since the two pure phases have in general different lattice parameters or lattice structures. The resulting growth patterns can then be highly anisotropic, reflecting the anisotropy of these elastic effects. Nevertheless, the essential feature that emerges from our simple description, i.e. the existence of a characteristic length scale that grows algebraically with time, is very robust. Some controlled porous materials (vycor) are formed by using such a spinodal decomposition pattern. After a controlled decomposition time, the system is quenched so that both pure phases become glassy and no more diffusion takes place. One of the components is then removed by chemical etching, leaving a well defined porous structure (made essentially of silica glass).

Further reading

R. Balian, *From Microphysics to Macrophysics*, Springer-Verlag, Berlin, 1991, gives a very clear account on the general structure of the phenomenological laws, with application to the derivation of general hydrodynamic equations.

A.J. Bray, *Adv. Phys.* **43**, 357 (1994), describes in detail the theory of spinodal decomposition, including consideration of exactly solvable models and models with more than one order parameter, which are useful for describing liquid crystalline systems.

J. Crank, *The Mathematics of Diffusion*, Clarendon Press, Oxford, 1975, describes in detail the methods of solution of the diffusion equation and gives a number of analytical solutions for various boundary conditions.

10 Brownian motion, diffusion and the Langevin equation

10.1 The Langevin equation

10.1.1 The model

The description of transport processes in the previous section was purely phenomenological, and independent of any microscopic model. Generally speaking, a more microscopic description of transport processes in the liquid state is a very difficult many-body problem. Although some exact formal expressions for the transport coefficients can be derived (see e.g. equations (11.7) and (12.15)), they are seldom amenable to analytical evaluation. The *Langevin equation*, originally devised to describe the *Brownian motion* of large (micrometric) colloidal particles in a solvent, provides an intermediate level, semi-empirical description with some microscopic ingredients. The basic equation proposed by Langevin reads

$$M\frac{\mathrm{d}\mathbf{V}}{\mathrm{d}t} = -\zeta M\mathbf{V} + \mathbf{F}(\mathbf{R}) + \theta(t) \tag{10.1}$$

Here $\mathbf{V}$ and $\mathbf{R}$ are the solute particle velocity and position, $-\zeta M\mathbf{V}$ is the 'systematic' contribution to the force that acts on this particle (i.e. the average force the solvent exerts on a particle with constant velocity $\mathbf{V}$; $M\zeta$ is often described as a friction coefficient [1]), $\mathbf{F}(\mathbf{R})$ is an external force (e.g. a gravitational field) and θ is the so-called 'random force'. This random force is supposed to describe the many collisions experienced by the solute particle. The essential assumption in the Langevin approach is that the typical time scale over which these collisions take place is very short compared to the time scale for the evolution of the velocity of the solute. Formally, the corresponding assumption is that the random force in equation (10.1) has the following statistical properties

$$\langle \theta_\alpha(t) \rangle = 0 \tag{10.2}$$

$$\langle \theta_\alpha(t)\theta_\beta(t') \rangle = 2\Theta_0\delta_{\alpha\beta}\delta(t - t') \tag{10.3}$$

---

[1] With this notation, the mass M conveniently factors out; ζ has the dimension of inverse time, and $1/\zeta$ is the time it would take for the velocity to relax to zero in the absence of the random force.

Here the Greek indices refer to the three directions in space, and the angular brackets correspond to a statistical average over the 'realizations' of the random force, a notion which will be made more precise below. Equation (10.3) simply expresses the fact that the average effect of the solvent has been taken into account in the first, 'systematic' term on the right-hand side of equation (10.1). Equation (10.3) reflects, in a mathematically idealized fashion, the uncorrelated character of the collisions. On the time scale of interest, the forces exerted at times t and t' by the solvent on the solutes are modelled as *independent random variables*.

Equation (10.1) is a stochastic differential equation, a mathematical object that may not be very familiar to all readers. As is usual in statistical physics, the quantities of interest are not the exact velocities or positions of the particles, but rather their *probability distribution*, which for non-stationary situations is time dependent. In this context, all the quantities appearing in (10.1) must be interpreted as random variables whose probability distribution corresponds to the probability distribution of the solute particle coordinates. Equation (10.1) provides a link between these different random variables, and allows one to compute the probability distribution of $\mathbf{V}$ and $\mathbf{R}$ from those of θ, which are assumed to be known. Hence the angular brackets in (10.2) and (10.3) correspond, in fact, to an average over the *time-independent* probability distribution of θ, and the Langevin equation can also be interpreted as an evolution equation for the probability distribution of $\mathbf{V}$, as we shall see in section 10.2.

10.1.2 The fluctuation–dissipation relation

In the absence of an external force, the general solution of equation (10.1) can be written as

$$\mathbf{V}(t) = \mathbf{V}(0) \exp(-\zeta t) + \frac{1}{M} \int_0^t ds \exp(-\zeta (t - s)) \, \theta(s) \tag{10.4}$$

Again, the quantities of interest are averages. For a system at equilibrium, the average value of $\mathbf{V}(t)$ vanishes. Its mean squared value, however, is non-zero and can be easily computed as

$$\langle \mathbf{V}(t)^2 \rangle = \langle \mathbf{V}(0)^2 \rangle \exp(-2\zeta t) + \frac{1}{M^2} \int_0^t \int_0^t ds \, ds' \exp(-\zeta(2t - s - s')) \langle \theta(s) \cdot \theta(s') \rangle \tag{10.5}$$

The cross terms $\langle \theta(s) \cdot \mathbf{V}(0) \rangle$ have vanished, since the velocity at $t = 0$ depends only on the previous force history, and is therefore decorrelated from the force at later times. Now, using equation (10.3), it is easily seen that

$$\lim_{t \to \infty} \langle \mathbf{V}(t)^2 \rangle = \frac{3 \Theta_0}{M^2 \zeta} \tag{10.6}$$

From equilibrium statistical mechanics and the equipartition theorem, we know that the left-hand side in equation (10.6) is simply $3k_B T / M$. Hence we have obtained the *fluctuation–dissipation* relation

$$M\zeta k_B T = \Theta_0 \tag{10.7}$$

between the strength Θ_0 of the *fluctuating* random force and the viscous, *dissipative* friction ζ. This very important relation exemplifies the intimate connection between the microscopic scale (the random force) and the macroscopic, or mesoscopic scale (the viscous friction on a mesoscopic particle can be obtained from purely macroscopic, hydrodynamic equations; see equation (9.11)). We shall see other examples of this fundamental connection below.

10.1.3 Velocity correlations and diffusion

The time-dependent, equilibrium correlation function of the particle velocity is easily obtained by multiplying equation (10.1) by the velocity at t' and averaging, as

$$\langle \mathbf{V}(t).\mathbf{V}(t') \rangle = \frac{3k_B T}{M} \exp(-\zeta |t - t'|) \tag{10.8}$$

From this it is a simple exercise to obtain the mean squared displacement of the particle

$$\langle (\mathbf{R}(t) - \mathbf{R}(t'))^2 \rangle = \frac{6k_B T}{M\zeta}(|t - t'| + \zeta^{-1}(\exp(-\zeta |t - t'|) - 1)) \tag{10.9}$$

Both equations are rich in physical information. From equation (10.8), it is clear that the time scale over which the particle keeps a memory of its initial velocity is ζ^{-1}, which defines a 'velocity relaxation time'. From equation (10.9), it appears that the mean squared displacement of the particle grows linearly with time for $t > \zeta^{-1}$. This *diffusive* behaviour is characterized by the diffusion constant $D = k_B T / M\zeta$. For shorter times, however, the mean squared displacement is proportional to t^2. This corresponds to ballistic motion, before the interactions with the bath (i.e. the random force) cause the velocity to become decorrelated from its initial value. The situation is summarized in figure 10.1. The Langevin equation is of course a very crude model, in which all the 'microscopic' aspects have been lumped into the 'random force', which is assumed to have a vanishing correlation time. Physically, this corresponds to the fact that the interactions with the solvent molecules take place on a time scale much shorter that ζ^{-1}. In a more microscopic description, this time scale should be considered and the part of the curve close to $t = 0$ in figure 10.1 would reveal a much richer structure, corresponding to the very short time correlations in the solvent–particle interactions. These correlations are quite irrelevant for a study that considers

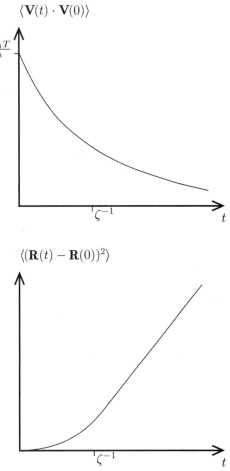

Figure 10.1. Velocity autocorrelation and mean squared displacement for a Brownian particle described by the Langevin equation. The cross-over from ballistic to diffusive motion takes place on the time scale ζ^{-1}.

mesoscopic time and length scales. They become essential, however, if the particle size is of the order of the molecular size, in which case the Langevin description is inappropriate [2].

10.2 The Fokker–Planck description

The Langevin description is physically appealing but, being based on a stochastic differential equation, not always easy to deal with. In this section we introduce an equivalent description which deals directly with the *probability distribution*

---

[2] In fact, the Langevin equation for colloidal particles is rigorously valid only in the (unpractical) limit where the mass density of the large particle is much larger than the mass density of the fluid. For equal mass densities, the particle motion tends to be correlated with the hydrodynamic flow it induces in the solvent over long time scales, and the separation of time scales between particle and solvent does not, rigorously speaking, hold any longer.

function (pdf) $P(\mathbf{R}, \mathbf{V}, t)$ of the stochastic variables. For the sake of simplicity we will limit our study to a unidimensional situation (the generalization to three dimensions is fairly obvious). In the first paragraph we discuss the evolution of the pdf for the velocity in the case where no external force is applied. The derivation of the evolution equation for this pdf (namely the Fokker–Planck equation) is somewhat technical, so that the reader who is not interested in mathematical developments may, in a first approach, skip the corresponding section.

10.2.1 Evolution of the velocity distribution

To study the evolution of the velocity distribution function, we first consider the stochastic variable $\xi(t, \Delta t)$ defined as the velocity increment between t and $t + \Delta t$. For small Δt, integration of (10.1) (in the absence of an external force, $F = 0$) yields

$$\xi(t, \Delta t) = -\zeta V(t)\Delta t + \frac{1}{M} \int_t^{t+\Delta t} \theta(s)\, ds \tag{10.10}$$

For a given $V(t)$, the distribution of the velocity increment ξ will be determined by the values taken by the random force $\theta(s)$ in the interval $[t, t + \Delta t]$. Since the random force has a vanishing correlation time, the integral on the right-hand side of (10.10) can be considered as an infinite sum of uncorrelated contributions. The central limit theorem then states that this sum has a Gaussian distribution, entirely characterized by the knowledge of its first and second moments. As a result, the distribution function of ξ, *given* $V(t)$, $P(\xi; V(t))$, can be written as

$$\Pi(\xi; V(t)) = \left(\frac{M^2}{2\pi\Theta_0\Delta t}\right)^{1/2} \exp\left(-\frac{M^2\,(\xi + \zeta V(t)\Delta t)^2}{2\Theta_0\Delta t}\right) \tag{10.11}$$

The next step is to obtain the distribution function for V. We start from the *convolution* equation

$$P(V, t + \Delta t) = \int d\xi\, P(V - \xi, t)\Pi(\xi; V - \xi) \tag{10.12}$$

which expresses that in order to have velocity V at time $t + \Delta t$, the particle must have velocity $V - \xi$ at time t *and* a velocity increment ξ in the interval $[t, t + \Delta t]$. The trick is then to expand $P(V - \xi, t)\Pi(\xi; V - \xi)$ in a Taylor series with respect to the increment ξ appearing in the $V - \xi$ argument of the functions, and $P(V, t + \Delta t)$ in a Taylor series with respect to Δt. This yields

$$P(V, t) + \Delta t\frac{\partial P}{\partial t} + \cdots = P(V, t)\int d\xi\, \Pi(\xi; V) - \frac{\partial}{\partial V}\int d\xi\, \xi P(V, t)\Pi(\xi; V)$$
$$+ \frac{1}{2}\frac{\partial^2}{\partial V^2}\int d\xi\, \xi^2 P(V, t)\Pi(\xi; V) + \cdots \tag{10.13}$$

Using equation (10.11) and identifying the two expansions to order Δt one ends up with the *Fokker–Planck* equation

$$\frac{\partial P}{\partial t} = \frac{\partial}{\partial V}\left(\zeta V P(V,t) + \frac{\partial}{\partial V}\left(\frac{\Theta_0}{M^2}P(V,t)\right)\right) \tag{10.14}$$

Equation (10.14) can be interpreted as an equation of conservation for the probability density. The 'probability current' is the sum of a *drift* (or convection) term $\zeta V P(V,t)$ and of a *diffusive* term, $\frac{\partial}{\partial V}(\frac{\Theta_0}{M^2}P(V,t))$. The fluctuation–dissipation relation (10.7) is recovered by expressing that, when the pdf is equal to the Maxwell distribution, $P(V) = (M/2\pi k_B T)^{1/2}\exp(-MV^2/2k_B T)$, the total probability current must vanish. That this is the case is of course necessary for the model to describe equilibrium systems properly.

The structure of equation (10.14) is quite similar to that of the Schrödinger equation, so that the general methods known from quantum mechanics can be used for its solution. We leave it as an exercise for the reader to prove

- that the function $\psi(V,t)$ defined by $\psi(V,t) = P(V,t)\exp(MV^2/4k_B T)$ obeys the Schrödinger equation for a quantum harmonic oscillator, with the time replaced by an imaginary time it.
- that, as a consequence, the solution of (10.14), starting from any initial condition $P(V,0)$, evolves towards the Maxwellian distribution over a time scale ζ^{-1}.

The Fokker–Planck equation provides hence a description of the approach to equilibrium for a 'free' (no external potential) Brownian particle. This approach always takes place over a time scale ζ^{-1} which, as we have seen above, is the relevant time scale for velocity correlations. We will use this result in the next section to treat the (physically more interesting!) case of a Brownian particle in an external potential.

10.2.2 Kramers and Smoluchovski equations

The *Kramers* equation is obtained by extending the method used in section 10.2.1 for $P(V,t)$ to the probability distribution function $P(X,V,t)$ of the position and velocity variables. Starting from the coupled differential equations

$$\frac{dX}{dt} = V \tag{10.15}$$

$$\frac{dV}{dt} = -\zeta V + \frac{1}{M}F_{\text{ext}}(X) + \frac{1}{M}\theta(t) \tag{10.16}$$

one ends up with an equation of conservation for the probability density in the form

$$\frac{\partial P}{\partial t} = -\frac{\partial}{\partial X}V P(X,V,t) + \frac{\partial}{\partial V}\left[\left(\zeta V - \frac{F_{\text{ext}}(X)}{M}\right)P(X,V,t)\right.$$
$$\left. + \frac{\partial}{\partial V}\left(\frac{\Theta_0}{M^2}P(X,V,t)\right)\right] \tag{10.17}$$

The structure of the Kramers equation (10.17) is similar to that of the Fokker–Planck equation. The absence of a diffusion term relative to the variable X simply reflects the absence of noise in equation (10.15). However, the added variable and the coupling term involving $F_{ext}(X)$ make this equation much more difficult to solve than the Fokker–Planck equation. Here we only note that when the force F_{ext} is conservative, of the form $F_{ext} = -dU_{ext}/dX$, the Maxwell–Boltzmann distribution $\exp[-(MV^2/2 + U(X))/k_B T]$ is an equilibrium solution of (10.17), provided the fluctuation–dissipation relation (10.7) is obeyed. Again, the fluctuation–dissipation relation is necessary to obtain a correct description of a system at equilibrium.

The Kramers equation can be considerably simplified in the limit of large friction, $\zeta \to \infty$, in which case it reduces to the diffusion equation studied in application 1 of chapter 9. In this context, this equation is usually known as the *Smoluchovski* equation. A rigorous derivation of the Smoluchovski equation from the Kramers equation necessitates the use of mathematically involved techniques such as the multiple time scale method[3], which we avoid here by a 'quick and dirty' heuristic derivation. We define the particle density and current, respectively, by

$$\rho(X, t) = \int dV \, P(X, V, t) \tag{10.18}$$

$$J(X, t) = \int dV \, V P(X, V, t) \tag{10.19}$$

Integrating equation (10.17) over velocities one easily obtains[4] the continuity equation

$$\frac{\partial \rho(X, t)}{\partial t} + \frac{\partial J(X, t)}{\partial X} = 0 \tag{10.20}$$

In the same way, a multiplication by V followed by an integration over V yields an equation of motion for J in the form

$$\frac{\partial J(X, t)}{\partial t} + \frac{\partial}{\partial X} \left(\int dV \, V^2 P(X, V, t) \right) - \frac{F_{ext}(X)}{M} \rho(X, t) = -\zeta J(X, t) \tag{10.21}$$

The problem is now to obtain a closed equation for $\rho(X, t)$ from some physically sensible approximation. In the limit of large friction, we assume that the velocity distribution function equilibrates much faster than the position distribution function. This assumption is of course motivated by the result obtained earlier for a free particle, that the velocity correlations relax on a time scale ζ^{-1}. After a time ζ^{-1}, it should therefore be legitimate to replace the distribution function $P(X, V, t)$ by a 'local equilibrium' distribution, $\rho(X, t) \times (M/2\pi k_B T)^{1/2} \exp(-MV^2/2k_B T)$,

---

[3] L. Bocquet, High friction limit of the Kramers equation: the multiple time-scale approach, *Am. J. Phys.* **65**, 140 (1997).

[4] Upon integrating by parts, the contribution from the last term in (10.17) vanishes.

to compute the second moment appearing in (10.21). With this approximation (10.21) becomes

$$\frac{\partial J(X, t)}{\partial t} + \frac{\partial}{\partial X}\left(\frac{k_B T}{M}\rho(X, t)\right) - \frac{F_{ext}(X)}{M}\rho(X, t) = -\zeta J(X, t) \tag{10.22}$$

A further approximation, again justified in the large ζ limit, is to neglect the inertial term $\partial J(X, t)/\partial t$ compared to the friction term $-\zeta J(X, t)$. Indeed, if we study phenomena that take place over a time scale t_0 much larger than ζ^{-1}, $\partial J(X, t)/\partial t$ is of order $J/t_0 \ll \zeta J$. Equation (10.22) is then replaced by

$$J(X, t) = \zeta^{-1}\left(\frac{F_{ext}(X)}{M}\rho(X, t) - \frac{\partial}{\partial X}\left(\frac{k_B T}{M}\rho(X, t)\right)\right) \tag{10.23}$$

which corresponds exactly to equation (9.7) with a mobility $(M\zeta)^{-1}$ and a diffusion constant $k_B T/M\zeta$, obeying the fluctuation–dissipation relation (9.10). The Smoluchovski equation is obtained by combining (10.20) and (10.23) into

$$\frac{\partial \rho(X, t)}{\partial t} + \frac{\partial}{\partial X}\left[\zeta^{-1}\left(\frac{F_{ext}(X)}{M}\rho(X, t) - \frac{\partial}{\partial X}\left(\frac{k_B T}{M}\rho(X, t)\right)\right)\right] = 0 \tag{10.24}$$

We can now try to understand more physically and to quantify the notion of a 'high friction' limit. The essential assumption is the existence of a time scale t_0 over which the velocity distribution of the particle equilibrates, while the position probability distribution function does not evolve much. Further evolution of the position pdf is then governed by equation (10.24). The time scale t_0 should obey the inequalities

$$\zeta^{-1} \ll t_0 \ll \frac{\sigma^2}{(k_B T/M\zeta)} \tag{10.25}$$

where the upper bound corresponds to the diffusion of the particle over its own size σ. Obviously both inequalities can be satisfied, and the approach is meaningful, if $\zeta \gg (k_B T/M\sigma^2)^{1/2}$.

The Smoluchovski equation is often taken as the starting point for studying the dynamics of colloidal particles, or other problems in which the *inertial term in the Langevin equation*, which governs the dynamics on the fast time scale ζ^{-1}, can be ignored. This is in principle valid when the dynamics governed by the Langevin equation is strongly overdamped. In practice, however, this reduced approach is often used for simplicity, and because there is little interest, in the problem under consideration, in studying the motion on very short time scales.

10.3 Application 1: rotational diffusion

In this chapter, we have up to now been interested in the description of translational dynamics for a large, slow particle. A completely parallel description is possible for the rotational degrees of freedom of a non-spherical particle. For simplicity, let us consider a rod-like particle (molecule). The orientation of the particle is described by a unit vector $\mathbf{u}(t)$, the orientation of the particle axis. Equations very similar to the Langevin equation may be written for the time evolution of $\mathbf{u}(t)$ and of the angular velocity $\boldsymbol{\omega}$, namely

$$\frac{d\mathbf{u}}{dt} = \boldsymbol{\omega} \wedge \mathbf{u} \qquad (10.26)$$

$$J\frac{d\boldsymbol{\omega}}{dt} = -\zeta J\boldsymbol{\omega} + \mathbf{C}_{\text{ext}}(\mathbf{u}) + \mathbf{C}_{\text{rand}}(t) \qquad (10.27)$$

Here $J = ML^2/12$ for a rod of length L is the moment of inertia of the particle, $\mathbf{C}_{\text{ext}}$ is an external torque and $\mathbf{C}_{\text{rand}}(t)$ is a random torque, with properties similar to the random force in (10.1). For an external torque that can be derived from an external potential V_{ext}, $\mathbf{C}_{\text{ext}}(\mathbf{u}) = -\mathbf{u} \wedge \partial V_{\text{ext}}/\partial \mathbf{u}$, where the derivative with respect to $\mathbf{u}$ is the angular part of the gradient operator.

It is then possible to carry out an analysis similar to that yielding (10.24). In the absence of an external torque, the resulting *angular diffusion* equation simply reads

$$\frac{\partial P(\mathbf{u}, t)}{\partial t} = D_{\text{r}}\frac{\partial^2 P(\mathbf{u}, t)}{\partial \mathbf{u}^2} \qquad (10.28)$$

with $D_{\text{r}} = k_{\text{B}}T/J\zeta$ the rotational diffusion constant. The differential operator on the right-hand side is the angular part of the Laplacian operator, which in spherical coordinates θ, ϕ reads

$$\frac{\partial^2}{\partial \mathbf{u}^2} = \frac{1}{\sin\theta^2}\left(\sin\theta\frac{\partial}{\partial\theta}\sin\theta\frac{\partial}{\partial\theta}\right) + \frac{\partial^2}{\partial\phi^2} \qquad (10.29)$$

Equation (10.28) is, in fact, the diffusion equation for a 'particle' (representing the end of the vector $\mathbf{u}$) diffusing on the surface of the unit sphere. As this equation is identical to the Schrödinger equation for a rigid rotator, the solution can be written in terms of the eigenfunctions of the Laplacian, i.e. the spherical harmonics $Y_l^m(\mathbf{u})$. For an initial condition $P(\mathbf{u}, t = 0) = \delta(\mathbf{u} - \mathbf{u}_0)$, the solution is

$$P(\mathbf{u}, t) = \sum_{l,m} \exp\left(-l(l+1)D_{\text{r}}t\right) Y_l^m(\mathbf{u})Y_l^{-m}(\mathbf{u}_0) \qquad (10.30)$$

which shows that the orientations become isotropically distributed over a time scale D_{r}^{-1}.

If we now include the possibility of an external torque, the diffusion equation becomes of the Smoluchovski type

$$\frac{\partial P(\mathbf{u}, t)}{\partial t} = \frac{\partial}{\partial \mathbf{u}}D_{\text{r}}\left(\frac{\partial P(\mathbf{u}, t)}{\partial \mathbf{u}} + \frac{1}{k_{\text{B}}T}P(\mathbf{u}, t)\mathbf{C}_{\text{ext}}(\mathbf{u})\right) \qquad (10.31)$$

which shows that the angular mobility μ_{r}, relating the current to the torque, is given by $\mu_{\text{r}} = D_{\text{r}}/k_{\text{B}}T$, which is the counterpart of equation (9.10) in the case of rotational diffusion. The angular mobility can be assumed to be inversely proportional to the viscosity of the fluid, which implies the *Debye–Einstein* relationship $D_r \sim k_B T / \eta$.

If the molecule has a dipole moment, the external torque may correspond to the action of an external electric field, and equation (10.31) will then allow a description of dielectric relaxation phenomena. In the following, we give an elementary description for the case of two-dimensional systems, in which the orientation of the molecules is restricted to the unit circle. In that case, the energy of a molecule with dipole moment $\mathbf{p}$ in a field $\mathbf{E}$ parallel to the x axis is

$$V_{ext}(\phi) = -\mathbf{p} \cdot \mathbf{E} = -pE \cos \phi \qquad (10.32)$$

where ϕ is the angle between the dipole and the x axis. For a two-dimensional system, the appropriate Smoluchovski equation for rotational diffusion is simply equation (10.24) describing one-dimensional Brownian motion on the unit circle, $0 \le \phi \le 2\pi$, with the appropriate periodic boundary conditions,

$$\frac{\partial P(\phi, t)}{\partial t} = \frac{\partial}{\partial \phi} \left(D_r \left(\frac{\partial P(\phi, t)}{\partial \phi} + \frac{1}{k_B T} P(\phi, t) \frac{\partial V_{ext}(\phi)}{\partial \phi} \right) \right) \qquad (10.33)$$

The equilibrium solution for a constant electric field is obviously the Boltzmann distribution, $P_{eq}(\phi) = A \exp(pE \cos \phi / k_B T)$, with A a normalization constant. Dielectric relaxation experiments, however, involve the study of the response to a weak, time dependent electric field. If the applied electric field is such that $pE \ll k_B T$, the external potential can be considered as a perturbation. Writing the time dependent distribution as $P(\phi, t) = (2\pi)^{-1} + \delta P(\phi, t)$ and linearizing equation (10.33) (i.e. neglecting terms of order $V_{ext} \delta P$), we obtain, for a time dependent field $E(t)$,

$$\frac{\partial \delta P(\phi, t)}{\partial t} = \frac{\partial}{\partial \phi} D_r \left(\frac{\partial \delta P(\phi, t)}{\partial \phi} + \frac{pE(t)}{2\pi k_B T} \cos \phi \right) \qquad (10.34)$$

If $E(t) = E_0 f(t)$, with $f(t)$ a dimensionless function, a solution for $\delta P(\phi, t)$ can be found in the form $(pE/k_B T)(g(t)/2\pi) \cos \phi$. Equation (10.34) reduces to a simple differential equation for $g(t)$,

$$\frac{dg(t)}{dt} = -D_r g(t) + f(t) \qquad (10.35)$$

Two cases are of interest here, that of a step function ($f(t) = 1$ for $t \le 0$, $f(t) = 0$ for $t \ge 0$), and that of a sinusoidal field. For a step function, one obtains

$$g(t) = \exp(-D_r t) \qquad \text{for} \qquad t \ge 0 \qquad (10.36)$$

Hence the dipole moment will equilibrate over a time scale D_r^{-1} after the electric field has been switched off. For a sinusoidal function $f(t) = \exp(-i\omega t)$, a forced oscillation of the dipole moment will follow after a transient, i.e.

$$g(t) = \frac{D_r}{2\pi(D_r - i\omega)} \exp(-i\omega t) = (g' + ig'') \exp(-i\omega t) \qquad (10.37)$$

Figure 10.2. Real (dashed line) and imaginary (full line) parts of the Debye response function, as a function of frequency. In a log–linear plot such as this one, which is often used to account for the broad frequency range used in experiments, the width at half maximum of the absorption peak in the imaginary part is 1.14 decades.

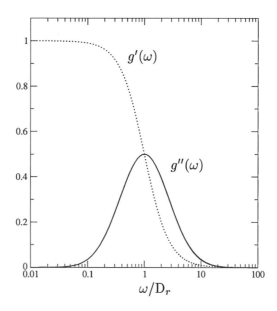

The real part $g'(\omega) = D_r^2/2\pi(D_r^2 + \omega^2)$ is the response in phase with the electric field. The imaginary part, $g''(\omega) = \omega D_r/2\pi(D_r^2 + \omega^2)$, represents the component of the response in quadrature with the field. The latter is associated with the dielectric loss in the liquid, which is proportional to the average of $E \cdot dP/dt$. Both components can be measured in a dielectric relaxation experiment, in which the capacitance and loss angle of a capacitor filled with the polarizable liquid are studied as a function of frequency. Figure 10.2 displays g' and g'' as a function of frequency. The behaviour of the response function illustrated in figure 10.2 is known as a *Debye response*, and is of course directly related to the simple exponential relaxation following a step function perturbation, equation (10.36). The Debye response is characteristic of an ensemble of non-interacting dipoles, in which case the relaxation time can be related to the rotational diffusion constant of a single dipole. This simple behaviour, however, is commonly found for the low frequency dielectric response of molecular liquids at high temperatures, even in the presence of interactions between dipoles. At high frequencies, other phenomena such as electronic and atomic polarizabilities come into play. At low temperatures, interactions become important and deviations from the simple Debye behaviour are observed. Dielectric relaxation is then a very useful tool for characterizing the collective behaviour in dense, cold molecular liquids. Deviations from the Debye behaviour are conveniently displayed in the so-called Cole–Cole plots, in which a parametric curve in the plane g',g'' is drawn as a function of ω. Deviations from a circular shape imply non-Debye relaxation (see figure 10.3).

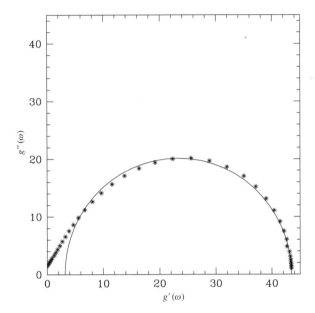

Figure 10.3. Cole–Cole representation of the frequency dependent dielectric constant of glycerol, $g(\omega) = g'(\omega) + ig''(\omega)$, at $T = 218$ K (courtesy of C. Alba, Orsay). The data points are taken for frequencies between 100 kHz and 10 Hz. Deviation from the circular, Debye behaviour (full line) is observable at high frequencies, near the origin.

10.4 Application 2: polymer dynamics, Rouse and reptation models

The Rouse model

The Rouse model of polymer dynamics, proposed by P.E. Rouse in 1955, can be seen as a dynamical extension of the Gaussian chain model (see section 1.5). The dynamics of each 'bead' is described by a Langevin equation [5] in the large friction limit. The position $\mathbf{R}_i(t)$ of bead i obeys

$$\zeta \frac{d\mathbf{R}_i}{dt} = \frac{3k_B T}{b^2}(\mathbf{R}_{i+1} + \mathbf{R}_{i-1} - 2\mathbf{R}_i) + \Theta_i(t) \tag{10.38}$$

The first term on the right-hand side is the 'elastic' force between beads of the Gaussian chain. The second term is a random force with the usual properties (equations (10.2), (10.3)), and which is uncorrelated for different beads. This 'Rouse dynamics' does not take into account the excluded volume interactions between beads, or the fact that chain crossings are forbidden. The chain is often described as a 'phantom chain'. The use of a Langevin equation for each bead clearly corresponds to a coarse-grained approach, in which each bead is much larger than a solvent molecule, and hence consists of several monomers. This is consistent with the use of an entropic elastic force, as in the usual Gaussian chain model (see section 1.5).

---

[5] Traditionally, the friction in the Rouse model is written ζ rather than $M\zeta$ as used in section 10.1, since the particle mass is irrelevant in the large friction limit.

Equation (10.38) is usually replaced by its continuum limit, where the position along the chain is described by a function $\mathbf{R}(s, t)$ with the coordinate s varying from 0 to N, the chain length. In this continuum limit the elastic force is replaced by a second derivative, and equation (10.38) becomes

$$\zeta \frac{\partial \mathbf{R}(s, t)}{\partial t} = \frac{3k_B T}{b^2} \frac{\partial^2 \mathbf{R}(s, t)}{\partial s^2} + \Theta(s, t) \tag{10.39}$$

The absence of correlation between different monomers is expressed by $\langle \Theta_\alpha(s, t)\Theta_\beta(s', t')\rangle = \delta_{\alpha\beta}\delta(s - s')\delta(t - t')(2k_B T\zeta)$. The appropriate boundary condition for the partial differential equation (10.39) is

$$\left. \frac{\partial \mathbf{R}(s, t)}{\partial s} \right|_{s=0, N} = 0 \tag{10.40}$$

This boundary condition expresses the fact that no external forces are applied to the first and last monomer, so that the tension of the chain vanishes at both extremities. The Rouse equations (10.39) and (10.40) are easily solved by introducing a discrete set of generalized coordinates known as the *Rouse modes*, defined by

$$\mathbf{X}_p(t) = \frac{1}{N} \int_0^N ds \cos\left(\frac{p\pi s}{N}\right) \mathbf{R}(s, t) \tag{10.41}$$

Inversely $\mathbf{R}(s, t)$ can be expressed as a Fourier series in terms of these Rouse modes

$$\mathbf{R}(s, t) = \mathbf{X}_0 + 2\sum_{p>0} \mathbf{X}_p(t) \cos\left(\frac{p\pi s}{N}\right) \tag{10.42}$$

The Rouse mode $\mathbf{X}_0(t)$ is simply the position of the chain's centre of mass. The first mode can be considered as defining some sort of 'dipole', with the first half of the monomers ($0 < s < N/2$) having a positive weight and the second half a negative weight. The evolution of this mode will be related to changes in the general chain orientation. Higher order modes are associated with internal motion of growing complexity. Multiplying equation (10.39) by $\cos(p\pi s/N)$ and integrating between 0 and N, it is easily shown that the Rouse modes obey

$$N\zeta \frac{d\mathbf{X}_0(t)}{dt} = \Theta_0(t) \tag{10.43}$$

$$2N\zeta \frac{d\mathbf{X}_p(t)}{dt} = -2N\left(\frac{p\pi}{N}\right)^2 \frac{3k_B T}{\pi^2} \mathbf{X}_p(t) + \Theta_p(t) \qquad \text{for} \qquad p > 0 \tag{10.44}$$

The new 'random forces' appearing in equation (10.44) are decorrelated from each other and satisfy the fluctuation–dissipation relation with the appropriate friction coefficient,

$$\langle \theta_{0,\alpha}(t)\theta_{0,\beta}(t')\rangle = (N\zeta)(2k_B T)\delta_{\alpha\beta} \tag{10.45}$$

$$\langle \theta_{p,\alpha}(t)\theta_{p,\beta}(t')\rangle = (2N\zeta)(2k_B T)\delta_{\alpha\beta} \tag{10.46}$$

These relations ensure that the equilibrium probability distribution function of the Rouse modes will be the Gaussian distribution

$$P(\mathbf{X}_0, \mathbf{X}_1, ...) \sim \exp\left(-\sum_{p>0} \frac{3\pi^2 p^2}{Nb^2} \mathbf{X}_p^2\right) \tag{10.47}$$

which is simply a rewriting of the equilibrium Gaussian distribution (1.27) in terms of the Rouse coordinates. More interesting are the dynamical properties that can be deduced from equations (10.43) and (10.44). Equation (10.43) shows that the centre of mass of the chain is undergoing simple Brownian motion under the effect of the random forces. The friction on the centre of mass is the sum of the frictions on the different monomers, and the corresponding diffusion constant, $D = k_B T / N\zeta$, scales as the inverse of the molecular weight. Each Rouse mode $\mathbf{X}_p$, $p \geq 1$, performs a Brownian motion within a harmonic potential well. At equilibrium, this motion is characterized by the correlation function

$$\langle \mathbf{X}_p(t) \cdot \mathbf{X}_p(0) \rangle = \frac{Nb^2}{6\pi^2 p^2} \exp\left(-\frac{p^2 t}{\tau_r}\right) \tag{10.48}$$

where

$$\tau_r = \frac{N^2 \zeta b^2}{3\pi^2 k_B T} \sim \frac{R_G^2}{D} \tag{10.49}$$

is the Rouse time. Note that in accordance with the general philosophy underlying the use of coarse-grained models, this Rouse time can be expressed in terms of other parameters characteristic of the chain, and does not involve explicitly the underlying microscopic parameters b and ζ. The Rouse time is the slowest relaxation time for the internal motion of the chain, and scales as the molecular weight squared.

From equations (10.48) and (10.49), it is seen that the internal motion of the chain is associated with a rich *spectrum of relaxation times*. Any experiment probing this motion is likely to reveal correlation functions which will be a combination of the correlation functions (10.48), and hence will be a complex sum of exponentials. This complex response is indeed characteristic of polymeric systems, as probed for example in dielectric or rheological experiments. A typical response curve is shown in figure 10.4, and compared to a single exponential response.

The Rouse model was originally proposed as a description for dilute solutions of polymer chains. It turns out, however, that the model is not appropriate in this case. The essential reason lies in the neglect of the *hydrodynamic interactions*. Briefly, the hydrodynamic interactions describe the fact that a given bead of the chain is experiencing a flow of the surrounding solvent, caused by the motion of the other beads. This results in a *long-range interaction*, enhancing the correlations between the motion of distant monomers. A discussion of the corresponding formalism goes beyond the scope of this chapter. The essential result, however, is

Figure 10.4. A correlation function formed by adding up weighted contributions of all Rouse modes, $g(t) = \sum_p p^{-2} \exp(-t/p^2 \tau_r)$ (full line) is compared to the single exponential $\exp(-t/\tau_r)$ (dashed line). Summing over modes implies a more stretched relaxation. The function $g(t)$ can be shown to describe the viscoelastic response of the Rouse model.

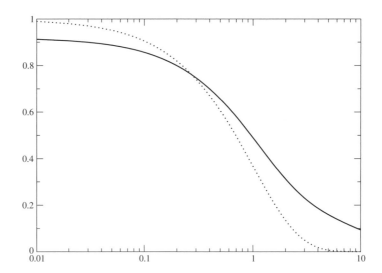

that the polymer chains in the solvent tend to behave essentially as hard spheres of diameter R_G, with a diffusion constant scaling as $k_B T / R_G \eta$, with η the solvent viscosity. The solvent molecules within the volume occupied by the chain are dragged along with the chain, and screened from the external flow. Another consequence of taking into account the hydrodynamic interactions correctly is that the longest relaxation time for internal motions of the chain scales as $N^{3/2}$ rather than N^2. Relaxation is enhanced by the long-range hydrodynamic interactions between monomers.

In spite of its failure to describe the dynamics of dilute chains, the Rouse model turns out to be quite successful for the description of polymer melts, at least for low molecular weights. In this situation, the surrounding chains behave as an effective medium that creates the 'noise' in equation (10.38). The hydrodynamic interactions become much smaller, due to the high viscosity of the medium, and the Rouse model becomes appropriate. Hence the diffusion coefficient of a short chain in a melt scales as N^{-1}, and the Rouse spectrum of relaxation times provides a reasonable description of the response function of the melt.

The reptation model

Experimentally, it is found that the dynamical properties of polymer melts are adequately described by the Rouse model only for short chains, with molecular weight N smaller than some cross-over value denoted by N_e. For $N \geq N_e$, the N dependence of the diffusion constant changes from the N^{-1} behaviour typical of the Rouse model to a N^{-2} dependence (see figure 10.5). Similar changes are found for the N dependence of the viscosity of the melt. These changes around

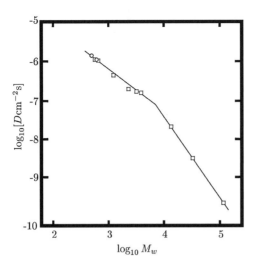

Figure 10.5. Diffusion constant of polyethylene chains of increasing molecular weight, obtained from pulsed field gradient NMR experiments (see section 11.7). The two straight lines have slopes equal to -1 (corresponding to the Rouse description) and -2 (reptation model). The cross-over molecular weight corresponds to a few hundred monomeric units. After data published in D.S. Pearson, G. Ver Strate, E. von Meerwall and F. Schilling, *Macromolecules* **20**, 1133 (1987).

N_e have been interpreted by Edwards and de Gennes as signalling the emergence of important topological constraints in the dynamics. Above N_e, the constraint that a chain cannot cross the other chains, which was ignored in the Rouse model, becomes a dominant dynamical feature. Obviously, the effect of these constraints on chain motion is a very complex many-body problem, and even their description in terms of static correlation functions is far from obvious. The problem was, to a certain extent, solved by the development of a simple, mean field like model known as the tube or reptation model. The basic idea is that the *effective* consequence of the topological constraints imposed by all other chains on the motion of a tagged chain is to confine its motion to a *tube*-like region of space. The chain then performs a curvilinear diffusive motion along the contour of this tube, described as *reptation* [6].

We emphasize that the tube is an effective concept, resulting from an average over the interactions with all other chains and the resulting topological constraints. Hence no particular spatial structures or correlations are associated with this tube. Nevertheless, it is meaningful to speak about a tube diameter and length, defined in a dynamical sense. The tube diameter d_T will be the typical lateral size of the region that can be explored by the chain without feeling the lateral confinement

[6] After latin *reptare*, which describes the crawling motion of a snake.

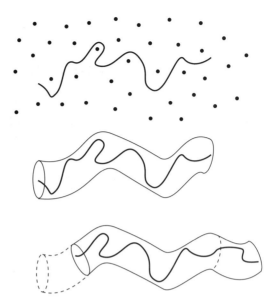

induced by the other chains. The length L_T of the tube is the curvilinear length of the chain, coarse-grained on a scale d_T, and is of course proportional to the molecular weight (figure 10.6).

The notion of a cross-over in molecular weight at which the motion becomes of the reptation type is, in this model, intimately related to the notion of tube diameter. For a chain whose radius of gyration is smaller than d_T, the motion remains completely isotropic and the tube does not really make sense. Hence we expect $N_e \sim (d_T/b)^2$. Longer chains with $N \geq N_e$ can be divided in subunits of size d_T. The tube length is given by $L_T \sim (N/N_e)d_T$. The radius of gyration of the chain is of course $R_G^2 = Nb^2 = (L_T/d_T)d_T^2$.

As the tube can be divided into sections of length d_T containing N_e monomers, the reptation model is often described in terms of *entanglements* that would restrict the chain motion every N_e monomers. d_T is then the distance between these entanglements. N_e is often described as the entanglement number. Care must be taken, however, that the entanglement, like the tube, is simply an effective description of the influence of all the other chains acting on N_e monomers of the tagged chain. The entanglement is not associated with a particular configuration of the chains in space, and the distance between entanglements is not detectable by experiments that probe the static structure of the melt.

The diffusion coefficient of the chain is computed by assuming that the chain performs a one-dimensional diffusive motion along the tube. When a section of size d_T has been vacated by one end of the chain a new section immediately appears at the other end. The tube will be entirely renewed by this process after a typical time $t_{rept} \sim L_T^2/D_c$, where D_c is the diffusion constant for the one-dimensional motion along the tube. Summing up the contribution of the Rouse frictions on each segment, we have

$$
D_c \sim \frac{N_e \, k_B T}{N \, N_e \zeta} \sim \frac{N_e}{N} D_r(N_e) \tag{10.50}
$$

where $D_r(N_e)$ is the Rouse diffusion constant for a chain of N_e monomers. Hence the reptation time is

$$
t_{rept} \sim \frac{L_T^2 N}{N_e D_r(N_e)} \sim \left(\frac{N}{N_e} \right)^3 \tau_r(N_e) \tag{10.51}
$$

This relaxation time of the chain scales like the cube of the molecular weight, and smoothly crosses over to the Rouse relaxation time for $N = N_e$. The diffusion constant is obtained by expressing that a complete tube renewal corresponds to a spatial displacement of the chain over a length of order R_G. Hence the diffusion constant is

$$
D_{rept} \sim \frac{R_G^2}{t_{rept}} \sim \left(\frac{N_e}{N} \right)^2 D_r(N_e) \tag{10.52}
$$

As observed in experiments, D_{rept} scales like the inverse of the molecular weight squared. Note that reptation theory involves very few adjustable parameters, which makes its quantitative success quite remarkable. All the physical quantities in equations (10.52) and (10.51) are expressed in terms of other quantities that can be experimentally measured unambiguously, essentially the ratio N/N_e (where N_e is by definition the cross-over molecular weight) and the relaxation time of chains with molecular weight N_e. For flexible polymers, typical values for N_e are in the range of 100 to 1000 Kuhn segments.

Although the reptation theory is remarkably successful in view of its great simplicity, it should be noted that it does not account perfectly for all the rheological properties of polymer melts. In particular, the viscosity is predicted within the theory to scale as $\eta \sim t_{rept} \sim N^3$, while experiments point towards a $N^{3.4}$ dependence. The reasons for this discrepancy are not well understood. Several improvements to the simple theory, involving for example tube length fluctuations or 'double reptation' by which the entanglements acquire a finite lifetime, have been proposed to improve the agreement with experiment. However, such improvements are always obtained at the expense of the introduction of new adjustable parameters. Reptation theory is then most often used in its simplest form described in this chapter, and remains, in spite of its obvious weaknesses, the most efficient description of the dynamics in melts of high molecular weight chains. It has also been adapted successfully to describe the dynamics of semi-dilute solutions of long chains.

10.5 Application 3: barrier crossing

Consider a Brownian particle subject to one of the external potentials $V(x)$ shown in figure 10.7. A particle thermalized at a temperature T such that $k_B T \ll V_B - V_A$, the barrier height, will remain trapped in the metastable position A until a favourable fluctuation allows a *barrier crossing* event. The particle will then overcome the potential barrier at B, and reach the second minimum at A′ (case 1) or leave the potential well forever (case 2). The crucial question then is that of the *escape time* from a potential well. How long must one wait, on average, for the thermal fluctuation that provides the particle with enough kinetic energy to overcome the barrier? This question can be addressed using the Kramers and Smoluchovski equations, (10.17) and (10.24).

Before discussing in detail the solution of this barrier crossing problem, let us mention that the applications extend far beyond the description of the dynamics of one colloidal particle in an external potential. Many physical or chemical problems involve the description of a 'slow' degree of freedom trapped in some

Figure 10.7. Two types of external potential for the barrier crossing problem in one dimension. Case 1: particles leaving the well A fall into the well A′. Case 2: particles leaving A fall into a much deeper part of the potential and never return.

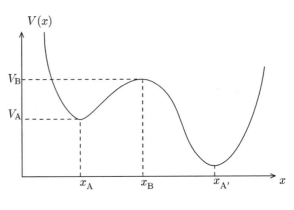

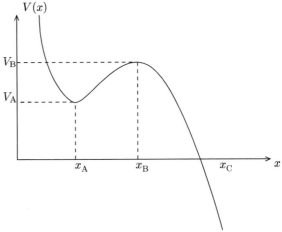

metastable position and coupled to many 'fast' degrees of freedom. Examples of such 'barrier crossing' problems include, among many others, chemical reactions in solution (the simplest examples being isomerization reactions), motion of defects in solids (crystalline or amorphous), the nucleation theory described in application 4 of this chapter, or the motion of a polymer across an interface. Overcoming an energy (or free energy) barrier is then the rate-limiting step in the dynamics of the system. Of course, the 'particle coordinate' x must in many cases be replaced by some fictitious 'reaction coordinate', which does not necessarily correspond to a particle position. The one-dimensional picture developed in this chapter is also an oversimplification, since the 'reaction coordinate' may involve several degrees of freedom evolving in a multidimensional energy landscape. In such cases, the definition of an appropriate 'reaction path' allows one to make use of the main results obtained for one-dimensional systems.

The large friction limit

We begin by considering the case of the strong friction limit, in which the dynamics of the particle is correctly described by the Smoluchovski equation (10.24). In this case, the escape time can be computed easily by considering the following situation. A large number (ensemble) of independent particles is trapped in the potential well A. As soon as a particle succeeds in escaping and reaches point C [7] an external operator removes it from the ensemble and introduces a new particle in the bottom region of well A. A stationary situation follows, in which an average current $j(x)$ flows from A to C. In a stationary state the continuity equation implies that $j(x) = J$ is independent of x. The particle density $\rho(x)$ then obeys

$$J = -\frac{k_B T}{M \zeta} \frac{\partial \rho(x)}{\partial x} - \frac{1}{M \zeta} \frac{\partial V}{\partial x} \rho(x) \tag{10.53}$$

This equation can be solved with the appropriate *absorbing boundary condition* at point C, which expresses that the probability density is continuous at x_C [8],

$$\rho(x_C^-) = 0 \tag{10.54}$$

Defining

$$\psi(x) = \exp\left(V(x)/k_B T\right) \rho(x) \tag{10.55}$$

we find

$$\frac{\partial \psi}{\partial x} = -\frac{M \zeta J}{k_B T} \exp\left(V(x)/k_B T\right) \tag{10.56}$$

---

[7] We focus here on case 2 in figure 10.7, a similar argument can of course be built for case 1.
[8] The density vanishes for $x = x_C^+$, since particles are removed at this point. As ρ is a continuous function it also vanishes for $x = x_C^-$.

and

$$\psi(x) = \frac{M\zeta J}{k_B T} \int_x^{x_C} \exp\left(V(x')/k_B T\right) dx' \tag{10.57}$$

The solution for $\rho(x)$ follows immediately. The escape rate (i.e. the inverse of the time needed to escape the well) can be expressed as

$$\tau^{-1} = \frac{J}{N} = \frac{J}{\int_{-\infty}^{x_B} \rho(x)\, dx} \tag{10.58}$$

where the denominator is simply the average number of particles trapped in the potential well. From equations (10.55), (10.57) and (10.58), it is seen that the problem of computing the escape rate has been reduced to quadratures. These quadratures can be considerably simplified when $V_B - V_A \gg k_B T$. In the bottom of the potential well (i.e. for x such that $V(x) - V_A$ is comparable to $k_B T$), the integral appearing in equation (10.57) is nearly independent of x, and is completely dominated by the vicinity of the barrier, where $V(x)$ is maximum. This implies that the choice of x_C is, fortunately, irrelevant, provided $V_C \ll V_B$. It also allows one to to estimate the integral using the quadratic approximation for the potential in the vicinity of x_B. Writing

$$V(x) = V_B - \frac{1}{2} M\omega_B^2 (x - x_B)^2 \tag{10.59}$$

for x close to x_B and extending the integration limits to infinity so that the Gaussian integral is easily computed, we obtain for $\psi(x)$ in the vicinity of x_A the expression

$$\psi(x) \simeq \frac{M\zeta J}{k_B T} \exp(V_B/k_B T) \left(\frac{2\pi k_B T}{M\omega_B^2}\right)^{1/2} \tag{10.60}$$

Now $\rho(x) = \psi(x) \exp(-V(x)/k_B T)$ is largest in the bottom of the well, so that the integral appearing in the denominator of equation (10.58) is dominated by the neighbourhood of A. Using a quadratic approximation similar to equation (10.59) for the potential near A, $V(x) = V_A + \frac{1}{2} M\omega_A^2 (x - x_A)^2$, and the same trick of extending the integration limits to infinity for the computation of the integral, we have

$$\frac{1}{\tau} \simeq \frac{\omega_A \omega_B}{2\pi\zeta} \exp\left(\frac{V_A - V_B}{k_B T}\right) \tag{10.61}$$

This approximate result shows that the escape rate depends only on the parameters of the potential in the vicinity of the potential well and at the barrier top, and on the friction coefficient. The dominant contribution is obviously the exponential, *Arrhenius factor* $\exp((V_A - V_B)/k_B T)$. The existence of this term implies that plotting $\log(\tau)$ as a function of $1/T$ will often produce a straight line, as the T dependence of the other parameters is generally weak. This T dependence is characteristic of so-called activated processes, and the slope $V_B - V_A$ of the

Arrhenius plot is termed the *activation energy*. Other parameters also influence the escape rate. A larger friction logically involves a smaller rate. A narrow barrier (large ω_B) will be easier to cross than a wide one, and a narrow well (large ω_A) is also easier to escape.

The small friction limit

It is easily seen, however, that the result (10.61) cannot be valid in the limit of zero friction. In this limit, the dynamics of a particle is described by the usual equations of Newtonian dynamics, so that a particle trapped at an energy smaller than V_B will oscillate for ever and never escape the well. The occurrence of thermal fluctuations is indeed intimately linked to the existence of a non-vanishing friction, as can be seen from the fluctuation–dissipation relation (10.7). For low frictions, the thermal fluctuations will be weak, so that the escape rate should decrease, rather than increase, as predicted by equation (10.61). To describe this *low friction limit*, one must resort to the full Kramers equation (10.17), which as indicated above is much more complex than the Smoluchovski equation. The solution of the problem initially proposed by Kramers involves rather complex mathematical developments, and we present here a very simplified derivation. The basic idea is that for weak friction, a trapped particle will oscillate for long times at constant energy. Hence the relevant distribution function is the probability distribution function of the particle energies, $f(E, t)$, rather than the full distribution function $P(x, v, t)$. If we assume that the escape rate is small enough, it is reasonable to infer that, within the well, this probability distribution is a quasi-equilibrium one,

$$f(E) = C \exp(-E/k_B T) \tag{10.62}$$

where C is a normalization constant. Although this distribution is stationary, the particles move up and down in energy under the combined effect of the friction and of the random force. To estimate how fast this motion takes place on the energy axis, we estimate the energy loss for a particle of energy E, oscillating in the well at a frequency ω_A during one oscillation period. An order of magnitude of this loss $\delta(E)$ is given by

$$\delta(E) = \int_0^{2\pi/\omega_A} (M\zeta v(t) \cdot v(t))\, dt \simeq \zeta(E - V_A)/\omega_A \tag{10.63}$$

where the second equality is obtained by replacing Mv^2 by its value for a particle performing harmonic oscillations at energy E, i.e. $Mv^2(t) = 2(E - V_A)\cos^2(\omega_A t)$. Of course, the stationarity of the quasi-equilibrium distribution implies that the particle can undergo a gain in energy of similar magnitude with equal probability over the same period of oscillation. Within the well, these energy changes only result in thermal equilibration. For the particles that are close enough to the top of the barrier, however, a gain in energy will allow them to escape the potential well. Typically, half of the particles that are within a range $\delta(V_B)$ from

Figure 10.8. Escape rate as a function of friction, for a given activation energy. The maximum corresponds to the 'Kramers turnover' phenomenon.

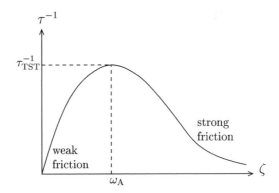

the top will escape during one period. Hence the escape rate is approximately given by

$$\tau^{-1} \sim \frac{\delta(V_B)\omega_A f(V_B)}{\int_{V_A}^{V_B} f(E)\, dE} \sim \zeta \left(\frac{V_B - V_A}{k_B T} \right) \exp \left(\frac{V_B - V_A}{k_B T} \right) \tag{10.64}$$

In view of the crudeness of the derivation, we have omitted factors of order unity. The dominant factor is again the Arrhenius factor. However, the influence of the friction is completely different in the low friction and high friction limits. According to equation (10.64), the escape rate increases with the friction and, as expected, vanishes for zero friction.

According to equations (10.61) and (10.64), the escape rate is a non-monotonous function of the friction ζ, with an increase at low friction and a decrease at large frictions. This behaviour is illustrated schematically in figure 10.8. The maximum of the escape rate is often described as the *Kramers turnover* region, and corresponds to values of ζ of order ω_A. For such values of the friction, the correlation time of the particle velocity ζ^{-1} is comparable to the oscillation period. Hence neither the large friction limit (which corresponds to a particle that does not move much on the time scale ζ^{-1}) nor the weak friction approach (which corresponds to many oscillations at constant energy before being affected by the thermal fluctuations) are applicable. In this turnover region, the escape rate depends only weakly on the friction (since $d\tau/d\zeta = 0$) and can be safely estimated using the so-called transition state theory (TST) briefly exposed in the next paragraph. The existence of the Kramers turnover has been beautifully demonstrated in experiments where the dynamics of a chemical isomerization reaction is studied (using time resolved spectroscopic methods) in solvents of various viscosities, higher viscosity corresponding of course to higher friction[9].

[9] M.Y. Lee, G.R. Holtom, and R.M. Hochstrasser, *Chem. Phys. Lett.* **118**, 359 (1985).

The transition state theory

The transition state theory (also known as Eyring's activated rate theory) is most easily described for case 1 of figure 10.7. The calculation is in fact a purely equilibrium one, resting on two assumptions. (i) The distribution $P(x, v, t)$ is the equilibrium distribution

$$\exp\left(-\left(V(x) + \frac{Mv^2}{2}\right)/k_B T\right)$$

(ii) The escape rate from well A is obtained by counting the particles that cross the barrier at B from left to right, per unit time, and dividing by the number of particles in the well. In other words, there are no barrier recrossings, and a particle that reaches B with a positive velocity is said to have crossed the barrier and will end up in well A' for a long time.

From these two assumptions the escape rate is easily obtained as

$$\tau_{\text{TST}}^{-1} = \frac{\int_0^\infty dv\, v\, \exp(-V_B/k_B T - Mv^2/2k_B T)}{\int_{-\infty}^{+\infty} dv \int_{-\infty}^{x_B} dx\, \exp(-V(x)/k_B T - Mv^2/2k_B T)} \tag{10.65}$$

Using the same quadratic approximation as in the large friction limit, and carrying out the Gaussian integrals, we obtain

$$\tau_{\text{TST}}^{-1} = \frac{\omega_A}{2\pi} \exp\left(\frac{V_B - V_A}{k_B T}\right) \tag{10.66}$$

Again, the Arrhenius dependence on the barrier height is found, with a prefactor that does not depend on the friction.

The essential approximation in this theory is that the barrier recrossings are neglected. As these recrossings will always tend to lower the escape rate, τ_{TST}^{-1} will always be an upper bound. It is also understandable that the approach is reasonable close to the Kramers turnover, since the occurrence of recrossings will be strongly dependent on the friction, and by construction the friction is irrelevant in the vicinity of the turnover. Note that the transition state calculation is very easily generalized to the crossing of a saddle point on a multidimensional energy surface, and is a rather robust way of estimating escape rates in systems for which the friction force is not known in detail.

10.6 Application 4: classical nucleation theory

Nucleation is the process by which a system quenched into a metastable state starts to phase separate. The situation is very different here from that of spinodal decomposition, in which small fluctuations undergo exponential growth due to the instability of the system. A metastable phase is indeed stable against the small, long wavelength fluctuations that trigger spinodal decomposition. A different route has to be found to reach the minimum of the free energy, and involves two steps. The first step, called nucleation, is the creation of small, localized droplets or 'nuclei' of the stable phase within the homogeneous metastable system. The second step is the growth and coalescence of these droplets. As this second step is in many respects similar to the domain growth phenomenon discussed in section 9.3, we discuss here only the first step of nucleus formation.

Note that in most first-order phase transitions, nucleation and growth are the only route by which the phase change takes place. This is the case especially for crystallization, for which it is generally believed that no spinodal line (or metastability limit of the liquid state) exists. Other examples, among many, include crystal or liquid formation from a vapour (the condensation phenomenon that gives rise to the formation of clouds from cold vapour), and phase separation in a binary mixture quenched in the vicinity of the binodal line.

The relationship of this nucleation phenomenon to barrier crossing problems (section 10.5) becomes obvious when considering the free energy of a droplet as a function of its size. For a spherical droplet of radius R, this free energy is the sum of volume and surface terms,

$$\Delta F(R) = \frac{4}{3}\pi R^3 \rho \Delta\mu + 4\pi R^2 \gamma \tag{10.67}$$

Here $\Delta\mu < 0$ is the difference in chemical potential between the stable phase and the metastable one, ρ is the density of the stable phase and γ is the surface free energy associated with the interface between the two phases. A schematic plot of ΔF as a function of R is shown in figure 10.9. For small R, the unfavourable surface term is dominant while, for large enough R, it becomes negligible. Hence the free energy goes through a maximum as the radius increases. This maximum in the free energy is the nucleation barrier,

$$\Delta F^* = \frac{16\pi}{3} \frac{\gamma^3}{\Delta\mu^2} \tag{10.68}$$

The corresponding radius $R^* = -2\gamma/\Delta\mu$ is called the critical radius. In order to minimize its free energy, a system has first to create at least one critical nucleus, from which the stable phase will grow.

The kinetic process by which the creation of a critical nucleus takes place is described by the classical phenomenological nucleation theory. This theory

Figure 10.9. Schematic plot of ΔF as a function of the radius of the stable phase droplet.

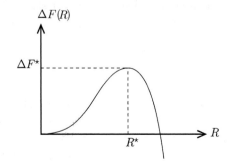

does not attempt to tackle directly the many-body problem of nucleus formation from individual atoms. Rather, the phenomenon is described as the evolution of a 'reaction coordinate', the number of atoms in the nucleus $p = 4\pi\rho R^3/3$ evolving in a 'potential' $\Delta F(p)$. This reaction coordinate is treated like the position of a Brownian particle in the high friction limit, which is reasonable since the growth of the nucleus is not expected to involve inertial effects. The diffusion constant, however, is taken to be dependent on the reaction coordinate, since the number of atoms that can be incorporated into the nucleus per unit time is proportional to its surface. Hence a reasonable guess for the diffusion constant $D(p)$ is

$$D(p) \sim p^{2/3}/\tau_0 \qquad (10.69)$$

where τ_0 is a characteristic time of atomic motion in the liquid, typically $1/\tau_0 \sim Da^2$ where D is the true diffusion coefficient in the liquid and a the distance between molecules. Hence the theory can be summarized in a Smoluchovski type equation (see equation (10.24)) for the reaction coordinate p, treated as a continuous variable

$$\frac{\partial P(p,t)}{\partial t} = \frac{\partial}{\partial p}D(p)\left(\frac{\partial}{\partial p}P(p,t) + \frac{1}{k_B T}P(p,t)\frac{\partial \Delta F(p)}{\partial p}\right) \qquad (10.70)$$

Here $P(p,t)$ is the probability that a nucleus has a size of p atoms at time t. The ensemble of nuclei over which this probability is calculated is, in principle, made up of all the atoms in the sample, each of which can serve as a starting point for the nucleus. Hence the number of nuclei of p atoms at time t is $NP(p,t)$, where N is the total number of atoms in the sample. This is of course a slight overestimate, since an atom that has been incorporated into a given nucleus cannot serve as a starting point for a new one. The number of atoms involved in nuclei during the nucleation period is however small, so that this approximation is well justified.

The nucleation rate is the number of nuclei per unit time and per unit volume of the liquid which reach the critical size. The procedure to estimate this rate parallels the one described for the barrier crossing problem (section 10.5) in the high friction limit. One imagines creating a stationary non-equilibrium state by having some 'Maxwell daemon' in the system, whose role is to destroy the nuclei that reach a size p_{max} somewhat larger than the critical size $p* = 4\pi\rho R^{*3}/3$. The corresponding atoms are returned to the $p = 1$ state. In this way a steady state is created, with a current J of nuclei that cross the barrier per unit time. The associated probability distribution is then,

$$P(p) = \frac{J}{N}\exp(-\Delta F(p)/k_B T)\int_p^{p_{max}} \exp(\Delta F(p')/k_B T)D(p')^{-1}\,dp' \qquad (10.71)$$

This is similar to equation (10.57), except that the p-dependence of D has been taken into account. In the vicinity of $p = 0$, the integral is dominated by the top of the nucleation barrier, and is therefore a constant. The escape rate for one atom is obtained by dividing the current J by the number $NP(1)$ of atoms in the $p = 1$

state, which gives

$$\frac{1}{\tau} = \frac{1}{\int_1^{p_{max}} \exp(\Delta F(p')/k_B T) D(p')^{-1}\, dp'} \tag{10.72}$$

The nucleation rate is $I = \rho/\tau$, where ρ is the number density. Note that, since one nucleus in the whole sample is sufficient to start the phase transformation, the relevant quantity is the nucleation rate *multiplied* by the volume of the sample V. After a time $V/I = \tau/N$, a nucleus will certainly have appeared in the sample, so that the phase transformation will begin.

The formula (10.72) can be simplified by using a quadratic approximation for $\Delta F(p)$ in the vicinity of p^*. The resulting integral is then evaluated by extending the bounds to infinity, and neglecting the slow variation of $D(p)$. With these approximations, one obtains

$$I = \rho D(p^*) \left(-\frac{1}{2\pi k_B T} \frac{d^2 F}{dp^2}\Big|_{p^*} \right)^{1/2} \exp\left(-\Delta F^*/k_B T\right) \tag{10.73}$$

The approximations used in establishing equation (10.73) make its practical application problematic. In fact, experimental estimates of nucleation rates for e.g. the liquid crystal transition can differ by orders of magnitude from the values that would be obtained using equation (10.73), and the resolution of such discrepancies is still an open subject. With this caveat, classical nucleation theory is still widely used to provide a qualitative description of the phase transformation rate, both from the gas phase (crystal or liquid formation from a vapour) and in condensed matter systems (liquid crystal transition or phase separation in mixtures).

The appealing aspects of equation (10.73) are its simplicity, and the fact that all the ingredients can be rather easily obtained from available experimental data. The essential feature of this formula is that the nucleation rate is expressed as the product of a kinetic factor, $D(p^*)$, and a thermodynamic, Arrhenius-like factor $\exp(-\Delta F^*/k_B T)$. In spite of the formal similarity, this factor is very different from the standard Arrhenius form, with a constant activation energy. The nucleation barrier ΔF^* (equation (10.68)) has a strong temperature dependence, and diverges at the temperature of phase coexistence, T_{coex}, where $\Delta\mu = 0$. Hence, in the vicinity of T_{coex}, this factor tends to increase very rapidly, as temperature is decreased and the nucleation barrier drops. In condensed phases, this increase is often counterbalanced by the decrease of the kinetic factor, which is proportional to the diffusion coefficient. In liquids this diffusion coefficient can be very strongly temperature dependent (see section 12.4). For simplicity we assume Arrhenius dependence of $D(T)$, i.e. $D(T) \sim \exp(-E_d/k_B T)$, where E_d is an activation energy for the diffusion process (see section 10.5) and use for $\Delta\mu(T)$

the expression, valid near T_{coex},

$$\Delta\mu(T) = \frac{\ell}{T_{coex}}(T - T_{coex}) \qquad (10.74)$$

where ℓ is the latent heat per particle released in the phase transformation. We obtain for the *logarithm* of the nucleation rate

$$\log I(T) = C(T) - \frac{A}{T} - \frac{B}{T(T - T_{coex})^2} \qquad (10.75)$$

where A, B are positive constants and $C(T)$ is a very slowly varying function of T. It is easily checked that this expression, which diverges at $T = 0$ and $T = T_{coex}$, has a pronounced maximum for some $T_{max} < T_{coex}$. The corresponding maximum in $I(T)$ is of course amplified by the exponentiation, so that the nucleation rate varies by orders of magnitude, going through a maximum, when temperature is decreased below T_{coex}. This very fast variation explains why nucleation appears as a rather sudden phenomenon when a liquid is cooled in a metastable state. It has important implications for the phenomenon of glass formation, as will be discussed in section 12.4.

Finally, we mention that we have limited our discussion in this section to *homogeneous nucleation*, which takes place in a perfectly pure system. A much more common phenomenon, which can be described along similar lines, is *heterogeneous nucleation*. In heterogeneous nucleation, the nucleus is formed on some 'impurity' present in the system (which could be the wall of the container, or a dust particle). If the impurity is locally flat, the nucleus will assume a spherical cap shape, with a contact angle θ that can be derived from Young's law.

Exercise: Show that the nucleation barrier for a spherical cap nucleus on a planar surface is decreased by a factor

$$f(\theta) = \frac{1}{4}(2 + \cos\theta)(1 - \cos\theta)^2 \qquad (10.76)$$

Equation (10.76) shows that the nucleation rate can be considerably increased by the presence of walls or dust. Unfortunately, the properties of impurities are generally unknown, so that heterogeneous nucleation is not a well controlled phenomenon. However, very careful experiments have to be designed in order to observe truly homogeneous nucleation.

Further reading

H.A. Kramers, Brownian motion in a field of force and the diffusion model of chemical reactions, *Physica* 7, 284 (1940), is the original reference on the 'Kramers problem' of barrier crossing. It is still a very readable reference, with a complete treatment of both the high and low friction limits. A recent reference on barrier crossing is the review paper by P. Hänggi, P. Talkner, and M. Borkovec, *Rev. Mod. Phys.* **62**, 251 (1990).

S. Chandrasekhar, Stochastic problems in physics and astronomy, *Rev. Mod. Phys.* **15**, 1 (1943) (reprinted in, N. Wax (ed.), *Noise and Stochastic Processes*, Dover, New York, 1954, p. 65) is a very interesting reference on Brownian motion, with applications that range from colloid physics to astrophysics.

A very complete review on the Fokker–Planck formalism, with detailed description of solution methods, is contained in H. Risken's book *The Fokker Planck Equation*, Springer, Berlin, 1984. Another general book of interest in this context is N. van Kampen's *Stochastic Processes in Physics and Chemistry*, North Holland, Amsterdam, 1981. Note that the term 'Fokker-Planck' is taken here in a much more general sense, including also the multivariate (Kramers) case and the high friction (Smoluschovski) limit.

Rotational diffusion and its application to dielectric experiments are discussed in the book by Berne and Pecora, *Dynamic Light Scattering* (see chapter 3).

Dynamics of polymeric systems is very clearly exposed in the classical books by de Gennes and Doi and Edwards (see chapter 3).

Nucleation theory is described in many treatises on materials science and metallurgy, usually in a way that is adapted to the particular topic under consideration. A good and relatively recent general reference are the lectures by J. Langer in *Solids far from Equilibrium*, ed. C. Godrèche, Cambridge University Press, Cambridge, 2001.

11 Response and correlation functions

11.1 Probing dynamical properties at equilibrium

By definition, the macroscopic properties of a system at equilibrium do not change with time. This does not mean, however, that the system is dynamically inert. On the contrary, the equilibrium state is associated with permanent motion at the molecular level. Although this motion is sometimes described as random 'thermal noise', it is in fact quite well organized, and reflects the microscopic processes that govern the dynamics of the system. As these same processes will also determine the system response to an external perturbation, understanding their organization and time evolution is of primary importance for the determination of the material response, and of its relationship to the microscopic structure.

As a simple illustration of how seemingly random dynamical processes are organized in a coherent fashion, one may consider the case of vibrations in a crystalline solid. Each atom oscillates around its equilibrium position, in a way which is apparently very random. The well known harmonic analysis, however, shows that this motion is really caused by the superposition of well defined sound waves, the *phonons*, with different phases and directions of propagation [1]. This organization of the atomic motions into excitations of well defined spatial and temporal structure determines many thermodynamic and transport properties of the crystal.

The harmonic crystal is of course particularly simple, since it is possible in that case to deduce analytically from the interaction potential the structure of the coherent excitations. In more complex, disordered systems, an analytical treatment is usually out of reach. Nevertheless, the information on the way atomic motions are organized is encoded in the correlation functions of atomic positions.

---

[1] See for example N. Ashcroft and D. Mermin, *Solid State Physics*, Holt Saunders International, Cornell University Press, Ithaca, NY, 1976.

11.2   Time dependent correlation functions

Generally speaking, a time dependent equilibrium correlation function is defined as

$$C_{AB}(t, t') = \langle A(t)B(t') \rangle \tag{11.1}$$

where $A(t)$ (respectively $B(t')$) refers to the value of some microscopic observable at time t (respectively t'). A microscopic observable is, by definition, a function of the atomic coordinates and momenta. Hence the time dependence in $A(t)$ is actually shorthand for indicating the time dependence of these coordinates, $A(t) \equiv A(\{r_i(t), p_i(t)\})$. The angular brackets in (11.1) refer, as usual, to an equilibrium average over an ensemble of different initial conditions.

By definition an equilibrium system is invariant under time translation. Hence the correlation function defined in equation (11.1) depends only on the time difference $\tau = t' - t$. Taking advantage of this time translation invariance, one can transform equation (11.1) into

$$C_{AB}(\tau) = \langle A(\tau)B(0) \rangle = \lim_{T \to \infty} \frac{1}{T} \int_0^T A(t + \tau)B(t)\, dt \tag{11.2}$$

The last equality in (11.2) is a consequence of the ergodic hypothesis, that averaging over a single trajectory of the system is equivalent to an ensemble average.

A simple time dependent correlation function is the so-called van Hove function which correlates the positions of a tracer particle at two successive times. Formally, this correlation function is written as

$$G_s(\mathbf{r}, \mathbf{r}', t, t') = V \langle \delta(\mathbf{R}(t') - \mathbf{r}')\delta(\mathbf{R}(t) - \mathbf{r}) \rangle \tag{11.3}$$

where $\mathbf{R}(t)$ is the position of the tracer particle at time t, and V is the volume of the system. Here the two observables A and B are $\delta(\mathbf{R} - \mathbf{r})$ and $\delta(\mathbf{R} - \mathbf{r}')$, respectively. The term between brackets is the joint probability density for finding the particle at point $\mathbf{r}$ at time t, and at point $\mathbf{r}'$ at time t'. This probability can be expressed as the product of the probability of finding the tracer at time t at point $\mathbf{r}$, times the probability that it moves from $\mathbf{r}$ to $\mathbf{r}'$ between t and t'. In a uniform system, at equilibrium, the first factor is a constant equal to $1/V$, and the second factor depends only on the position and time differences $\mathbf{r} - \mathbf{r}'$ and $t' - t$. If the system is, moreover, isotropic, the van Hove correlation function is of the form

$$G_s(\mathbf{r}, \mathbf{r}', t, t') = g_s(|\mathbf{r}' - \mathbf{r}|, t' - t) \tag{11.4}$$

The function $4\pi r^2 g_s(r, t)\, dr$ gives the probability for finding the tracer at a distance r (within dr) from the origin, *knowing* that this tracer was located at the origin at time 0.

With this interpretation, it is clear that the time dependent correlation function G_s contains information on the diffusion coefficient of the tracer, a property that was defined from a macroscopic viewpoint in chapter 9. Indeed, if the tracer is observed to obey the diffusion equation on long time scales, the direct implication is that the van Hove correlation function $g_s(r, t)$ must, for long times, behave as $g_s(r, t) = (4\pi Dt)^{-3/2} \exp(-r^2/4Dt)$. This relationship between a correlation function and a macroscopic transport coefficient is easily understood. The diffusive motion that takes place in an equilibrium system is exactly identical to the motion leading to a uniform distribution in response to a concentration gradient.

The same behaviour can be characterized by studying the incoherent scattering function $F_s(\mathbf{k}, t)$ which is simply the Fourier transform of $g_s(r, t)$, and can be measured in scattering experiments (section 11.6). Formally, $F_s(\mathbf{k}, t)$ is defined as

$$F_s(\mathbf{k}, t) = \langle \exp(i\mathbf{k} \cdot (\mathbf{R}(t) - \mathbf{R}(0))) \rangle = \int d\mathbf{r} \, g_s(r, t) \exp(i\mathbf{k} \cdot \mathbf{r}) \qquad (11.5)$$

In the practical case where many independent tracers are present, an average over the different particles must be performed. In the long time limit, the Gaussian shape of $g_s(r, t)$ implies that $F_s(k, t) = \exp(-k^2 Dt)$. Identifying with the small k expansion in equation (11.5), one recovers the well known Einstein formula (see equation (10.9))

$$\lim_{t \to \infty} \frac{1}{t} \langle (\mathbf{R}(t) - \mathbf{R}(0))^2 \rangle \sim 6D \qquad (11.6)$$

which again expresses that the motion of the particles in the equilibrium system is described by the same diffusion constant that describes a non-equilibrium phenomenon. As a byproduct, the following *explicit* formula is obtained for the diffusion constant in terms of the time dependent correlation function of the particle velocity,

$$D = \frac{1}{3} \int_0^\infty dt \, \langle \mathbf{V}(t) \cdot \mathbf{V}(0) \rangle \qquad (11.7)$$

Exercise: Prove equation (11.7).
Hint: Express $\mathbf{R}(t) - \mathbf{R}(0)$ as $\int_0^t \mathbf{V}(s) \, ds$. Use then the fact that the double integral $\int_0^t \int_0^t \langle \mathbf{V}(s) \cdot \mathbf{V}(s') \rangle \, ds' \, ds$ is, for large t, proportional to $2t \int_0^\infty dt \, \langle \mathbf{V}(t) \cdot \mathbf{V}(0) \rangle$.

11.3 Response functions

Time dependent correlation functions can be directly obtained from inelastic scattering experiments (see section 11.6), or from numerical simulations in which the coordinates of the particles are explicitly known. In many cases, however, the simplest way to investigate the dynamics of a system is to perturb it with an external field and to monitor the response of some observable. Generally, both the field and the observable are macroscopic in nature, i.e. correspond to a wavelength much larger than atomic or molecular dimensions. For liquids, typical experiments of this type are provided by dielectric relaxation, in which the perturbing field is an electric field and the observable is the electric dipole moment of the sample, as was briefly described in section 10.3. In this case, although the field is always homogeneous (or corresponds to a large wavelength), its frequency can be very high, so that information on microscopic processes is gained from a macroscopic experiment.

Formally, the perturbation to the system can be written in the form of a change to its Hamiltonian,

$$\Delta H = -f(t)A \qquad (11.8)$$

where $f(t)$ is some prescribed function. In the case of the dielectric experiment, $f(t) = E(t)$ is the applied field, and the observable A is simply the component of the total dipole moment in the direction of the field. The observable whose response is monitored, B (which may in some cases be identical to A), is written as

$$\langle B(t)\rangle_{\text{pert}} = \langle B(t)\rangle_0 + \int_{-\infty}^{t} \chi_{BA}(t,s)f(s)\,ds + \mathcal{O}(f^2) \qquad (11.9)$$

Equation (11.9) corresponds to a Taylor expansion of the response in terms of the perturbation. The notation $\langle\rangle_{\text{pert}}$ denotes an ensemble average over perturbed systems, while $\langle\rangle_0$ indicates an unperturbed average. The second term in equation (11.9) corresponds to the *linear response* of the system to the external field, which for weak enough perturbations is the observable response. Whether the linear term provides a sufficient description of the response has to be checked explicitly by comparing the response obtained for different field intensities. It will be seen in the following that, in principle, this *linear response* approximation can be expected to work only if the perturbing Hamiltonian is, on the average, not larger than the thermal energy $k_B T$ for each degree of freedom. However, the approximation sometimes proves to be accurate far beyond this theoretical limit.

Two important points are apparent in equation (11.9). First, the integral that defines the linear response only involves times $s \leq t$, where t is the time at which the measurement is made. This restriction is an expression of the causality principle, namely that the observable at t can only be affected by earlier history. Secondly, the definition of linear response implies that the *response function* $\chi_{BA}(t,s)$ does not have any dependence on the perturbation (such a dependence would appear as as a non-linear contribution to the response). This means that χ_{BA} is a property of the unperturbed system. In particular, in a system at equilibrium, the time translation invariance implies that $\chi_{BA}(t,s)$ is a function of the time difference only, $\chi_{BA}(t-s)$.

In the next section, the function $\chi_{BA}(t)$ will be shown to be intimately related to the correlation function of the two observables A and B. Some important general properties of χ_{BA} are however independent of this relationship. They are most easily expressed in terms of the Fourier transform of $\chi_{AB}(t)$, which naturally arises when a sinusoidal excitation $f(t) = f_0 \exp(-i\omega t)$ is considered. In the forced sinusoidal regime, the response is of the form

$$\langle B(t)\rangle_{\mathrm{pert}} = f_0(\chi'_{AB}(\omega) + i\chi''_{AB}(\omega)) \exp(-i\omega t) \qquad (11.10)$$

where

$$\chi'_{AB}(\omega) = \int_0^\infty \chi_{AB}(t) \cos \omega t \, dt \qquad \chi''_{AB}(\omega) = \int_0^\infty \chi_{AB}(t) \sin \omega t \, dt \qquad (11.11)$$

From a practical standpoint, the most important property is probably the relationship between $\chi''_{AA}(\omega)$ and the absorption of energy by the system. If the perturbation is of the form $-f(t)A$, the function $f(t)$ can be interpreted as a 'force' acting on the 'coordinate' $A(t)$. The total work done on the system by the perturbing 'force' is of the form

$$W = \int_{-\infty}^\infty dt \, f(t) \frac{d\langle A(t)\rangle}{dt} \qquad (11.12)$$

(note that the function $f(t)$ may be non-zero over a finite time interval only). Transforming this relation through an integration by parts, and using the convolution theorem one arrives at

$$W = \frac{1}{2\pi} \int_{-\infty}^\infty d\omega \, \omega \chi''_{AA}(\omega) |\hat{f}(\omega)|^2 \qquad (11.13)$$

where $\hat{f}(\omega) = \int_{-\infty}^\infty dt \, f(t) \exp(i\omega t)$. Hence $\omega \chi''_{AA}(\omega)$ is related to the power absorbed by the system for an input signal with a spectral density concentrated near ω. This implies in particular that $\omega \chi''_{AA}(\omega)$ is always a positive quantity. Otherwise, it would be possible to extract work from the system by imposing a sinusoidal perturbation for a finite time. The system would depart from and eventually return to equilibrium when the perturbation is turned off, and the external operator would receive work ($W < 0$), in contradiction with the second law of thermodynamics.

11.4 Fluctuation–dissipation theorem

We now turn to the connection between the fluctuations in an equilibrium system, characterized by their time dependent correlation functions, and the response of this system to an external perturbation. The existence of such a connection was first shown by Onsager. His famous principle of the regression of fluctuations was already illustrated in the case of tracer diffusion in section 11.2. There we saw that the van Hove correlation function coincides, for long times, with the solution of the diffusion equation. The diffusion coefficient obtained from the study of this correlation function is identical to that obtained from studying, on a macroscopic level, the relaxation of an initial density gradient. The evolution of

Figure 11.1. Left: $A(t)$ in an equilibrium system. The perturbation is zero at all times, the non-zero value at $t = 0$ results from a fluctuation. The restricted average of $B(t)$ is proportional to $C_{AB}(t)$. Right: a perturbation is imposed for $t < 0$, and creates the non-zero value of A at $t = 0$. The decay of B for $t > 0$ is the same as in the first case.

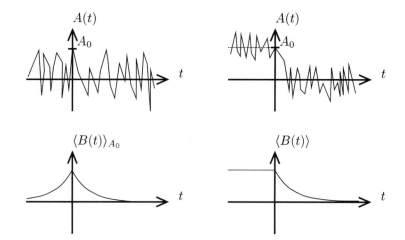

a spontaneous fluctuation is identical to that induced by an external disturbance. This is the essence of Onsager's principle.

To express it in a slightly more general fashion, let us consider a system in which, at time $t = 0$, the observable A deviates by A_0 from its equilibrium value $\langle A \rangle_{\text{eq}} = 0$. This deviation can either be due to a spontaneous fluctuation or caused by an external disturbance. The two situations are illustrated schematically in figure 11.1 In the case of a spontaneous fluctuation, we can guess what the average value of an observable $B(t)$ will be at time t. If the averaging is restricted to those systems in which $A(0) = A_0$ (a restriction which is indicated by $\langle B(t) \rangle_{A_0}$), one expects

$$\langle B(t) \rangle_{A_0} = \frac{C_{BA}(t)}{C_{AA}(0)} A_0 \tag{11.14}$$

so that when all possible values of A_0 are summed over with the appropriate equilibrium distribution $p(A_0)$, the correct correlation function $C_{BA}(t)$ is recovered, i.e.

$$\int \langle B(t) \rangle_{A_0} A_0 p(A_0) \, dA_0 = C_{BA}(t) \tag{11.15}$$

Consider now the case in which A_0 has been created externally by imposing an external perturbation $-fA$ between $t = -\infty$ and $t = 0$. The perturbation is switched off at $t = 0$. From the definition of the response functions one obtains

$$A_0 = f \int_{-\infty}^{0} \chi_{AA}(-s) \, ds \tag{11.16}$$

and

$$\langle B(t) \rangle_{A_0} = f \int_{-\infty}^{0} \chi_{BA}(t - s) \, ds \tag{11.17}$$

Comparing (11.17) and (11.16) with (11.14), a mathematical formulation of the regression hypothesis is obtained in the form

$$\frac{\int_{-\infty}^{0} \chi_{BA}(t-s)\,ds}{\int_{-\infty}^{0} \chi_{AA}(-s)\,ds} = \frac{C_{BA}(t)}{C_{AA}(0)} \tag{11.18}$$

This implies that the ratio $\int_{t}^{\infty} \chi_{BA}(s)\,ds/C_{BA}(t)$ is independent of the pair of observables under consideration. A complete calculation, described below, shows that this ratio (which dimensionally is the inverse of an energy) is equal to $-1/k_{B}T$. Equation (11.18) can therefore be reformulated as

$$\chi_{BA}(t) = -\frac{1}{k_{B}T}\frac{dC_{BA}(t)}{dt} \tag{11.19}$$

which constitutes the fluctuation–dissipation theorem.

In order to establish the theorem completely, one needs to compute the ratio $X = (\int_{-\infty}^{0} \chi_{AA}(-s)\,ds)/C_{AA}(0)$. This is done by assuming that at $t = 0$ the perturbed system, described by the Hamiltonian $H_{0} - fA$, is at equilibrium. Hence (see for example the analogous calculation presented in section 3.6, equation (3.66))

$$A_{0} = \frac{\text{Tr}A \exp -\beta(H_{0} - fA)}{\text{Tr} \exp -\beta(H_{0} - fA)} = -\beta f\langle AA \rangle + \mathcal{O}(f^{2}) \tag{11.20}$$

This proves that the ratio is indeed $X = -1/k_{B}T$, since A_{0} is also given by equation (11.16).

The fluctuation–dissipation theorem, equation (11.19), provides the route by which the response of a system to a small perturbation can be inferred from the study of time dependent correlation functions. The relationship holds only for the response to small perturbations, since the theorem was proved within the hypothesis of linear response. It implies that the same physics rules both the correlation functions, which can be measured by nonintrusive scattering techniques, and the response functions. In particular, the same time scales will be involved in the response and in the spontaneous fluctuations.

11.5 Application 1: collective modes

In this chapter, we have studied, using the example of particle diffusion, how the macroscopic, phenomenological equations could be used to determine the behaviour of correlation functions for long times and large distances (or small wavevectors). The basic idea is that the long wavelength components of the observable densities relax according to the phenomenological transport equations. In the case of diffusion, the relevant density is the tagged particle density, $\rho_{t}(\mathbf{r}, t) = \delta(\mathbf{r} - \mathbf{R}(t))$. Its Fourier component is $\rho_{t}(\mathbf{k}, t) = \exp(i\mathbf{k} \cdot \mathbf{R}(t))$, and obeys the diffusion equation in the small k limit. This yields the long time, large wavelength limit of $F_{s}(\mathbf{k}, t) = \exp(-k^{2}Dt)$, which is conveniently expressed in terms of its Laplace transform

$$\tilde{F}_s(\mathbf{k}, z) = \int_0^\infty F_s(\mathbf{k}, t) e^{-zt}\, dt = \frac{1}{z + Dk^2} \tag{11.21}$$

The experimentally accessible quantity (see sections 11.6 and 11.7) is the Fourier transform of $F_s(\mathbf{k}, t)$, $S_s(\mathbf{k}, \omega) = 2Dk^2 / (\omega^2 + (Dk^2)^2)$. When plotted as a function of ω for a given $\mathbf{k}$, this function exhibits a central peak at $\omega = 0$, with a width proportional to Dk^2. Such a peak is usually assigned to the existence of a *diffusive mode* in the system, in analogy to the resonance peaks that are observed in vibrating systems when the frequency equals that of an eigenmode. In this section, we will see how this notion can be usefully generalized to classify the dynamical behaviour of complex systems.

As seen in chapter 9, a system can generally be characterized, on some mesoscopic, coarse-grained scale, by the densities relative to a small number, p, of thermodynamic variables. These densities may describe *conserved quantities*, such as mass, momentum, energy, or species concentration, or non-conserved quantities, for example the order parameter of a transition, or the polarization density. In the long wavelength limit, the time evolution of these densities will be described by linear response equations of the type described in chapter 9. In general, the relaxation of the system will involve a coupled description of the relaxations associated with relevant densities. The analysis of this chapter implies that the fluctuations of the corresponding quantities will be governed by the same phenomenological equations. Therefore, we expect that if we form a column vector $A(\mathbf{k}, t)$ with the Fourier components of the fluctuations for the p relevant quantities, this vector will obey an equation of motion of the form

$$\frac{\partial A(\mathbf{k}, t)}{\partial t} = -[M(\mathbf{k})]\, A(\mathbf{k}, t) \tag{11.22}$$

Here $[M(\mathbf{k})]$ is a $p \times p$ square matrix involving phenomenological transport coefficients[2].

Equation (11.22) can be solved through diagonalization of $[M(\mathbf{k})]$. The corresponding eigenvectors are linear combinations of the original densities, and each of them exhibits an exponential relaxation, with relaxation rates $\lambda_\alpha(\mathbf{k})$ ($\alpha = 1, \ldots, p$), given by the eigenvalues of $[M(\mathbf{k})]$. Each of these eigenvectors describes a *mode* of the system, and a classification is made according to the behaviour of $\lambda_\alpha(\mathbf{k})$ as $k \to 0$, a limit which is called the *hydrodynamic limit*.

Three situations are commonly encountered. First, $\lambda_\alpha(\mathbf{k})$ may level off at a finite value λ_0 as $k \to 0$. if this is the case, the mode is called a purely relaxational mode. The corresponding fluctuations always decay in a finite time $\tau_0 = 1/\lambda_0$,

---

[2] The linearity of the phenomenological equations implies that the relaxation equations for different wavevectors are uncoupled.

whatever their wavelength. In a macroscopic description of the system, valid at long times and long wavelengths, they can always be considered to have relaxed and can be discarded from the description. An example of such fluctuations is given by the polarization fluctuations in a dipolar liquid such as water, which will relax on a molecular time scale. The reason why the description of chapter 9 is nevertheless extended to include such relaxational modes is that the associated relaxation times, although finite, may become large, and therefore relevant in macroscopic experiments in specific circumstances. Examples include critical phenomena and glass transitions, to be discussed in chapter 12. The second possibility is that $\lambda_\alpha(\mathbf{k})$ vanishes as $k \to 0$. The mode is then called a *hydro-dynamic mode*, and the corresponding relaxation time $\tau_\alpha(\mathbf{k}) = 1/\lambda_\alpha(\mathbf{k})$ diverges in the long wavelength limit. Again, two cases arise, depending on the small $\mathbf{k}$ behaviour. Either $\lambda_\alpha(k) = \pm ick - \Gamma k^2$, or $\lambda_\alpha(k) = -\Gamma k^2$ [3]. The first case corresponds to a *damped propagative* mode, while the second describes a *diffusive mode*. For both diffusive and propagative modes, the vanishing of τ at small k is related to the conservation laws in the system. Roughly speaking, the divergence operator in the conservation law becomes a factor k in Fourier space. Diffusive modes are associated with variables whose current is not a conserved quantity, e.g. the density of tracers. Propagative modes, however, are obtained when the current of a conserved variable is itself a conserved variable, as is the case for mass density, whose current is the momentum density.

To clarify the role of these modes, consider the correlation function of a variable such as the mass density $\rho(\mathbf{r}, t) = \sum_i m\delta(\mathbf{r} - \mathbf{r}_i(t))$ in a one-component fluid. In section 11.6, we will see that the associated *dynamical structure factor*

$$S(k, \omega) = \int_{-\infty}^{+\infty} dt \, \langle \rho(\mathbf{k}, t)\rho(-\mathbf{k}, 0) \rangle \exp(i\omega t) \tag{11.23}$$

is a measurable quantity. In a one-component fluid, the conserved quantities are mass, momentum and energy. Equation (11.22) involves a 3×3 matrix, so that three eigenmodes can be identified. It can be shown that two of these modes are propagative – they correspond to the propagation of longitudinal sound waves – while one is diffusive, and corresponds to heat diffusion. All three modes will contribute to the density–density correlation function, which will therefore be a sum of terms $\exp(-t/\tau_\alpha(k))$. Varying the frequency at fixed k, one observes a peak at finite frequency $\omega = ck$ (with width Γk^2) for each propagative mode, and a peak centred around $\omega = 0$ (with width Δk^2) for a diffusive mode. The damping coefficients Γ and Δ are related to the volume and shear viscosities of

---

[3] A behaviour of the form $\lambda_\alpha(\mathbf{k}) = +Dk^2$ is of course excluded, since it would result in an exponential growth of fluctuations. In the same way, $\lambda_\alpha(\mathbf{k}) = -ck$ is impossible, since it would imply an exponential growth with rate $+ck$ for the fluctuations of wavevector $-\mathbf{k}$.

Figure 11.2. The dynamical structure factor of a one-component liquid in the long wavelength, low frequency limit. The central peak is related to heat diffusion, the two lateral peaks to propagation of longitudinal sound waves.

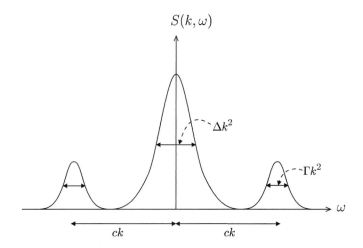

the fluid and to its thermal conductivity. The corresponding dynamical structure factor for a one-component fluid is shown in figure 11.2[4].

Summarizing, the long wavelength fluctuations in a fluid are organized in eigenmodes that can be identified by diagonalizing the linearized phenomenological equations describing the relaxation of inhomogeneities. There are at least as many modes as there are conserved variables in the system, and extra modes associated with the slow relaxation of an order parameter may also play an important role in specific situations. Experimentally, modes are identified by plotting the spectrum of the correlation functions as a function of frequency. Each mode is then associated with a peak. For relaxational modes, the peak is at $\omega = 0$, and its width is independent of the wavevector. For propagative modes, the peak is at finite ω, and has a linear dispersion relation. For diffusive modes the peak is at $\omega = 0$, and its width is proportional to k^2.

An interesting byproduct of this analysis is the proof of the Onsager reciprocity relations, which were alluded to in chapter 9. A schematic proof of these relations proceeds as follows. Consider the binary system of chapter 9. Forming a column vector A with the number and energy densities, ρ_N and ρ_E (which are related to the currents $\mathbf{j}_N$ and $\mathbf{j}_E$ introduced in section 9.1 by a continuity equation), the phenomenological equations can be written in Fourier space in the form

$$\frac{\partial A(\mathbf{k}, t)}{\partial t} = -k^2 [L][\Gamma] A(\mathbf{k}, t) \tag{11.24}$$

where $[L]$ is the 2×2 matrix of the Onsager coefficients, and $[\Gamma]$ is a symmetric matrix with elements $\Gamma_{ij} = \frac{\partial \gamma_i}{\partial \rho_j} = \frac{\partial^2 S}{\partial \rho_j \partial \rho_i}$. Defining the 2×2 matrix of correlation functions $C_{ij}(k, t) = \langle \rho_i(\mathbf{k}, t) \rho_j(-\mathbf{k}, 0) \rangle$, it can be shown from equation (11.24)

[4] For an analytical expression of this structure factor, see e.g. J.P. Hansen and I.R. McDonald, *Theory of Simple Liquids*, Academic Press, New York, 1986.

that

$$\lim_{\omega \to 0} \lim_{k \to 0} \frac{\omega^2}{k^2} [\tilde{C}(k, i\omega)] = [L][\Gamma][C(k, t = 0)] \qquad (11.25)$$

where $\tilde{C}$ denotes a Laplace transform. The theory of fluctuations presented in section 3.3 implies that $[C(k, t = 0)]$ is proportional to $[\Gamma]^{-1}$. Hence the symmetry of $[L]$ is related to the symmetry of $[C]$, i.e. to the fact that $\langle \rho_1(t)\rho_2(0) \rangle = \langle \rho_2(t)\rho_1(0) \rangle$, where for convenience we have dropped the wavevector dependence. Invoking microscopic reversibility [5] in the form $\langle \rho_2(t)\rho_1(0) \rangle = \langle \rho_2(0)\rho_1(-t) \rangle = \langle \rho_1(t)\rho_2(0) \rangle$ completes the proof.

11.6 Application 2: inelastic scattering of neutrons and light

In section 3.8, we have seen how the scattering of radiation from a liquid sample could be used to understand its spatial structure. The essential point is that the *amplitude* $A(\mathbf{k}, t)$ of the radiation scattered in the direction that corresponds to a scattering vector $\mathbf{k}$ is proportional to the Fourier component $\rho(\mathbf{k}, t)$ of the density of scatterers at time t. The nature of the scatterers, and the associated cross-section, depend on the type of radiation.

If one is interested in the static structure only, it is sufficient to measure the total *intensity* of the scattered radiation, $I(\mathbf{k}) = \langle |A(\mathbf{k}, t)|^2 \rangle$. To gain information on the dynamics, it is necessary to measure the correlation function of the amplitude, $C(\tau) = \langle A(\mathbf{k}, t + \tau)A^*(\mathbf{k}, t) \rangle$, which is proportional to the time correlation function of the density of scatterers. A direct measurement of such a correlation function is difficult, since detectors are sensitive to the intensity rather than to the amplitude [6]. The solution is to analyse the frequency power spectrum of the scattered signal. A frequency filter is interposed before the detector, so that only the Fourier components of the signal $A(t)$ within a window $d\omega$ are detected (for simplicity, we have dropped the $\mathbf{k}$ dependence). If the signal is recorded during a time interval $[-T, T]$, its Fourier decomposition is of the form (with $\omega_n = 2\pi n/T$)

$$A(t) = \frac{1}{2T} \sum_n \exp(i\omega_n t)A_n \qquad \text{for} \qquad -T < t < T \qquad (11.26)$$

With a filter that isolates the Fourier component around frequency ω, the intensity reaching the detector is proportional to $|A_n|^2$, with $2\pi n/T = \omega$. Letting $T \to \infty$

---

[5] That is, the fact that the evolution from 0 to $-t$ is identical to the evolution from 0 to t.

[6] In light scattering, a direct measurement is nevertheless possible by adding to the scattered light a reference, unscattered signal (obtained by splitting the original laser beam). The *intensity* of the resulting signal contains a cross term proportional to $A(\mathbf{k}, t)$, and can be autocorrelated to give $C(\tau)$. This is the photon correlation spectroscopy technique, particularly suited for studying long time dynamics, in particular the Brownian dynamics of colloidal particles.

at fixed ω, the average detected intensity is

$$\lim_{T \to \infty} \frac{1}{2T} |A_n|^2 = \lim_{T \to \infty} \frac{1}{2T} \int_{-T}^{T} dt \int_{-T}^{T} dt' \, A(t) A^*(t') \exp(i\omega t) \exp(-i\omega t')$$

$$= \int_{-\infty}^{\infty} \langle A(\tau) A^*(0) \rangle \exp(i\omega \tau) \tag{11.27}$$

where in the second line we have made use of the ergodic assumption, equation (11.2). The essential result here is that, after frequency selection, the scattered intensity yields information on the Fourier transform of the time correlation function of the density of scatterers.

Like the cross-section, the way the frequency selection is done is strongly dependent on the type of radiation which is scattered. For visible light, appropriate filters are Fabry–Perot type interferometers. For neutrons, a quantum description is necessary, and the frequency selection in fact corresponds to an energy selection, which can be made e.g. with a time of flight apparatus. In the quantum description, the change of energy between the incoming and outgoing particles is described as an energy exchange between the scattered particle and a thermal excitation in the sample.

We now turn to the correlation functions that can be measured using neutron or light scattering. In condensed matter studies, 'thermal' neutrons from a nuclear reactor with kinetic energies of order $k_B T$ are used, so that the energy exchange with the system is easily detectable. The wavelength of these neutrons is of the order of a few angstroms, which makes them well suited for structural investigations. The neutrons interact only with the atomic nuclei, and not with the surrounding electrons. The interaction of the neutron with a given nucleus is usually characterized by the scattering length b of this nucleus, which quantifies the strength of the nucleus–neutron coupling. On the atomic scale, this interaction can be appropriately represented by a Fermi potential $(2\pi\hbar^2 b / m_n)\delta(\mathbf{r} - \mathbf{R})$, where m_n is the neutron mass, while $\mathbf{r}$ and $\mathbf{R}$ are the neutron and nucleus positions. As a result, the density that is 'seen' by the neutrons and scatters them is $\sum_i b_i \delta(\mathbf{r} - \mathbf{R}_i)$, where the sum runs over all nuclei in the sample. The corresponding scattering amplitude for a wavevector $\mathbf{k}$ is

$$A_n(\mathbf{k}, t) = \sum_i b_i \exp(i\mathbf{k} \cdot \mathbf{R}_i(t)) \tag{11.28}$$

If the nuclei are all identical, the information obtained from the frequency spectrum of the signal is therefore proportional to the Fourier transform of the density–density correlation function, the dynamical structure factor introduced in section 11.5. More generally, the scattering length depends on the nature of the nuclei. Even in a monoatomic system, different isotopes with different values of b_i are usually present. Since there is no correlation between the isotopic nature of the nucleus and its position, it is legitimate to perform an average over the isotopic distribution. The correlation function that is actually measured is therefore

$$C(\mathbf{k}, \tau) \sim \sum_{i,j} \overline{b_i b_j} \langle \exp[i\mathbf{k} \cdot (\mathbf{R}_i(t + \tau) - \mathbf{R}_j(t))] \rangle \tag{11.29}$$

where $\overline{b_i b_j}$ denotes the average over the isotope distribution. This average can be written as

$$\overline{b_i b_j} = \delta_{ij}(\overline{b^2} - \bar{b}^2) + \bar{b}^2 \tag{11.30}$$

where $\bar{b}$ is the average scattering length. The total signal therefore contains a so-called incoherent part proportional to $(\overline{b^2} - \bar{b}^2)\sum_i \exp(i\mathbf{k} \cdot (\mathbf{R}_i(t + \tau) - \mathbf{R}_i(t)))$, which is the van Hove function defined in section 11.2. By varying the isotope content of a sample appropriately, this incoherent contribution, characteristic of single particle motion, can be isolated from the signal. Generally speaking, iso-topic substitution (which in complex liquids is most often the hydrogen/deuterium substitution) is a powerful technique for sorting out contributions from different correlation functions in the neutron scattering amplitude.

Light scattering operates in a completely different wavelength domain, with a radiation wavelength of typically 500 nm and an energy 100 times larger than thermal energies. Optical methods are sufficiently accurate, however, that the frequency shifts induced by the interaction with the sample are easily detectable. It is customary to distinguish between Rayleigh (or quasi-elastic) scattering, in which the relative frequency shift is small (typically $\hbar\omega \simeq k_B T$) and Raman, or inelastic, scattering, which involves larger frequency changes caused by internal transitions in molecules, or by localized vibrations similar to optical phonons in solids. In the following we focus on the simplest case of Rayleigh scattering by isotropic molecules. In view of the large wavelength of the incident radia-tion, a macroscopic description in terms of an inhomogeneous, time-dependent dielectric permittivity $\epsilon(\mathbf{r}, t)$ is appropriate here. Each volume element acts like a small dipole which re-emits the light, the scattering being caused by the inho-mogeneities in the dielectric constant. The measured correlation function, for a scattering vector $\mathbf{k}$, is then $\langle \epsilon(\mathbf{k}, t)\epsilon(-\mathbf{k}, 0)\rangle$, with prefactors that can be obtained from the classical electromagnetic theory of dipolar radiation.

In a simple fluid, the inhomogeneities in $\epsilon(\mathbf{r}, t)$ arise from local temperature or density fluctuations. The measured correlation function will therefore contain contributions from temperature–temperature, density–temperature and density–density correlations. Fortunately, the dielectric constant is much more sensitive to density variations than to temperature variations, so that the density–density contribution is normally the dominant one. This means that quasielastic light scattering allows a direct measurement of the dynamical structure factor described in section 11.5, the so-called Rayleigh–Brillouin spectrum (figure 11.2).

In complex fluids, especially in colloidal suspensions, the scattering is mainly due to the difference in refractive index between the particles and the solvent. The scattering in this case is therefore related to the fluctuations in the concentration of the large particles. The situation is very similar to that encountered in neutron scattering, with the large particles playing the role of atomic nuclei. Isotopic substitution is replaced by the use of particles with a slightly different refractive index. An important practical difference, however, is that multiple scattering is a much more sensitive issue for light than it is for neutrons, due to the much larger scattering cross-section. For simple fluids, this multiple scattering becomes important when the density fluctuations increase close to a critical point, and is responsible for the critical opalescence phenomenon. In colloidal solutions, light scattering studies are possible in concentrated solutions only when the index of the solution closely matches that of the particles.

11.7   Application 3: pulsed field gradient nuclear magnetic resonance

Many atomic nuclei carry a magnetic moment proportional to their spin. These magnetic moments can be oriented in an external magnetic field $\mathbf{B}_0$. At equilibrium, the magnetic moments are, on average, aligned with the field. Each moment actually precesses around the field direction at the *Larmor frequency* $\omega_L = \gamma B_0$. Because of this precession, the moments are sensitive to electromagnetic radiation at the frequency ω_L. In the classical picture, a rotating field B_1 orthogonal to B_0 and with frequency ω_L induces, in the rotating frame, a precession of the total magnetic moment around its direction. Applying such a field during an appropriate time window, it is then possible to tilt the direction of the average magnetization by any desired angle. The so-called $\pi/2$ pulses use such a rotating field to tilt the average magnetization by $\pi/2$[7]. By studying how the transverse magnetization created in this way returns to its equilibrium value, information can be gained on the local dynamics of the spins, and therefore – via some modelling of the interactions with the surroundings – on the dynamics of the liquid. This is the basis of the nuclear magnetic resonance (NMR) techniques, which provide a powerful tool for investigating atomic dynamics in liquids and solids. Conceptually, the simplest of these techniques is probably the pulsed field gradient technique, which has the advantage that it does not involve any assumption concerning the interactions of the spins with the surrounding medium. This technique is used to investigate diffusion on long time and length scales, and is therefore complementary to incoherent neutron scattering techniques.

In a pulsed field gradient experiment, the first step consists in applying a $\pi/2$ pulse to the sample, creating a transverse magnetization M_0. Next, a small position dependent field, $\mathbf{B}_i = \alpha z \mathbf{e}_z$, is superimposed on the magnetic field $\mathbf{B}_0 = B_0 \mathbf{e}_z$.

---

[7] In the more accurate quantum description, this corresponds to preparing the system in a linear combination of the 'up' and 'down' states.

The duration of this gradient pulse is δ. When this inhomogeneous field is present, the Larmor frequency becomes position dependent. As a result, the transverse magnetizations at different values of z rotate by different amounts during this pulse, and the total transverse magnetization of the sample decays rapidly.

After a time lag Δ, a second gradient pulse is applied to the sample. This pulse is identical to the first in duration, but the sign of the inhomogeneous field is now reversed, $B_1 = -\alpha z \mathbf{e}_z$. If the atoms were immobile, the dephasing caused by this second pulse would exactly compensate the dephasing caused by the first one. Hence after the second pulse all the transverse magnetization at different positions would be in phase, and the total transverse magnetization would recover its initial value M_0. Atomic motion during the time interval Δ will result in different phase shifts during the first and second pulse, so that the initial transverse magnetization will not be recovered. Quantitatively, an atom which is at z during the first pulse and at z' during the second one will have its transverse magnetization rotated by $\gamma\alpha(z - z')\delta$. The total transverse magnetization after the second gradient pulse will therefore be

$$M(\Delta) = M_0 \int dz\, P(z, z', \Delta) \cos(\gamma\alpha(z - z')\delta) \qquad (11.31)$$

where $P(z, z', \Delta)$ is the probability that an atom initially at z has moved to z' in the time interval Δ. For a diffusive motion characterized by a diffusion constant D, $P(z, z', t) = (4\pi Dt)^{-1/2} \exp(-(z' - z)^2/4Dt)$. More generally, $P(z, z', t)$ is related to the van Hove function introduced in equation (11.3). Therefore measuring the transverse magnetization after the second gradient pulse gives direct access to the Fourier transform of the van Hove correlation function. The result is analogous to that obtained for inelastic neutron scattering, with the quantity $\gamma\alpha\delta$ playing the role of the momentum transfer[8]. There are, however, important differences between the two techniques, as shown in the following.

Two implicit assumptions have been made in deriving equation (11.31). First, the diffusion constant has been assumed to be small enough, so that atomic motion can be neglected during the gradient pulse of duration δ. This turns out not to be a serious limitation for measuring diffusion constants in liquids, but makes the method inappropriate for measuring the short time behaviour of $P(z, z', t)$. The second assumption is that all the relaxation processes that contribute to the decay of the transverse magnetization could be neglected, so that in the absence of diffusive motion a perfect 'spin echo', $M(\Delta) = M_0$, would be obtained. In fact, the transverse magnetization after a $\pi/2$ pulse returns slowly to its equilibrium value, over a time scale usually denoted by T_2. Equation (11.31) must therefore be corrected by replacing M_0 with $M_0 \exp(-\Delta/T_2)$, where the exponential factor

---

[8] Here the field gradient was assumed to be parallel to the field $\mathbf{B}_0$, so that the projection of the van Hove function in this direction is observed. Information on diffusion in other directions can be obtained by using appropriately directed field gradients.

accounts for the spontaneous decay of the transverse magnetization. As a result, the values that can be used for the time lag Δ between the gradient pulses are limited to typically $\Delta < T_2$, since for longer times the signal vanishes rapidly. However, Δ must be large enough so that appreciable motion takes place between the two pulses. Typically the condition is that $(D\Delta)^{1/2}\gamma\alpha\delta > 1$ for $\Delta \simeq T_2$. This puts a lower limit on the measurable values of D, this limit being strongly dependent on T_2. Unfortunately, systems with low diffusion constants have also small values of T_2 (faster nuclear relaxation), so that this limitation is a serious one. Nevertheless, refinements of the method allow measurements of diffusion constants in the range of 10^{-14} m^2/s.

Further reading

D. Chandler's *Introduction to Modern Statistical Mechanics*, Oxford University Press, Oxford, 1987, is probably the only elementary textbook that deals with time dependent phenomena, response and correlation functions. His last chapter will be a very useful introduction to this topic.

At a much more advanced level, D. Forster, *Hydrodynamic Fluctuations, Broken Symmetry, and Correlation Functions*, Addison-Wesley, New York, 1983, deals with general properties of correlation functions and collective modes, with applications to a large variety of systems, including simple fluids, superfluids and liquid crystals. At a comparable level, J-P. Hansen and I.R. McDonald, *Theory of Simple Liquids*, Academic Press, New York, 1986, gives a thorough treatment of correlation functions in simple fluids and mixtures. Collective modes in various systems are also discussed in P. Chaikin and T. Lubensky, *Principles of Condensed Matter Physics*, Cambridge University Press, Cambridge, 1995.

Dynamic light scattering is discussed in great detail in the book by Berne and Pecora (see chapter 3) and neutron scattering in S.W. Lovesey, *Theory of Neutron Scattering from Condensed Matter*, Clarendon Press, Oxford, 1984.

Finally, pulsed field gradient NMR, which has undergone fast developments relatively recently, is reviewed in the paper by J. Kärger, H. Pfeifer and W. Heink, *Adv. Magn. Reson.* **12** (1988).

12 Slow relaxations

12.1 Memory and viscoelastic effects

Implicit in all studies involving the Langevin equation (10.1) is the assumption of a separation of time scales. The degrees of freedom described by the equation are assumed to evolve much more slowly than those giving rise to the random force. In many interesting physical chemistry problems, this separation of time scales is, however, not achieved. An example is the reorientation of a polar molecule in a solvent, in which the mass and size of the polar molecule are in the same range as those of the solvent molecules. It is therefore often desirable to consider a generalization of the Langevin equation that allows for *correlation* effects in the random force, i.e. equation (10.3) is replaced by

$$\langle \theta_\alpha(t)\theta_\beta(t')\rangle = \Theta_0(t-t')\delta_{\alpha\beta} \tag{12.1}$$

If the solvent has an intrinsic time scale, its average response (the friction term) cannot be instantaneous, and the generalization of the Langevin equation therefore also includes a *memory* kernel,

$$M\dot{\mathbf{V}} = -\int_{-\infty}^{t} M\zeta(t-s)\mathbf{V}(s)\,\mathrm{d}s + \theta(t) \tag{12.2}$$

The intuitive link between the existence of correlations in the random force and that of a memory in the friction term is expressed mathematically in a fluctuation–dissipation relation that generalizes equation (10.7), and reads

$$M\mathrm{Re}\left(\int_0^\infty \zeta(t)\exp(\mathrm{i}\omega t)\,\mathrm{d}t\right) = \frac{1}{k_B T}\int_{-\infty}^{+\infty}\Theta_0(t)\exp(\mathrm{i}\omega t)\,\mathrm{d}t \tag{12.3}$$

273

Exercise (difficult): Prove (12.3).

Hints: By adding a time dependent sinusoidal force in equation (12.2), show that a frequency dependent mobility can be defined in the form $\lambda(\omega) = 1/M(\zeta(\omega) + i\omega)$. Show next, using the fluctuation–dissipation relation (11.19), that for a system in thermal equilibrium, the mobility is related to the autocorrelation function of the velocity through $\lambda(\omega) = \frac{1}{3k_B T} \int_{-\infty}^{\infty} \langle \mathbf{V}(t) \cdot \mathbf{V}(0) \rangle \exp(-i\omega t)\, dt = \frac{1}{k_B T} Z(\omega)$. (Consider the case where both observables A and B are a cartesian coordinate of the particle.) Use equation (12.2) to show that $|\zeta(\omega) + i\omega|^2 Z(\omega) = \int_{-\infty}^{\infty} \Theta_0(t) \exp(i\omega t)\, dt$.

There is an interesting formal analogy between equation (12.2) and the equations describing the dynamical behaviour of *viscoelastic* fluids. As pointed out in section 1.1, viscous fluids are characterized by a stress σ which is proportional to the strain rate $\dot{u}$, while elastic solids have a stress proportional to the strain u. An intermediate situation frequently arises, which may be described by

$$\sigma(t) = \int_{-\infty}^{t} G(t - s)\dot{u}(s)\, ds \tag{12.4}$$

where for simplicity we have treated σ and u as scalar quantities. The stress in (12.4) is very similar to the systematic part of the force appearing in the right-hand side of equation (12.2). The extreme cases of the elastic solid and of the viscous liquid are recovered by taking $G = $ constant and $G(t) = \eta \delta(t)$, respectively. A well known intermediate model is the so-called Maxwell model, in which $G(t) = G_0 \exp(-t/\tau)$. In this model, liquid-like behaviour is obtained on time scales larger than τ, with a viscosity $\eta = G_0 \tau$. The system behaves as an elastic solid with modulus G_0 on shorter time scales. This cross-over can be illustrated by considering the response to an oscillatory strain, $u(t) = u_0 \exp(i\omega t)$. For $\omega\tau \ll 1$ one finds $\sigma(t) = i\omega G_0 \tau u(t) = G_0 \tau \dot{u}(t)$, i.e. a viscous response. For $\omega\tau \gg 1$ on the contrary, one has $\sigma(t) = G_0 u(t)$, as in an elastic solid.

Equation (12.2) can be interpreted as a microscopic equivalent of (12.4), including the thermal fluctuations that are important when dealing with microscopic degrees of freedom. The force on the tagged particle is elastic (proportional to displacement) on short time scales, and viscous (proportional to the velocity) on longer time scales. In simple fluids, the time scale τ involved in the equivalent of the Maxwell model, $\zeta(t) = \zeta_0 \exp(-t/\tau)$, is generally small, typically a few picoseconds. In complex fluids which display viscoelastic behaviour, it can be much longer and enters the region of macroscopic times. The memory effect is then related to the slow relaxation of large molecules, such as the polymers described in section 10.4.

As long as the memory function ζ remains rapidly decaying (typically exponentially), the existence of memory effects simply implies that the relaxation will be more complex that the simple Debye form (10.37). Nevertheless, if the memory function ζ is a rapidly decaying function of time, the correlation function of **V** will decay on a similar time scale. Typically it will be a sum of exponentials rather than a single exponential as in equation (10.8). In some cases, however, memory effects induce much more dramatic changes in the behaviour of correlation functions. This occurs whenever the 'slow' variable described by the generalized Langevin equation is coupled to a continuum of other slow variables, resulting in a non-trivial collective behaviour. The slow variables can be associated with the hydrodynamic behaviour, or with order parameters in phase transitions. Specific examples are considered in the following sections.

12.2 Coupling to hydrodynamic modes and long time tails

A beautiful illustration of how hydrodynamic variables can couple to microscopic observables to produce long term memory is provided by the phenomenon of 'long time tails', discovered in 1967 by Alder and Wainwright [1] using molecular dynamics simulations. These showed that the velocity autocorrelation of a tagged particle in a dense fluid of similar particles, $Z(t) = \langle \mathbf{v}(t) \cdot \mathbf{v}(0) \rangle$, displays a 'tail' that decays much more slowly than an exponential. Before these simulations, it was generally believed that, due to the large number of uncorrelated collisions experienced by the tagged particle, the decay of $Z(t)$ would be fast, similar to what is predicted for a particle obeying the Langevin equation (see equation (10.8)). The failure in this reasoning lies in the assumption of 'uncorrelated' collisions. The organization of the microscopic motions in hydrodynamic modes (section 11.5) implies that the collisions are indeed *correlated* over long times, so that a more careful analysis is required.

The first step is to introduce a coarse-grained velocity field,

$$\mathbf{v}(\mathbf{r}, t) = \sum_{i=1}^{N} \mathbf{v}_i(t) f(\mathbf{r} - \mathbf{r}_i(t)) \tag{12.5}$$

where f is a slowly varying function of r, typically a normalized Gaussian with a width equal to the interparticle distance. This coarse-grained field is expected to obey, for large distances and long times, the equations of hydrodynamics. The velocity autocorrelation function $Z(t)$ is then approximated as

$$Z(t) \simeq \langle \mathbf{v}(\mathbf{r}_1(t), t) \cdot \mathbf{v}(\mathbf{r}_1(0), 0) \rangle = \int \frac{d^3 k}{(2\pi)^3} \frac{d^3 k'}{(2\pi)^3} \langle \mathbf{v}(\mathbf{k}, t) \cdot \mathbf{v}(\mathbf{k}', 0) \rangle$$
$$\exp(i\mathbf{k} \cdot \mathbf{r}_1(t)) \exp(i\mathbf{k}' \cdot \mathbf{r}_1(0)) \rangle \tag{12.6}$$

where we have introduced the Fourier components $\mathbf{v}(\mathbf{k}, t)$ of the velocity field. The next step is to assume that for long times, the position of the particle becomes decorrelated from the velocity field. This assumption allows one to express the correlation function appearing in (12.6) as a product of two correlation functions.

[1] B.J. Alder and T.E. Wainwright, *Phys. Rev. Lett.* **18**, 988 (1967).

The final expression for $Z(t)$ is then

$$Z(t) \simeq V^{-1} \int \frac{d^3k}{(2\pi)^3} C_{VV}(\mathbf{k}, t) F_s(k, t) \qquad (12.7)$$

where V is the volume, C_{VV} is the autocorrelation of the velocity field $\mathbf{v}(\mathbf{k}, t)$, and $F_s(k, t)$ is the incoherent scattering function introduced in equation (11.5).

The velocity field in a fluid obeys the Stokes equation (valid for the small fluctuations that are of concern here)

$$\frac{\partial \mathbf{v}}{\partial t} = \nu \nabla^2 \mathbf{v} - \nabla P / \rho \qquad (12.8)$$

where $\nu = \eta/\rho$ is the kinematic viscosity and P is the pressure. The *transverse* part of $\mathbf{v}(\mathbf{k}, t)$ (i.e. the component of $\mathbf{v}(\mathbf{k}, t)$ which is orthogonal to the wavevector $\mathbf{k}$) therefore obeys the diffusion-like equation [2]

$$\frac{\partial \mathbf{v}_\perp}{\partial t} = \nu \nabla^2 \mathbf{v}_\perp \qquad (12.9)$$

In analogy with the results obtained in section 11.2, we conclude that the transverse part of C_{VV} is

$$C_\perp(\mathbf{k}, t) = (N k_B T / m \rho^2) \exp(-\nu k^2 t) \qquad (12.10)$$

The transverse part of the momentum (or velocity) field is therefore characterized by a diffusive behaviour. This behaviour is characteristic of the fluid state which, in contrast to elastic solids or viscoelatic fluids, cannot sustain propagative transverse (shear) waves. A completely different behaviour is obtained for the longitudinal part of the C_{VV} (i.e. the correlation function of the component parallel to the wavevector). Since longitudinal sound waves can propagate in the system, this part is typical of a propagative mode,

$$C_\parallel(\mathbf{k}, t) = (N k_B T / m \rho^2) \text{Re} \exp(-ickt - \Gamma k^2 t) \qquad (12.11)$$

where c is the velocity of sound and Γ is the sound damping coefficient (see section 11.5). These expressions are then inserted into equation (12.7), using $C_{VV} = C_\parallel + 2 C_\perp$ [3] and $F_s(k, t) = \exp(-Dk^2 t)$. Integrating over $\mathbf{k}$, one finds that the longitudinal part contributes a term which decays exponentially fast with time. In contrast, the contribution from the transverse part decays much more slowly with time, like

$$Z(t) \sim (k_B T / m \rho) ((D + \nu) t)^{-3/2} \qquad (12.12)$$

or more generally $Z(t) \sim t^{-d/2}$ in a d-dimensional space.

---

[2] The pressure gradient appearing in equation (12.8) is a longitudinal field, i.e. its Fourier components are parallel to the wavevector. Therefore this term does not appear in the equation of motion for the *transverse* velocity field.

[3] The coefficient 2 accounts for the two possible polarizations of the transverse motion, in three dimensions.

A qualitative understanding of the algebraic decay described by equation (12.12) can be obtained from the following reasoning. Consider a tagged particle located at the origin at $t = 0$. This particle has an initial velocity $\mathbf{v}_0$. Due to the interactions, the corresponding initial momentum will soon be distributed over a surrounding region of the fluid. The longitudinal part of this momentum field propagates away in the form of sound waves, and does not affect the long time behaviour. The transverse part, however, diffuses away and is, after a time t, distributed over a sphere of radius $R(t) = (vt)^{1/2}$. This sphere will have an overall velocity of order $\mathbf{v}_0/(\rho R(t)^3)$, and the tagged molecule will be carried away by the corresponding flow. From this simple argument one predicts that the average velocity of the tagged particle at time t should be of order $\mathbf{v}_0/(\rho R(t)^3)$, so that $Z(t) \sim t^{-3/2}$ as in equation (12.12).

12.3 Critical slowing down

At the level of static properties, critical phenomena are characterized by the appearance of a diverging length scale, the correlation length (chapter 5). This length can be interpreted as the size of coherent domains (e.g. concentration domains in a binary mixture close to a demixing point). These coherent domains are fluctuating entities, and their formation and breakup will involve large time scales. On a purely dimensional basis, one may expect this time scale to be of order ξ^2/D, where ξ is the correlation length and D a molecular diffusion constant. This kind of behaviour is indeed confirmed by equation (9.17). Close to the critical point, and within mean field theory (see section 5.1) $\frac{d^2 f}{d\rho^2}$ is proportional to $(T - T_c) \sim \xi^{-2}$, so that the collective diffusion coefficient behaves as $L_{NN}/T\xi^2$. This is the so-called critical slowing down phenomenon, which occurs because of the softening in the thermodynamic driving force near a critical point.

It is interesting to note that the critical slowing down phenomenon will take place even in the case where the order parameter is not associated with a hydrodynamic mode. If the dynamics of the order parameter is purely relaxational, one may write a phenomenological equation in the form

$$\frac{\partial M(\mathbf{r}, t)}{\partial t} = -\lambda \frac{\delta F}{\delta M(\mathbf{r})} \tag{12.13}$$

where F is the free energy expressed as a function of a generic order parameter M (e.g. the magnetization in a magnetic system, or the orientation of molecules in a liquid crystal). Equation (12.13) simply expresses that the order parameter should relax towards the minimum of $F(M)$ ($M = 0$ in the disordered phase), with a finite relaxation time. This relaxation time is easily obtained by using the appropriate free energy in terms of the Fourier components $M(\mathbf{k})$ of M, i.e.

$$F[\{M_{\mathbf{k}}\}] = F_0 + \frac{k_B T}{2} \sum_{\mathbf{k}} \frac{|M(\mathbf{k})|^2}{\langle |M(\mathbf{k})|^2 \rangle} \tag{12.14}$$

which is analogous to the corresponding expression for an inhomogeneous fluid obtained in chapter 7, equation (7.11). Using this expression, it is seen that the relaxation time of a Fourier component $M(\mathbf{k})$ is $\tau(\mathbf{k}) = \langle|M(\mathbf{k})|^2\rangle/\lambda$. Far from a phase transition, $\tau(\mathbf{k})$ is finite even at $\mathbf{k} = 0$, $M(\mathbf{k})$ has a purely relaxational behaviour. At a critical point, however, the susceptibility $\chi(k) = \langle|M(\mathbf{k})|^2\rangle/k_B T$ diverges as $|\mathbf{k}| \to 0$, and so does the relaxation time. Therefore the relaxation of the order parameter at the critical point becomes a hydrodynamic mode (see section 11.5). Strictly speaking, the mode is hydrodynamic only at the critical point. Close to the critical point, however, the evolution becomes so slow that it cannot be neglected when discussing hydrodynamic behaviour.

According to the 'classical' theory (also called van Hove theory) summarized by equations (12.13) and (12.14), one would expect that only the relaxation of the order parameter is affected by the approach to a critical point. However, this is not the case, as the order parameter relaxation couples in a non-trivial way to that of other, seemingly unrelated quantities. The treatment of such couplings is theoretically involved, and our discussion here will remain at a qualitative level. Let us consider, for example, the case of a binary liquid mixture close to a critical demixing point. Assuming that the two components have similar viscosities, one may expect that the viscosity of the mixture is not affected by the proximity of the critical point. Experimentally, however, it is observed that the viscosity of the mixture increases strongly when approaching the critical temperature. Qualitatively, one may interpret this increase as being related to the appearance of many fluctuating interfaces in the system, that separate the fluctuating domains of the two phases. An interface carries stress (the surface tension) so that the amount of stress in response to a fixed shear rate (i.e. the viscosity) will be increased by the presence of these fluctuating interfaces. A quantitative expression of this idea can be obtained starting from an exact expression for the viscosity in terms of the equilibrium stress–stress autocorrelation function, known as the Kubo formula [4]

$$\eta = \frac{V}{k_B T} \int_0^\infty \langle\sigma_{xy}(t)\sigma_{xy}(0)\rangle \, dt \tag{12.15}$$

The structure of equation (12.15) is quite reminiscent of that of equations (11.7) and (9.10) for the mobility. It expresses in a similar way the relationship between the response of the system to an external force and the time evolution of spontaneous fluctuations. From this formula, it is possible to express mathematically the role of the interfacial stresses by isolating the contribution of the stress associated with concentration fluctuations. If $S(k)$ is the concentration–concentration structure factor, and $c_\mathbf{k}$ a Fourier component of the concentration fluctuations, an approximate expression for the off-diagonal component of the stress tensor is

---

[4] See e.g. J-P. Hansen and I.R. McDonald, *Theory of Simple Liquids*, Academic Press, London, 1986.

given by

$$\sigma_{xy} = (k_B T/N) \int \frac{d^d k}{(2\pi)^d} k_x k_y c_k c_{-k} \frac{\partial}{\partial(k^2)} \frac{1}{S(k)} \tag{12.16}$$

Exercise: Prove equation (12.16).

Hints: Consider a sample in which a concentration fluctuation characterized by its Fourier component c_k is present. The free energy associated with this concentration fluctuation is given by

$$F[c] = \frac{k_B T}{2\rho V} \sum_k \frac{c_k c_{-k}}{S(k)} \tag{12.17}$$

Now assume the sample undergoes a sudden shear deformation, $x \to x + \epsilon y$, $k_x \to k_x + \epsilon k_y$. Compute the free energy in the deformed sample. Use the definition of the stress from elasticity theory

$$V \sigma_{xy} = \left(\frac{\partial F}{\partial \epsilon} \right)_{\epsilon=0} \tag{12.18}$$

to justify equation (12.16). Use the identity $\frac{1}{V} \sum_k \leftrightarrow \int \frac{d^d k}{(2\pi)^d}$ for an integration over reciprocal space in d dimensions.

When the formula (12.16) is inserted into equation (12.15), a four-point correlation function of the form $\langle c_k(t) c_{-k}(t) c_{k'}(0) c_{-k'}(0) \rangle$ appears. The usual practice is to approximate this function by a product of two correlation functions, in the form

$$\langle c_k(t) c_{-k}(t) c_{k'}(0) c_{-k'}(0) \rangle \simeq (\delta_{k+k'} + \delta_{k-k'}) \langle c_k(t) c_{-k}(0) \rangle \langle c_{k'}(t) c_{-k'}(0) \rangle \tag{12.19}$$

This approximation, combined with a hydrodynamic approximation of the form

$$\langle c_k(t) c_{-k}(0) \rangle \simeq N S(k) \exp(-k^2 D_c t) \tag{12.20}$$

for the time dependence of the concentration fluctuations, where D_c is a collective diffusion coefficient (see section 9.3), eventually yields the following expression for the interfacial contribution η^* to the viscosity

$$\eta^* = k_B T \int \frac{d^d k}{(2\pi)^d} k_x^2 k_y^2 \left(\frac{\partial}{\partial(k^2)} \frac{1}{S(k)} \right)^2 S(k)^2/(2 D_c k^2) \tag{12.21}$$

If the Ornstein–Zernike form $S(k) = S(0)/(1 + k^2 \xi^2)$ is used for the structure factor, the integral can be computed in the form

$$\eta^* \sim k_B T \xi^{2-d}/D_c \tag{12.22}$$

where we have omitted numerical constants. Equation (12.22) shows that the contribution to the viscosity from the concentration fluctuations results (in three dimensions) from a balance between a growing correlation length and a vanishing diffusion constant D_c. If the diffusion constant is written in the form (similar to equation (9.17)) $D_c = (L_{NN}/k_B T \xi^2)$, where L_{NN} is a kinetic coefficient, equation (12.22) implies

$$\eta^* \sim \xi^{4-d}/L_{NN} \tag{12.23}$$

Hence if the kinetic coefficient L_{NN} is assumed to be regular at the transition, a divergent viscosity $\eta^* \sim \xi$ is predicted in three dimensions. The situation is, however, more complicated, since the kinetic coefficient is affected by the slowing down of the viscous flow. A detailed calculation taking into account the hydrodynamic equations of transport coupled to the diffusion equation shows that the kinetic coefficient L_{NN} has a divergent contribution due to this viscous slowing down, and that the divergence of the viscosity itself is relatively mild, namely $\eta^* \sim \log \xi$.

12.4   The glass transition

Most liquids, as they are cooled down, undergo a first-order transition to a crystalline solid phase. In a limited number of systems, however, a different behaviour is observed. The transition to a crystalline phase does not take place, either because no such phase is thermodynamically stable or, more generally, because the nucleation rate is so low that the transition does not take place on any reasonable laboratory time scale. A classical example is a silica (SiO_2) melt which, when cooled during a laboratory experiment, never crystallizes into its equilibrium quartz structure, but becomes a 'glass', i.e. a system disordered on the atomic scale, yet behaving like a solid.

The transformation from a liquid to a glass takes place in a continuous manner. Its characteristic feature is the fact that the relaxation times of the liquid, and its shear viscosity, increase very rapidly as the temperature is lowered. This increase makes the nucleation of a crystalline phase impossible on the laboratory time scale, as discussed in section 10.6. Moreover, when the internal relaxation time of the system[5] becomes comparable to laboratory time scales (typically 10^2 s), the system becomes unable to relax an externally applied stress through viscous flow, and starts behaving as a solid. This defines the glass transition temperature T_g. In laboratory experiments, T_g is conventionally defined by $\eta(T_g) = 10^{13}$ Pa s.

The dependence of viscosity on temperature is shown in figure 12.1. The Arrhenius representation ($\ln(\eta)$ versus T_g/T) ensures that a system in which the relaxation can be described as resulting from a process with fixed activation energy (e.g. bond breaking) will be represented by a straight line (see section 10.5).

---

[5] The relaxation time can be approximately defined by using the Maxwell model, $\eta = G_0 \tau$ (see section 12.1) where the high frequency modulus G_0 is only weakly temperature dependent.

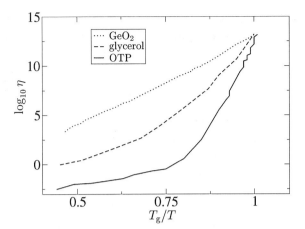

Figure 12.1. Viscosity (on a logarithmic scale) of several glass formers as a function of T_g/T. Data for 'strong' liquids like GeO_2 (dotted line, $T_g = 814\,K$) are almost straight lines in this Arrhenius representation, while data for 'fragile' liquids like orthoterphenyl (OTP, full line, $T_g = 242\,K$), display a marked curvature. Glycerol (dashed line, $T_g = 184.3\,K$) is an intermediate case.

This is indeed the case for some systems in which interactions can be described as having a rather strong covalent character, like silica. Such systems are described as 'strong', according to a classification proposed by C.A. Angell[6] 'Fragile' liquids, in contrast, display a strong curvature in this Arrhenius representation; typical fragile systems have interactions dominated by van der Waals or ionic forces.

An important feature displayed by glass-forming liquids is the 'stretching' of relaxation functions. Typical correlation or response functions (e.g. dielectric relaxation functions or density–density correlation functions) cannot be represented by a single exponential (a Lorentzian in frequency space, see section 10.3), but are rather of the so-called 'stretched exponential' form[7]. The intermediate scattering function $F(k, t)$ can be represented for long times by

$$F(k, t) = A(k)\exp-\left(\frac{t}{\tau(k, T)}\right)^{\beta} \qquad (12.24)$$

where the 'stretching exponent' β is smaller than 1. Equation (12.24) can be interpreted as resulting from a broad distribution of relaxation times, with a typical relaxation time $\tau(k, T)$ which scales with temperature like the viscosity. A typical correlation function is shown in figure 12.2.

Finally, one should also mention that when the glass transition is reached, i.e. when the relaxation time of the system becomes larger than the experimental time scale, anomalies are observed in the behaviour of static thermodynamic quantities, such as the volume or the energy. These anomalies, e.g. a sudden drop in thermal expansivity or in specific heat, are a consequence of the fact that the system falls out of equilibrium, and does not sample phase space efficiently below

[6] C.A. Angell, *J. Phys. Chem. Solids* **49**, 63 (1988).
[7] Also known as Kohlrausch–Williams–Watts or KWW form.

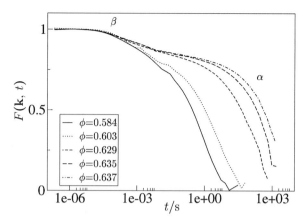

Figure 12.2. Intermediate scattering function $F(k, t)$, measured by light scattering, for a glass forming colloidal suspension, made of small polystyrene microgel beads in a good solvent. The different curves, from left to right, correspond to increasing colloid volume fractions. For the highest packing fractions, the two-step relaxation (β relaxation to a plateau value, followed by α or structural relaxation) is clearly visible. (Courtesy of E. Bartsch, University of Mainz.)

Figure 12.3. Constant pressure specific heat C_P of a fragile organic liquid, toluidine, as a function of temperature. The drop around $T = 180\,\text{K}$ corresponds to the glass transition; the dashed line is specific heat of the crystal, and stops at the melting temperature. (Courtesy of C. Alba, Orsay.)

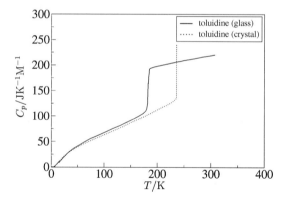

T_g. The loss of the corresponding 'configurational' degrees of freedom results in a decreased specific heat, as shown in figure 12.3. The glass transition can also be observed as a change of slope in the volume versus temperature curve at constant pressure (figure 12.3). The precise location of these anomalies, as well as their amplitude, depends strongly on the rate at which the system is cooled down. Faster quenches imply more rounded anomalies, at higher temperatures.

At present, there is no unified understanding of the mechanisms that give rise to the tremendous increase of relaxation times observed in glass-forming liquids. In the following we briefly discuss two very different – and probably complementary – approaches to this problem, the 'configuration entropy' and the 'mode-coupling' approach.

Entropic theories of the glass transition

Below T_g, the system cannot equilibrate any more, and does not obey equilibrium statistical mechanics. In particular, its properties (e.g. thermodynamic
quantities) will generally evolve in time, a feature known as *physical ageing*. The
glassy state is, by definition, a non-equilibrium state. The transition towards this
non-equilibrium state is generally accompanied by changes in the measured thermodynamic quantities, such as the specific heat or bulk modulus. The specific heat
decreases at the transition, while the bulk modulus displays a measurable increase.
For both quantities, this behaviour can be qualitatively explained through the loss
of configurational entropy which takes place at the transition. Configurational
entropy, a key concept in the discussion of glassy systems, is associated with the
possibility for a system of exploring many different disordered atomic configurations. A simple example is the residual entropy of ice at low temperature, which
results from the many possible organizations of the hydrogen bond network. In
low temperature amorphous systems, it is customary to assume that configuration
space can be partitioned into 'valleys' or 'states' which correspond to local energy minima. The system performs vibrational motion around a minimum, with
occasional hops between two minima. In this picture, the configurational entropy
is related to the number of minima by the usual Boltzmann formula. In the same
way as the entropy is an extensive function of the energy, the configurational
entropy is an extensive function of the energy of the minima. If $u = U/V$ is
the energy density of a minimum, one defines the configurational entropy density

$$s_c(u) = \frac{k_B}{V} \ln \Omega(u) \qquad (12.25)$$

where $\Omega(u)$ is the number of minima with energy density u. If one introduces the
average *vibrational* free energy of the minima with energy density u, $f_v(\beta, u)$,
the partition function of the system can be written

$$Z = \int du \exp\left(-\beta V(u + f_v(\beta, u) - T s_c(u))\right) \qquad (12.26)$$

The equilibrium energy of the minima populated by the system for a given temperature $k_B T = 1/\beta$ is obtained from a minimization of the argument in the
exponential of equation (12.26). Let us assume, for simplicity, that the vibrational free energy is independent of the energy minimum. f_v is then independent
of u, and the minimization yields

$$\frac{\partial s_c}{\partial u} = \beta \qquad (12.27)$$

An immediate problem arises if the function $s_c(u)$ vanishes for $u < u_{min}$ (meaning
that below this energy density there is no longer an exponential number of minima) with a finite slope $\alpha = 1/T_0$. If this is the case, $s_c(u)$ behaves in the vicinity
of u_{min} as $s_c(u) = (u - u_{min})/T_0$, and equation (12.27) cannot be satisfied for

$T < T_0$. A system which is cooled to a temperature below T_0 is therefore out of equilibrium, unless a phase transition takes place at T_0. This is the so-called 'entropic' explanation for the existence of glassy behaviour, in terms of a thermodynamic transition at a finite temperature. T_0[8].

Although this thermodynamic approach predicts a transition to a non-equilibrium situation at some finite temperature T_0, it does not make any prediction regarding the time scale for relaxation, which is the core of the glass transition problem. The thermodynamic approach is therefore usually supplemented by the Adams–Gibbs hypothesis, which relates the relaxation time of the system to the configurational entropy in the form

$$\tau(T) = \tau_0 \exp\left(\frac{A}{T s_c(T)}\right) \tag{12.28}$$

where τ_0 and A are two constants. If $s_c(T) \sim \alpha(T - T_0)$ in the vicinity of T_0, equation (12.28) implies a divergence of the relaxation time at T_0. The corresponding functional form, $\tau \propto \exp(B/(T - T_0))$, is often referred to as a Vogel–Fulcher law, and generally fits the experimental data well.

Briefly speaking, the rationale for the Adams–Gibbs hypothesis is the idea of partitioning the whole system into a collection of independently relaxing subsystems, the 'cooperatively relaxing regions'. The configuration entropy is, through simple statistical considerations, related to the number N_d of such regions, roughly as

$$s_c = N_d k_B / V \tag{12.29}$$

Assuming that the energy barrier for relaxation is proportional to the domain size V/N_d, and an Arrhenius law for the domain relaxation time, equation (12.28) is eventually obtained.

Many different versions of this heuristic approach, with various degrees of sophistication, are available in the literature on the glass transition. A drawback is the absence of a rigorous statistical mechanics framework. The idea of decomposing phase space into valleys in which the system performs vibrational motions, although physically appealing, has no general justification. Obviously the use of *free energy minima* would be more appropriate, but again there are no rigorous definitions for local free energy minima. Moreover, it can be shown that in a system in which point defects with finite energy can exist, the slope of $s_c(u)$ diverges at $s_c = 0$[9]. This raises doubts on the vanishing of

---

[8] The better known implementation of this approach is probably the Gibbs–DiMarzio theory (J.H. Gibbs and E.A. DiMarzio, *J. Chem. Phys.* **28**, 373 (1958)), which is based on an explicit evaluation for the configuration entropy of polymers on a lattice. This approach, however, was recently shown to be inaccurate (M. Wolfgardt, J. Baschnagel, W. Paul and K. Binder, *Phys. Rev. E* **54**, 1535 (1996)).
[9] This is related to the ideal mixing term $x \ln(x)$ in the entropy, where x is the defect concentration. See F.H. Stillinger, *J. Chem. Phys.* **88**, 7818 (1988).

the entropy with a finite slope, which is needed in order to obtain a true thermodynamic transition. Note that the latter argument may be rejected as practically irrelevant if the Adams–Gibbs expression for the relaxation time is accepted. The relaxation times then become much larger than macroscopic times before the potential thermodynamic phase transition is reached, so that assessing the existence, or non-existence, of this transition is out of reach of laboratory experiments.

Mode coupling theories

A completely different view of the viscous slowing down in liquids as a function of temperature is obtained within the framework of the so-called mode coupling theories. The essential idea in these theories is that the transition towards a glass is a purely dynamical one, driven by a non-linear feedback mechanism. In a first approach, one may consider the general Kubo formula (12.15) for the shear viscosity. In a simple, one-component system, the local stresses are related to the distortion in the configuration of neigbouring atoms, and therefore to the deformation of the local pair structure. In the same way as the stress–stress correlations could be expressed in terms of the concentration–concentration correlation functions in a mixture, they can be expressed in the present case in terms of the time dependent density–density correlation functions. These density correlation functions behave, for long times, in a diffusive manner. Therefore it is not unreasonable to assume a relationship of the form

$$\eta = \eta_0 + B(T)/D \qquad (12.30)$$

where η_0 and B are some temperature dependent functions, and D is the diffusion constant. If one makes use of the Stokes–Einstein relation $D \simeq k_B T/6\pi\eta\sigma$, one arrives at a relation of the form

$$\eta(T) = \eta_0/(1 - 6\pi B(T)/\sigma) \qquad (12.31)$$

which indicates a possible divergence in the viscosity at a finite temperature. This viscosity feedback mechanism [10] is of course largely oversimplified here; however, the idea of a feedback mechanism generating slow relaxations is at the heart of the much more sophisticated mode coupling theories developed in the 1980s to explain the sharp increase in the relaxation time of liquids at low temperature.

---

[10] T. Geszti, *J. Phys. C* **16**, 5805 (1984). For an accessible account of more recent theoretical developments, see W. Götze and L. Sjögren, *Rep. Prog. Phys.* **55**, 241 (1992).

Mode coupling theories concentrate on the normalized density–density correlation function at wavevector $\mathbf{k}$, $\phi(\mathbf{k}, t) = F(\mathbf{k}, t)/S(\mathbf{k})$ (see section 11.6). Using operator projection techniques that are beyond the scope of this book [11], it can be shown that these relaxation functions obey a non-linear equation of the form

$$\ddot{\phi}(\mathbf{k}, t) + \Omega_0(\mathbf{k})^2 \phi(\mathbf{k}, t) + \int_0^t M(\mathbf{k}, t - s)\dot{\phi}(\mathbf{k}, s)\, ds = 0 \qquad (12.32)$$

Here $\Omega_0(k)^2 = k^2 k_B T / m S(k)$ is the vibrational frequency of sound waves with wavevector $\mathbf{k}$[12]. Hence equation (12.32) is just the equation for a damped harmonic oscillator, with a non-local damping term containing a *memory function*. In mode coupling theory, this memory function is expanded in terms of products of pair correlations, and the essential term is a quadratic form of the correlation function:

$$M^{MC}(\mathbf{k}, t) = \int d^3\mathbf{k}' V(\mathbf{k}, \mathbf{k}')\phi(\mathbf{k}', t)\phi(\mathbf{k} - \mathbf{k}', t) \qquad (12.33)$$

This memory term is the mathematical expression of the feedback effect discussed above. A slower decay of the correlation function implies a longer memory and a weaker damping, which in turn slows down the relaxation. The 'vertices' $V(\mathbf{k}, \mathbf{k}')$ can be expressed in terms of the structure factor. The detailed expression is of no interest here, but the important feature is that they tend to increase with $S(\mathbf{k})$. Feedback effects are enhanced in strongly structured (low temperature or high density) systems.

A schematic version of equations (12.32) and (12.33) is obtained by selecting a special wavevector $\mathbf{k}_0$, corresponding to the main peak in the structure factor, and restricting the integral in equation (12.33) to the shell of wave-vectors around this peak. The resulting equation for $\phi(t) \equiv \phi(\mathbf{k}_0, t)$ reads

$$\ddot{\phi}(t) + \Omega_0^2\phi(t) + \lambda_2\Omega_0^2 \int_0^t \phi(t - s)^2 \dot{\phi}(s)\, ds = 0. \qquad (12.34)$$

Restriction to a single wavevector, although it misses some of the mathematical subtleties associated with the full equations (12.32) and (12.33), is physically reasonable since structural relaxation can be expected to be driven by wavevectors close to the peak in $S(\mathbf{k})$, which correspond to the most prominent feature of short-range order in the liquid. The gain in using equation 12.34 lies of course in its mathematical simplicity. Indeed, it is easily shown that when the coupling parameter λ_2 is increased, a transition from an ergodic state to a non-ergodic one

---

[11] See e.g. J-P. Hansen and I.R. McDonald. *Theory of Simple Liquids*, Academic Press, New York, 1986.

[12] Recall that $S(\mathbf{k})$ can be interpreted in terms of a generalized compressibility for wavevector $\mathbf{k}$, see section 3.6.

takes place. Mathematically:

$$\lim_{t\to\infty} \phi(t) = 0 \qquad \text{for} \qquad \lambda_2 < 4$$

$$(12.35)$$

$$\lim_{t\to\infty} \phi(t) = \phi_0 > 0 \qquad \text{for} \qquad \lambda_2 > 4$$

Since all the structure, temperature and density dependence has been cast into the single parameter λ_2, equation (12.35) illustrates the possibility of a *dynamical phase transition* taking place at some critical temperature (or density) in the more general description of equations (12.32), and (12.33). Indeed, a careful theoretical analysis performed on these non-linear integrodifferential equations has shown that they can describe such a transition, with many features reminiscent of the actual behaviour of supercooled liquids. Some of the main results that emerge from this analysis are the following

(i)   The existence of a sharp transition at some critical temperature T_c, with a typical relaxation time diverging like $\tau(T) = (T - T_c)^{-\gamma}$. The exponent γ is system dependent, and generally of order $\gamma \simeq 2$

(ii) Slightly above T_c, the correlation functions such as $\phi(\mathbf{k}, t)$ exhibit a slow, 'two-step' relaxation. In a first step, usually described as 'β' relaxation, the correlation appears to be reaching a plateau, as in a non-ergodic system. Eventually, on the longer time scale $\tau(T)$, the α or terminal relaxation takes place and the correlations relax to zero.

(iii) The α relaxation strongly departs from the single exponential, Debye behaviour. It can be described by the so-called stretched exponential[13] form (12.24), with *stretching exponents* β typically of order 0.6 for k vectors corresponding to the interatomic distance, which are the most important in describing the relaxation of the structure. Note that although the relaxation time in (12.24) is wavevector dependent, its temperature scaling does not depend on $\mathbf{k}$. Another important point is that equation (12.24) implies that the relaxation functions obey, close to T_c, the so-called time–temperature superposition property, i.e. the correlation can be expressed as $\hat{\phi}(t/\tau(\mathbf{k}))$, where $\hat{\phi}$ is a temperature independent 'master function'.

The problematic feature of this analysis is that the predicted dynamical transition at some critical temperature (or density) is not observed in actual systems[14]. Rather, careful experimental studies have shown that the mode coupling description holds relatively well only down to a temperature for which the relaxation time is in the range 10^{-8}–10^{-7}s. Below this temperature – and therefore in the temperature range where the experimental glass transition takes place – it becomes inappropriate. The reason for this failure is not perfectly understood, but is often qualitatively related to changes in the topology of the potential energy surface

[13] Also known as the Kohlrausch–Williams–Watts (KWW) form.

[14] This statement is certainly correct for all molecular glasses; the behaviour of hard-sphere like colloids, however, seems to be relatively well described by mode coupling theories.

visited by the system along its trajectory in phase space. Above T_c, the system has access to regions of phase space in which relaxation can take place through very collective rearrangements, involving only small energy barriers. These collective mechanisms, which are well accounted for by the mode coupling analysis, become inefficient below the critical temperature. They are then replaced by more localized 'hopping' or 'activated' mechanisms, involving energy barrier crossing of the form discussed in section 10.5, albeit in a high dimensional space. The theoretical description of these processes, however, has up to now escaped a first principles approach.

A unified viewpoint?

The two approaches described above seem, at first sight, very difficult to reconcile. The 'entropic' approach is entirely thermodynamic in nature, while the mode coupling one is purely kinetic. A very interesting discovery of the late 1980s [15] is that some exactly solvable theoretical models in fact exhibit a behaviour that includes both a dynamical phase transition of the mode coupling type at a temperature T_c, and an 'entropy crisis' at a lower temperature, $T_s < T_c$. The paradigm of these models is the so called '3-spin' model, which describes the motion of a 'particle' with N coordinates (each coordinate often being described as a 'continuous spin') $\sigma_1...\sigma_N$ in a high N-dimensional space, under the influence of a random potential [16] of the form [17]

$$H = \sum_{i_1,i_2,i_3} J_{i_1,i_2,i_3} \sigma_{i_1} \sigma_{i_2} \sigma_{i_3} \tag{12.36}$$

Here the J_{i_1,i_2,i_3} are random variables drawn from, e.g., a Gaussian distribution with zero mean. The sum in equation (12.36) runs over all possible triplets of coordinates, thus defining a '3-spin' model. The model can be solved in the 'mean field' limit, which amounts to taking the limit $N \to \infty$. Note that this limit is a very peculiar one, in the sense that the energy barriers separating various minima of the energy (12.36) are expected to be extensive, while they are finite in

---

[15] T.R. Kirkpatrick, P.G. Wolynes and D. Thirumalai, *Phys. Rev. A* **40**, 1045 (1989). J-P. Bouchaud, M. Mézard, J. Kurchan and L. Cugliandolo, *Physica A* **226**, 243 (1996).

[16] The existence of an imposed randomness in such models was first thought to imply a behaviour different from that of molecular systems, in which the Hamiltonian does not have a random component. A number of models without any quenched disorder have however been shown to exhibit the same type of behaviour. Rather than disorder, the important concept is that of 'frustration', i.e. the fact that energy minimization is hampered by a number of antagonist local constraints. The disorder present in molecular systems is sometimes described as 'self-induced' in contrast to the 'quenched' disorder of the p-spin model.

[17] The '3-spin' model has interactions between three different variables σ_i. The important feature that makes the model solvable is its infinite connectivity; each spin interacts with all others.

low dimensional systems[18]. In the mean field limit, the correlation functions $\langle \sigma_i(t)\sigma_i(0)\rangle$ are found to obey an integrodifferential equation similar to equation (12.34). This of course results in the existence of an ergodic–non-ergodic transition at some temperature T_c. Thermodynamic properties of the model, however, can be computed from the usual Gibbs approach at any temperature. The calculation is difficult, but results in the prediction of a phase transition, with a vanishing configuration entropy, at a 'static' transition temperature $T_s < T_c$.

Obviously such mean field systems, in the temperature range $T_s < T < T_c$ can never be at equilibrium, and hence the existence of a thermodynamic transition at T_s is in that sense irrelevant. The thermodynamic study of the mean field model is however interesting, in that it reveals a change in behaviour taking place at T_c. Below T_c, the phase space of the system breaks up into well separated 'ergodic components' or free energy valleys. It is therefore possible to define unambiguously the configurational entropy (equation (12.25)), and this entropy is found to vanish at T_s with a finite slope [19]. The conclusion is that both the 'mode coupling' and 'entropic' scenarios for glassy behaviour are realized in these models.

A reasonable speculation is that some remnants of this ideal, mean-field behaviour are observed in real, finite dimensional systems. This would account for the success of mode coupling approaches at relatively high temperatures. At the critical mode coupling temperature, a modification of the available phase space takes place. In mean field models, this corresponds to a breakup into perfectly separated minima, with infinite barriers between them. In real systems, the barriers are finite, and the system is not arrested. The dynamics at temperature lower than T_c will therefore be dominated by hopping between free energy minima, and the Adams–Gibbs theory could be applicable there (although a satisfactory theoretical framework for describing dynamical behaviour in this context is still missing). The existence, or not, of an actual entropic transition at a finite temperature in finite dimensional systems remains open, but one must emphasize that the question is rather academic, in the sense that low temperature systems cannot reach thermodynamic equilibrium for kinetic reasons.

Ageing and effective temperature

A very interesting outcome of these mean field theories is the emergence of a new theoretical framework for describing a class of non-equilibrium systems. In mean field theories, systems at temperatures between T_c and T_s are never at equilibrium.

---

[18] In the mean field system, going from one minimum to another through nucleation and growth is not possible, since the cost of nucleating a single defect is extensively large, as a result of the complete connectivity.

[19] In mean field models, this is possible since the energy cost of possible 'defects' is infinite.

Their dynamical properties can nevertheless be computed essentially exactly[20].
In this temperature range, the non-equilibrium character of the dynamics results
in correlation and response functions (C and χ in the notation of sections 11.2
and 11.3; for simplicity, we drop the subscripts A, B relative to the observables
in the present discussion) which are not invariant under time translation. $\chi(t, t')$
and $C(t, t')$ are functions of the two times at which the observables are measured,
rather than of the time difference only. This *ageing* property, which has been
experimentally observed for a long time in polymeric glasses as well as in spin
glasses, can be quantified by writing the correlation (or response) functions in
the form

$$C(t + \tau, t) = C_{short}(\tau) + C_{ageing}(t + \tau, t) \tag{12.37}$$

where the ageing function $C_{ageing}(\tau, t)$ is typically found to be of the form

$$C_{ageing}(t + \tau, t) = f(\tau/t) \tag{12.38}$$

A typical set of correlation functions, for different values of the *waiting time* t, is
shown in figure 12.4. The interpretation of such curves is the following. During
its non-equilibrium exploration of phase space, the system evolves on two well
separated time scales. On a short time scale, it explores the local neighbourhood
of the instantaneous configuration, through vibrational motions. Such short time
motions are in equilibrium with the external thermostat, even if the configuration
is not an equilibrium one[21]. On longer time scales, the system explores phase
space beyond the local neighbourhood, and the non-equilibrium aspect of the
dynamics becomes evident. For the correlation, this is described by the 'ageing'
part $C_{ageing}(\tau, t)$, which according to equation (12.38) becomes increasingly slow
as more time is spent in the nonequilibrium state.

The mean field '3-spin' model is not the only one that can exhibit 'ageing'
in its correlation functions. Another example of ageing, much simpler (although
probably not the most relevant in this case), is the 'domain growth' situation.
A system undergoing some type of spinodal decomposition is, indeed, out of
equilibrium and 'ages'. It can be shown that the correlation function in such a
system is of the form $C(t', t) = f(\xi(t')/\xi(t))$, where $\xi(t)$ is the domain size at
time t (see section 9.3), which reduces to the form (12.38) when the domain size
grows as a power law.

Other models for the description of ageing correlation functions include 'trap
models', in which the system is assumed to evolve in an energy landscape made of
many minima with a broad distribution of depths, $p(E) \sim \exp(-xE)$. Assuming
that the escape time from a minimum with depth E is proportional to $\exp(-\beta E)$,

---

[20] L.F. Cugliandolo and J. Kurchan, *Phys. Rev. Lett.* **73**, 171 (1993).
[21] In the same way, phonon-like vibrations in a solid are described by equilibrium statistical
mechanics, even if the solid is in a metastable crystalline form.

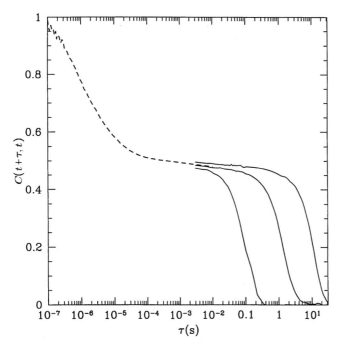

Figure 12.4. Correlation functions in an ageing system. The system is a dense colloidal suspension, and the correlation function is that of the intensity measured in a multiple light scattering experiment. The sample is 'quenched' from a state of high shear. The dashed line is the short time decay, independent of waiting time. The different full curves correspond, from right to left, to successive waiting times differing by one decade, $t = 0.7$ s, 7 s and 70 s. The 'simple ageing' relation (12.38) implies that on a logarithmic scale, the three ageing curves can be deduced from each other by shifts of 1 along the horizontal axis. (Courtesy of V. Viasnoff and F. Lequeux, ESPCI, Paris.)

it can be shown that for $\beta > x$ the system cannot equilibrate any more (the average escape time becomes infinite) and displays ageing in its correlation function. Such models are not easily related to a microscopic description, but are very useful in gaining some qualitative understanding of the ageing phenomenon.

The study of the 'ageing' correlation function has led to an interesting extension of the fluctuation–dissipation theorem, probably applicable to a number of non-equilibrium systems (including systems that are stationary, but out of equilibrium under the influence of some external driving force, e.g. shear flow). If the response function χ is separated into a short time part and an ageing part, in the same way as the correlation function (12.37), it is found that the short time contributions obey the equilibrium fluctuation–dissipation theorem (11.19), while the ageing contributions are related through

$$\chi(t', t) = \frac{1}{k_B T_{\text{eff}}} \frac{\partial C(t', t)}{\partial t} \tag{12.39}$$

which defines an *effective temperature* T_{eff}. The discussion of section 11.4 has shown that the existence of a well defined *fluctuation–dissipation ratio* (as defined in equation (11.18)) should be a rather robust property. Identification of this ratio as an effective temperature can be justified heuristically in the following way. For equilibrium systems, this identification was derived from the Boltzmann–Gibbs canonical distribution for a slightly perturbed system (see section 11.4). Therefore, equation (12.39) implies that a statistical ensemble of ageing systems responds, on long time scales, to external perturbations in the same way as an ensemble described by a canonical distribution with a temperature T_{eff}. The non-equilibrium situation can therefore be described as a 'superposition' of two quasi-equilibrium systems: a vibrational part, which responds on short time scales and has a temperature equal to the external temperature T, and a slowly evolving configurational part, which has an 'internal' temperature T_{eff}. Pushing the analogy with standard statistical thermodynamics further, one can infer that this effective temperature is indeed related to the configurational entropy of the system through

$$\frac{\partial s_c}{\partial u} = \frac{1}{k_B T_{eff}}. \qquad (12.40)$$

an equality that was verified through extensive numerical simulations of model systems [22]. Another property that makes T_{eff} similar to a thermodynamic temperature is the general property that a thermometer weakly coupled to the non-equilibrium system will, if its relaxation time is comparable to the slow time scale in the system, measure T_{eff} rather than the external temperature [23]. This notion of an effective temperature has appeared for a long time in the glass literature, but was only recently given such a clear meaning in terms of measurable correlation and response functions. The extension of equilibrium statistical mechanical concepts, such as temperature, to systems which are out of equilibrium, is still a very open problem, relevant to simple as well as complex fluids.

Further reading

Memory and viscoelastic effects, which are prominent in complex fluids, are discussed in many books dealing with soft condensed matter. In particular, interested readers should consult M. Doi and S.F. Edwards, *Theory of Polymer Dynamics*, Clarendon Press, Oxford, 1986, or R.G. Larson, *The Structure and Rheology of Complex Fluids*, Oxford University Press, Oxford, 1999.

For memory effects and long time tails in simple liquids, standard reference texts are J-P. Hansen and I.R. McDonald, *Theory of Simple Liquids*, Academic Press, London, 1986, and

---

[22] F. Sciortino and P. Tartaglia, *Phys. Rev. Lett.* **86**, 107 (2001).

[23] This idea, together with several other conceptual problems linked to the effective temperature concepts, are reviewed in the paper by L. Cugliandolo, J. Kurchan and L. Peliti, *Phys. Rev. E* **55**, 3898 (1997).

J.P. Boon and S. Yip, *Molecular Hydrodynamics*, McGraw Hill, New York, 1980 (reprinted by Dover, 1991).

Mode coupling effects and critical slowing down are discussed in review papers by B. Halperin and P.C. Hohenberg, *Rev. Mod. Phys.* **49**, 435, (1977) and K. Kawasaki in *Phase Transitions and Critical Phenomena*, ed. C. Domb and M.S. Green, Academic Press, New York, 1976.

The phenomenology of glass-forming liquids is beautifully described in S. Brawer, *Relaxation in Viscous Liquids, and Glasses*, American Ceramic Society, Columbus, OH, 1985. Another interesting, more recent, reference on the subject is the book by P. Debenedetti, *Metastable Liquids*, Princeton University Press, Princeton, NJ, 1996. Finally, a recent compilation of papers on glassy behaviour can be found in a special issue of *Science*, volume **267**, 1995. The mode coupling theory of glass formation is reviewed in W. Götze and L. Sjögren, *Rep. Prog. Phys.* **55**, 241 (1992).

A very recent update of theories of the glass transition is contained in the *Proceedings of the les Houches School 2002, Nonequilibrium Dynamics and Slow Relaxations in Condensed Matter*, ed. J.L. Barrat, J. Kurchan, M. Feigelman and J. Dalibard, Springer, Berlin, 2003.

Index